Distributed Sensor Networks: Technology and Applications

Distributed Sensor Networks: Technology and Applications

Edited by **Marvin Heather**

Willford Press
New York

Published by Willford Press,
118-35 Queens Blvd., Suite 400,
Forest Hills, NY 11375, USA
www.willfordpress.com

Distributed Sensor Networks: Technology and Applications
Edited by Marvin Heather

International Standard Book Number: 978-1-68285-183-8 (Hardback)

Printed in China.

Contents

Preface

Every book is a source of knowledge and this one is no exception. The idea that led to the conceptualization of this book was the fact that the world is advancing rapidly; which makes it crucial to document the progress in every field. I am aware that a lot of data is already available, yet, there is a lot more to learn. Hence, I accepted the responsibility of editing this book and contributing my knowledge to the community.

Distributed sensor networks are sensor networks that are regulated and manipulated by distributed control framework. This book deals with the various algorithms, architectures and concepts associated with distributed sensor networks (DSNs). The topics discussed herein are sensor networks, sensor devices and sensor arrays, different measurement techniques, etc. It is compiled in such a manner, that it will provide in-depth knowledge about distributed sensor networks and its applications to all the students, academicians and professionals engaged in this field.

While editing this book, I had multiple visions for it. Then I finally narrowed down to make every chapter a sole standing text explaining a particular topic, so that they can be used independently. However, the umbrella subject sinews them into a common theme. This makes the book a unique platform of knowledge.

I would like to give the major credit of this book to the experts from every corner of the world, who took the time to share their expertise with us. Also, I owe the completion of this book to the never-ending support of my family, who supported me throughout the project.

Editor

1

Lower Extremity Joint Angle Tracking with Wireless Ultrasonic Sensors during a Squat Exercise

Yongbin Qi, Cheong Boon Soh *, Erry Gunawan, Kay-Soon Low and Rijil Thomas

School of Electrical and Electronic Engineering, Nanyang Technological University,
50 Nanyang Avenue, 639798 Singapore; E-Mails: qiyo0001@e.ntu.edu.sg (Y.Q.);
egunawan@ntu.edu.sg (E.G.); ekslow@ntu.edu.sg (K.-S.L.); rijil001@e.ntu.edu.sg (R.T.)

* Author to whom correspondence should be addressed; E-Mail: ecbsoh@ntu.edu.sg

Academic Editor: Xiaoning Jiang

Abstract: This paper presents an unrestrained measurement system based on a wearable wireless ultrasonic sensor network to track the lower extremity joint and trunk kinematics during a squat exercise with only one ultrasonic sensor attached to the trunk. The system consists of an ultrasound transmitter (mobile) and multiple receivers (anchors) whose positions are known. The proposed system measures the horizontal and vertical displacement, together with known joint constraints, to estimate joint flexion/extension angles using an inverse kinematic model based on the damped least-squares technique. The performance of the proposed ultrasonic measurement system was validated against a camera-based tracking system on eight healthy subjects performing a planar squat exercise. Joint angles estimated from the ultrasonic system showed a root mean square error (RMSE) of $2.85° \pm 0.57°$ with the reference system. Statistical analysis indicated great agreements between these two systems with a Pearson's correlation coefficient (PCC) value larger than 0.99 for all joint angles' estimation. These results show that the proposed ultrasonic measurement system is useful for applications, such as rehabilitation and sports.

Keywords: ultrasound; wireless sensor network; joint angles; squat; rehabilitation; inverse kinematics

---

1. Introduction

Squat exercise is an effective training exercise for maintaining mobility and improving lower-limb muscle function [1]. Therefore, it has been included as part of athletic training and rehabilitation. Squat exercise can be investigated using a wide variety of measured and estimated parameters. In [2], the hip and knee torques combined with the position of the knee have been used for the assessment of barbell squat. Ankle, knee and hip joint kinematics have been used in [3] to quantitatively evaluate the motion of squat. In many of these applications, it is essential to monitor the squat exercise under a natural environment without hindrance. Therefore, this entails the need for an unrestrained low-cost motion tracking system.

The most accurate measurement system in capturing the squat motion or other lower trunk movement is the camera-based tracking system, which employs one or more cameras to capture the displacement of reflective markers placed at specific anatomical sites on limb segments [4]. However, such a system requires a complex and expensive experimental setup [5]. Furthermore, it is sensitive to changes in lighting, clutter and shadow. However, a recent development in computer gaming technology, the Microsoft Kinect, is inexpensive, portable and does not require markers to determine anatomical landmarks and, consequently, may overcome the limitations associated with laboratory-based movement analysis systems [6,7]. However, the limitation of the Microsoft Kinect system is the inability to assess internal/external joint rotations in the peripheral limbs [7].

The development of microelectromechanical system technology has produced many low -cost and small inertial sensors, which can be used in a human tracking system, such as an accelerometer and a gyroscope [8–10]. In [11], Brandes et al. use the accelerations measured at the lower trunk to extract spatial-temporal gait parameters with healthy children. Daniel *et al.* [12] proposed an ambulatory measurement system, using a triaxial accelerometer and a dual-axis gyroscope, to assess the vertical displacement of the foot during walking. Although these inertial sensors overcome the shortcomings of a camera-based tracking system, they require double integration to estimate the position and orientation of human motion [13]. Unfortunately, it is difficult to obtain accurate motion accelerations, because of the presence of sensor bias and measurement noise, thus making the displacement error increase exponentially over integration time [12]. This issue can be mitigated either by applying some techniques to correct it periodically, such as zero velocity update (ZUPT), or by applying a Kalman filter [14], or by combining with other sensors, such as imaging sensors, radio frequency identification (RFID) technology or ultra-wide band (UWB) techniques [15–18]. These mentioned hybrid motion tracking systems can improve the tracking accuracy, but with an increased cost, complexity of experiment installation and maintenance.

Since wireless healthcare has received significant interest in recent years, ultra-wideband technology has become a promising technique for general health-monitoring due to its high-temporal resolution and high ranging and positioning accuracy [19]. In [20], wearable UWB transceivers are used to measure some gait parameters (such as the base of support) during the gait. A joint flexion/extension angle measurement system has also been developed in [21,22]. The model is based on providing a high ranging accuracy (intersensor distance) between a pair of UWB transceivers placed on the adjacent segments of the joint center of rotation. The measured distance is then used to compute the joint angles based on the law of cosines. However, in an impulse radio (IR) UWB system, where a sub-nanosecond pulse is used, the required synchronization is extremely challenging. A highly accurate synchronization clock that provides precise timing information is required for tracking, which is difficult to achieve in low-cost systems [23]. Furthermore, it is difficult to sample the received signal in real time with current ADC technology, due to the large bandwidth of the IR-UWB pulse [24].

An alternative method to UWB technology is to use the ultrasound technique due to its low cost, safety, simplicity and high temporal resolution for low range measurement [25]. There are two types of ultrasonic transceivers: one is with an ultrasonic transmitter and receiver on the same circuit board as the one used in [26]. The distance measurement of such an ultrasonic sensor is the returned distance reflected from the ground/surface, and the orientation of the sensor is not taken into consideration. The other type is with an ultrasonic transmitter and receiver on separate boards [27]. However, there are only two receivers used in [27], which only measures one directional displacement.

In this paper, a wireless wearable sensor system based on ultrasound for monitoring squat exercise is proposed. Only one ultrasonic sensor (transmitter) is needed to be attached to the human trunk, which minimizes the discomfort for users and avoids complex calibration procedures and synchronization issues. Several receivers (anchors) with known positions are used to define a coordinate reference system. The system measures propagation delay from anchor to transmitter, thus computing the range in terms of the speed of sound. Then, the range information defines a circle centered at this anchor with a radius equal to the measured distance, and the position of the transmitter should be within the intersection of several such circles. This is called the spherical tracking technique [28]. The proposed approach makes use of the wireless sensor network concept with all of the mobile sensor nodes wirelessly communicating with the coordinator, thus eliminating the usage of wires. In addition, the sensor node attached to the human trunk is light and small. Furthermore, it is low cost compared to the camera system.

The paper is organized as follows: Section 2 gives a brief overview of the configuration of the wireless wearable system. The biomechanical model of the leg, the displacement measurement model and the inverse kinematic model using the damped least-squares technique are given in Section 3. Section 4 investigates the performance of our ultrasonic tracking system by comparing with the camera-based tracking reference system. Finally, the discussion and conclusion are given in Section 5.

2. Materials

Figure 1 shows the general configuration of the proposed system. The ultrasonic measurement system, used in our previous work [29,30], is comprised of one ultrasonic transmitter (mobile) and four ultrasonic receivers (anchors). The mobile node is composed of a circuit board (form factor: 4 cm × 6 cm) with a microcontroller and a radio frequency (RF) module, along with a separate circuit board (form factor: 4 cm × 3 cm × 1.6 cm) with an ultrasound generator. The displacements of the ultrasonic transmitter measured using the spherical tracking technique were expressed in a global coordinate system, which described trunk position relative to the ground. The *X*-axis was defined as the direction of anterior-posterior direction; the *Y*-axis was made to be in the vertical direction. The third axis of the coordinate system, the *Z*-axis, was determined in such a way, so as to form a right-handed coordinate system.

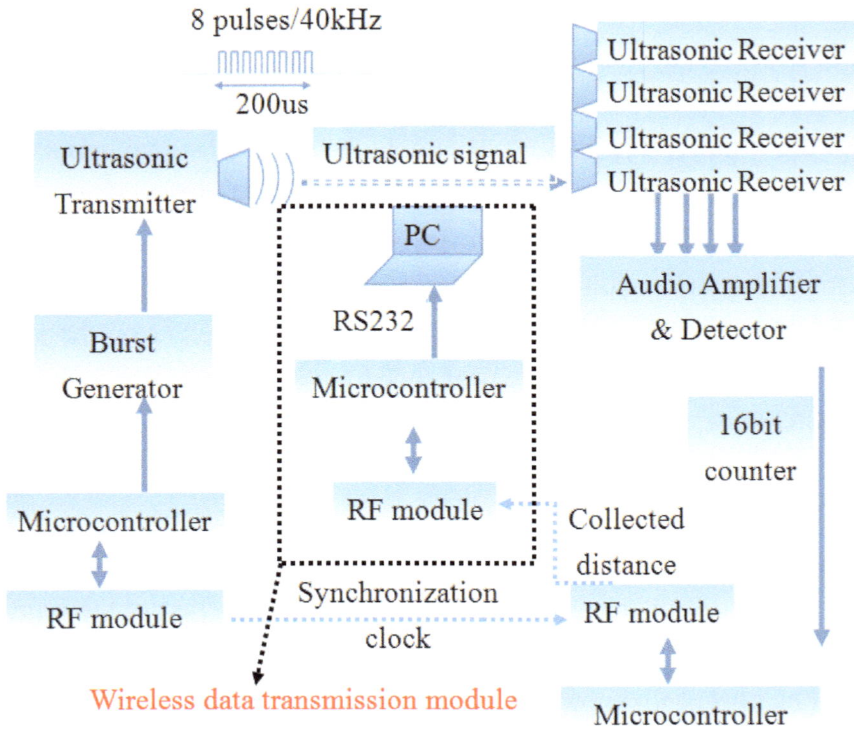

Figure 1. General configuration of the system.

In the proposed system, ultrasonic range measurements are initiated by a periodic trigger input with a pulse duration of 10 μs. Then, the ultrasound transmitter is triggered to produce an ultrasonic burst consisting of 8 pulses with a frequency of 40 kHz. Meanwhile, the RF module on the mobile node is triggered synchronously, thus sending out a data package with the timer starter command (TSC) using the broadcast address to notify the anchors that an ultrasound signal has been transmitted. Once the anchor receives the TSC, it will start its 16-bit counters to record the propagation delay from the mobile to the anchor. The transmission time of the RF signal from the mobile is negligible, since the speed of light is much faster than the speed of ultrasound. The 16-bit counter will stop counting immediately after each of the transmitted bursts is detected by the anchor. Then, the counted steps will be converted to propagation delay by multiplying the time resolution (instruction cycle) of the microcontroller. The device uses the STC12LE5612AD microprocessor with high crystal frequency (Fosc = 22.1184 MHz), which provides high time resolution, resulting in a theoretical resolution of 0.031 mm (340 m/s /(Fosc/2)). From this delay, the distance between the mobile and the anchor can be calculated using:

$$d = t \cdot v_s \tag{1}$$

where d is the distance in meters, t is the propagation delay in seconds and v_s is the speed of ultrasound in air. The ultrasound velocity can be approximated to:

$$v_s = 331.5 + 0.6T_c \tag{2}$$

where T_c is air temperature in degrees Celsius. After collecting all four distances from the transmitter to anchors, the coordinator node forwards all of the collected information to a wireless data transmission module, and then, it is transmitted to a personal computer through an RS232 cable for post-processing.

3. Methods

In this section, we discuss our proposed angle tracking algorithm. We first construct the leg kinematic model using a set of links connected by joints. Then, the Denavit–Hartenberg parameters of the leg model are computed. Horizontal and vertical displacement of the ultrasonic sensor are obtained from the data recorded with a single ultrasound transmitter placed at the center line of the torso using an extended Kalman filter. These displacements, together with known joint constraints during squatting, are then used as input to a damped least-squares-based inverse kinematic model, thus allowing the determination of the corresponding ankle, knee and hip joint flexion/extension angles.

3.1. Biomechanics of the Human Body

The commonly-used method for modeling a human rigid body is based on a sequence of links connected by joints. This model could better represent the movement of any parts on the human body [31]. In order to describe the position and orientation between adjacent links, Denavit and Hartenberg (D-H) in 1955 proposed a systematic notation for assigning right-handed orthogonal coordinate frames to each link [32]. The transformations between two adjacent frames can be described by a 4×4 homogeneous transformation matrix. Then, the transformation is described by four parameters associated with each link, known as D-H parameters ($\boldsymbol{L}_k = (\theta_k, d_k, a_k, \alpha_k)$). The first one is the joint angle θ_k, which is defined as the angle between x^{k-1} and x^k axes about the z^{k-1} axis. The second parameter is d_k, which is the distance between x^{k-1} and x^k axes along the z^{k-1} axis. The third one is a_k, which is defined as the distance between z^{k-1} and z^k axes along the x^k axis. The last parameter is α_k, which is the twist angle between z^{k-1} and z^k axes about the x^k axis.

Figure 2 and Table 1 show the biomechanical model of the human body used in our study and D-H parameters for the leg kinematics, respectively, which correspond to a kinematic chain comprised of three rigid segments (shank (a_1), thigh (a_2) and torso (a_3)).

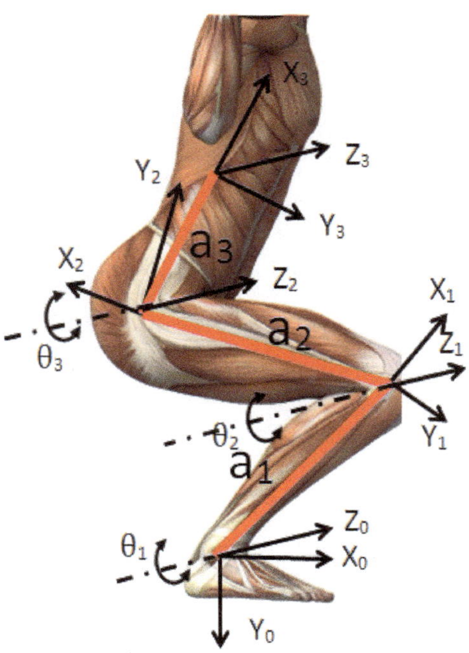

Figure 2. Three degrees-of-freedom kinematic model of the leg.

Table 1. Denavit–Hartenberg parameters of the leg model.

Joint	θ_k	d_k	a_k	α_k
Ankle (L_1)	θ_1	0	a_1	0
Knee (L_2)	θ_2	0	a_2	0
Hip (L_3)	θ_3	0	a_3	0

3.2. Determination of Orientation and Position

Given the D-H model and parameters presented before, we then proceed to investigate the orientation and position of the sensor point, where the ultrasonic sensor will be attached. The orientation R and position $\boldsymbol{p} = [x\ y\ z]^T$ of the torso (or ultrasonic sensor) can be calculated with respect to the ankle frame as follows:

$$\boldsymbol{R} = \begin{bmatrix} C\theta_{123} & -S\theta_{123} & 0 \\ S\theta_{123} & C\theta_{123} & 0 \\ 0 & 0 & 1 \end{bmatrix}$$
$$\boldsymbol{p} = \begin{bmatrix} a_3C\theta_{123} + a_2C\theta_{12} + a_1C\theta_1 \\ a_3S\theta_{123} + a_2S\theta_{12} + a_1S\theta_1 \\ 0 \end{bmatrix} \tag{3}$$

where the orientation angle $\phi = \theta_{123} = \theta_1 + \theta_2 + \theta_3$ and $\theta_{12} = \theta_1 + \theta_2$. θ_1, θ_2 and θ_3 are the ankle, knee, and hip joint angles, respectively.

3.3. Determination of the Ultrasonic Sensor Position

In most cases, it is expected to minimize the number of sensors attached to the human body, not only for unrestrained movement or financial cost, but also for simplifying the calibration process and synchronization issues [1,33]. Therefore, only one mobile target is used in our measurement system. As mentioned in Section 2, we assume that the mobile lying on the position $\boldsymbol{p}$ sends out ultrasound signals to the anchors during the discrete sampling time. The positions of the anchors are known to be $\boldsymbol{p_i} = [x_i\ y_i\ z_i]^T$, respectively. After receiving all of the measurements, the coordinator will compute an estimate of the state (position and velocity) of the mobile using extended Kalman filter (EKF) estimation. We will explain the tracking algorithm in more detail later.

3.3.1. Mobile Target Motion Model

We design a state vector with six components, three Cartesian coordinates $\boldsymbol{p} = [x\ y\ z]^T$ and three velocity components along the three coordinate axes $\dot{\boldsymbol{p}} = [\dot{x}\ \dot{y}\ \dot{z}]^T$, to describe the mobile moving in a 3-dimensional space. In other words, the state of the mobile target on time step k can be expressed as:

$$\boldsymbol{x}(k) = [x(k)\ y(k)\ z(k)\ \dot{x(k)}\ \dot{y(k)}\ \dot{z(k)}]^T \tag{4}$$

Then, the following model is used to describe the motion of the mobile:

$$\boldsymbol{x}(k) = A\boldsymbol{x}(k-1) + B\boldsymbol{\omega}(k-1) \tag{5}$$

where:

$$A = \begin{bmatrix} I_{3\times3} & T \cdot I_{3\times3} \\ O_{3\times3} & I_{3\times3} \end{bmatrix},\ B = \begin{bmatrix} \frac{T^2}{2} \cdot I_{3\times3} \\ T \cdot I_{3\times3} \end{bmatrix} \tag{6}$$

where $I_{3\times3}$ indicates the identity matrix and $O_{3\times3}$ indicates the zero matrix. T is the sampling interval between time step k and $k-1$. $\omega(k-1) = [\omega_x\ \omega_y\ \omega_z]^T$ is a white Gaussian noise with zero mean and covariance matrix $Q_\omega = diag(q_x^2, q_y^2, q_z^2)$. In most cases, q_x, q_y and q_z can be considered as standard deviations of the velocity noise along the *x*-, *y*- and *z*-axes, respectively.

3.3.2. Measurement Model

At time step k, let $d_i(k)$ denote the absolute distance measured at the i-th anchor using the following equation:

$$d_i = ||\boldsymbol{p} - \boldsymbol{p}_i|| + \tilde{d}_i \tag{7}$$

where $\tilde{d}_i$ is the range measurement error and $d_i(k)$ is simplified to d_i. Stacking all of the distance information, we have the measurement model expressed as:

$$D(k) = F(\boldsymbol{x}(k))\boldsymbol{x}(k) + \tilde{D}(k) \tag{8}$$

where:

$$\begin{aligned} D(k) &= \begin{bmatrix} d_1 & d_2 & \cdots & d_n \end{bmatrix}^T \\ F(\boldsymbol{x(k)}) &= \begin{bmatrix} F_1 & F_2 & \cdots & F_n \end{bmatrix}^T \\ F_i &= \begin{bmatrix} \frac{\partial d_i}{\partial x} & \frac{\partial d_i}{\partial y} & \frac{\partial d_i}{\partial z} & 0 & 0 & 0 \end{bmatrix} \\ \tilde{D}(k) &= \begin{bmatrix} \tilde{d}_1 & \tilde{d}_2 & \cdots & \tilde{d}_n \end{bmatrix}^T \end{aligned} \tag{9}$$

n is the number of anchors, and $\tilde{D}(k) \sim N(0, R(k))$ are the measurement errors. $R(k) = diag(e_1^2, e_2^2, \cdots, e_n^2)$ is the covariance matrix of measurement errors. e_i is always considered as the standard deviation of the measurement error of anchor i.

3.3.3. Tracking Using the Extended Kalman Filter

The basic idea of the extended Kalman filter is that the filter uses prior knowledge of all distance information to predict and produce an estimate of the position for where the mobile might be in the next time step. Once the next distance sample arrives, the filter first corrects the state based on the actual distance information [34,35]. In each iteration, the current state $\boldsymbol{x}(k-1)$ is used to predict the velocity and position at the time step $k-1$. The error covariance $\hat{P}(k)$ is also predicted using the state space model in Equation (5). This is followed by the computation of the Kalman filter gain $K(k)$. Once $K(k)$ is computed, the range measurements $D(k)$ are calibrated by the predicted state of the mobile and are used to update the mean and covariance of the state $\boldsymbol{x}(k)$ and $P(k)$, respectively. The standard expression is given below [36]:

- Prediction:

$$\begin{aligned} \hat{\boldsymbol{x}}(k) &= A\boldsymbol{x}(k-1) \\ \hat{P}(k) &= AP(k-1)A^T + BQ_\omega B^T \end{aligned} \tag{10}$$

- Update:

$$\begin{aligned} \Theta(k) &= D(k) - F(\hat{\boldsymbol{x}}(k))\hat{\boldsymbol{x}}(k) \\ S(k) &= F(\hat{\boldsymbol{x}}(k))\hat{P}(k)F^T(\hat{\boldsymbol{x}}(k)) + R(k) \\ K(k) &= \hat{P}(k)F^T(\hat{\boldsymbol{x}}(k))S(k)^{-1} \\ \boldsymbol{x}(k) &= \hat{\boldsymbol{x}}(k) + K(k)\Theta(k) \\ P(k) &= \hat{P}(k) - K(k)S(k)K(k)^T \end{aligned} \tag{11}$$

Table 2. Range of motion and constraints imposed in the system.

	Upper Limit (deg)	**Lower Limit (deg)**
θ_1 ankle joint	150	80
θ_2 knee joint	5	−120
θ_3 hip joint	110	−15

Once the state of the mobile $\boldsymbol{x}(k)$ is computed at time step k, the position information is used to obtain the joint angles when subjects are doing the squat exercise. Since the displacement along lateral direction does not change, the squat exercise is always planar. Therefore, in the rest of our paper, only horizontal and vertical displacement (Cartesian coordinates) are considered as input to the inverse kinematic model.

3.4. Inverse Kinematic Model

The direct kinematic mapping of interest for the system is given by:

$$\begin{aligned} \mathbf{p} &= f(\boldsymbol{\theta}) \\ \dot{\mathbf{p}} &= J(\boldsymbol{\theta})\dot{\boldsymbol{\theta}} \end{aligned} \tag{12}$$

where $\boldsymbol{\theta} = [\theta_1\ \theta_2\ \theta_3]^T$, $\dot{\boldsymbol{\theta}}$ and $\dot{\boldsymbol{p}}$ are joint velocities and Cartesian velocities, respectively, f is the nonlinear function described in Equation (3) and J is a Jacobian matrix defined as $J(\theta) = [\frac{\partial f}{\partial \theta_1}\ \frac{\partial f}{\partial \theta_2}\ \frac{\partial f}{\partial \theta_3}]^T$. The system under study is kinematically redundant during a planar squat exercise, since there is a 3-degrees-of-freedom (DoF) movement with only 2 displacements known.

A general solution in terms of a generalized inverse of the Jacobian matrix is [37,38]:

$$\dot{\boldsymbol{\theta}} = F(\boldsymbol{p} - \hat{\boldsymbol{p}}_e) + (I - FJ)\Phi \tag{13}$$

where $\hat{\boldsymbol{p}}_e$ is the estimated position obtained from the forward kinematic model and $\boldsymbol{p}$ is the Cartesian coordinates (target position) acquired by the ultrasonic sensor using EKF. F is a generalized inverse matrix of the Jacobian matrix, $(I - FJ)$ is the matrix that projects the vector Φ in the null space of J, $J(\theta)$ is simplified to J and:

$$\Phi = \xi_0 \left(\frac{\partial W(\boldsymbol{\theta})}{\partial \boldsymbol{\theta}} \right)^T \tag{14}$$

where $\xi_0 > 0$ and $W(\boldsymbol{\theta})$ is a scalar (objective) function of joint variables. Since the solution moves along the direction of the gradient of $W(\boldsymbol{\theta})$, it attempts to locally maximize subject to the kinematic constraint. The objective function is based on a limited joint range from lower $\boldsymbol{\theta}_{min}$ to upper $\boldsymbol{\theta}_{max}$ limits, shown in Table 2 [1]:

$$\begin{aligned} W(\boldsymbol{\theta}) &= -\frac{1}{6}\sum_{i=1}^{3} \left(\frac{\theta_i - \bar{\theta}_i}{\theta_{i\max} - \theta_{i\min}} \right)^2 \\ \bar{\theta}_i &= \frac{(\theta_{i\max} + \theta_{i\min})}{2} \end{aligned} \tag{15}$$

where $\bar{\theta}_i$ is the middle value of the joint range; thus, redundancy is exploited to keep the joint variables $\boldsymbol{\theta}$ as close as possible to the center of their ranges [39].

3.5. Damped Least-Squares Scheme

The system is close to a singular point near/at the beginning and ending of the squat cycle, since the Jacobian matrix does not have full rank [40]. The unavoidable measurement errors in displacement estimations using the EKF algorithm will produce some unreachable targets, which leads to oscillations and shaking of the inverse kinematic problem. An alternative solution overcoming the problem of inverting differential kinematics in the neighborhood of a singularity is damped least-squares inverse (DLS):

$$F = J^T(JJ^T + \kappa^2 I)^{-1} \tag{16}$$

where κ is a damping factor that ensures continuity and good shaping of the solution near/at singularity and for out-of-reach targets. The damping factor should be critically determined for obtaining good performance over the entire workspace as follows:

$$\kappa^2 = \begin{cases} 0 & when\ \sigma_{min} \geq \varepsilon \\ \left(1 - \left(\frac{\sigma_{min}}{\varepsilon}\right)^2\right)\kappa_{\max}^2 & otherwise \end{cases} \tag{17}$$

where κ_{max} is used to adjust the solution in the neighborhood of a singularity, σ_{min} is the estimate of the smallest singular value and ε is defined as the size of the singular region [41].

To gain more insight into the features of the DLS, the singular value decomposition [40,41] of the Jacobian matrix is introduced, which is:

$$J = \sum_{i=1}^{3} \sigma_i \boldsymbol{u_i} \boldsymbol{v_i^T} \tag{18}$$

where $\boldsymbol{u_i}$ and $\boldsymbol{v_i}$ are the output and input singular vectors and σ_i are the singular values. Thus, the solution of DLS can be expressed in the form of:

$$F = J^T(JJ^T + \kappa^2 I)^{-1} = \sum_{i=1}^{3} \frac{\sigma_i}{\sigma_i^2 + \kappa^2} \boldsymbol{v_i} \boldsymbol{u_i^T} \tag{19}$$

From Equation (19), the components for which $\sigma_i \gg \kappa$ are little affected by the damping factor, since for large σ_i, $\sigma_i/(\sigma_i^2 + \kappa^2) \approx 1/\sigma_i$. On the contrary, when σ_i is of a similar or even smaller magnitude as κ, then the associated component of the solution is driven to zero by the factor σ_i/κ^2. Therefore, the DLS method tends to effectively smooth out the performance in the neighborhood of singularities.

Once the solution of Equation (13) is found, the joint angles can be initially computed using previous joint angles and the estimated joint velocities:

$$\boldsymbol{\theta}_{k+1} = \boldsymbol{\theta}_k + \dot{\boldsymbol{\theta}} \tag{20}$$

The above joint angles estimation can be further refined iteratively: For $j = 0, 1, \cdots$, compute:

$$\begin{aligned} \boldsymbol{p}_{k+1}^{j+1} &= f(\boldsymbol{\theta}_{k+1}^{j}) \\ \dot{\boldsymbol{\theta}}_{k+1}^{j+1} &= F_{k+1}^{j}(\boldsymbol{p} - \hat{\boldsymbol{p}}_{k+1}^{j+1}) + (I - F_{k+1}^{j} J)\Phi_{k+1}^{j} \end{aligned} \tag{21}$$

until:

$$\|\boldsymbol{p} - \boldsymbol{p}_{k+1}^{j+1}\| \leq \xi \tag{22}$$

and $\xi \geq 0$ is a predefined threshold.

3.6. Parameters Identification

3.6.1. Noise Statistics in the Kalman Filter

Using the measurement model in Equation (7), we first conduct a static measurement to estimate the covariance matrix of measurement error, $R(k)$. It is reasonable to assume that all of the sensors have independent distributed noise. We take a sensor and run M tests with different distances between the receiver and transmitter. The actual distance for test i, r_i, is known, and there are N measurement samples m_i^j collected for each r_i, where $j = 1, \cdots, N$.

Therefore, straightforward calculations lead to the estimation of the mean and variance of the measurement errors:

$$\begin{aligned} u &= \frac{1}{MN} \sum_{i=1}^{M} \sum_{j=1}^{N} (m_i^j - r_i) \\ e^2 &= \frac{1}{M(N-1)} \sum_{i=1}^{M} \sum_{j=1}^{N} (m_i^j - u)^2 \end{aligned} \tag{23}$$

3.6.2. Parameter Identification of the Inverse Kinematic Model

The damping factor is computed in (17) with $\varepsilon = 0.04$ and $\kappa_{max} = 0.04$. These values were determined by calculating the smallest singular value among all of the near singular Jacobian matrices J, where the subjects under test were close to an upright posture.

4. Experimental Validation

4.1. Experimental Setup

To compare the performance of the proposed measurement system with a conventional measurement system, experiments were conducted in a motion analysis lab with eight high speed cameras in the School of Mechanical and Aerospace Engineering, Nanyang Technological University. Eight healthy subjects (mean age 24.5 ± 1.73 years, mean weight 61.3 ± 6.40 kg, mean height 171.8 ± 5.80 cm) volunteered in the experiments. All subjects were required to wear a ultrasound transmitter on the torso and five reflective markers, as shown in Figure 3, to perform an unrestrained squat exercise at a self-selected speed, keeping their feet flat on the ground. Each subject was asked to repeat the squat exercise five times. The segment length and joint angles with respect to the kinematic model in Figure 2 were calculated using the instantaneous positions of the above-mentioned five reference markers.

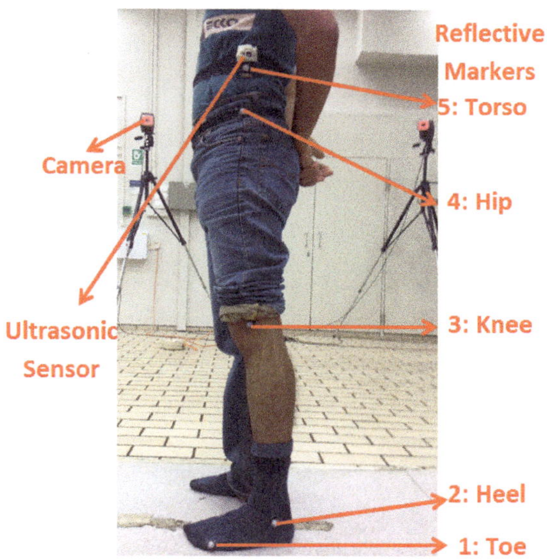

Figure 3. Wireless unit with embedded ultrasonic sensor placed on the lateral side of the trunk and five reflective markers fixed on the toe, heel, knee, hip and torso for the optical tracking reference system.

There were four anchors used in our experiment with positions ($\boldsymbol{p}_1 = [0\ 0\ 0]^T$, $\boldsymbol{p}_2 = [0.324m\ 0\ 0]^T$, $\boldsymbol{p}_3 = [0.324m\ 0.230m\ 0]^T$, $\boldsymbol{p}_4 = [0\ 0.230m\ 0]^T$). In our method, only one ultrasonic sensor (transmitter) is needed, which is to be attached to the foot, which minimizes the discomfort for users and avoids complex calibration procedures and synchronization issues. All of the data transmission between anchors, coordinator and transmitter are done wirelessly through the RF module. Therefore, it does not restrict the movement of the subjects.

The synchronization between the proposed ultrasonic and reference camera measurement system was done by maximizing the correlation between the horizontal displacements of the torso point estimated by both systems. The proposed ultrasonic sensor data were acquired at 50 Hz. Data from the reference system were acquired at 200 Hz. All data were low-pass filtered by a second order low-pass Butterworth filter at 5 Hz. The positions of ankle, knee, hip and torso markers and related joint angles estimated from the motion capture system were considered as reference data and used for the evaluation of the ultrasound-based system.

4.2. Data Analysis

To verify the joint angle estimation accuracy, which is implemented in the hardware platform, as described in Section 2, the mean and standard deviation value of the difference between the parameters extracted with the optical tracking reference system, together with root mean square error (RMSE), were computed. Pearson's correlation coefficient (PCC) is also used in gauging the degree of agreement between these two systems. Furthermore, a statistical method, Bland–Altman analysis [42], for assessing the level of agreement between the proposed and reference systems was applied. The limits of agreement refer to the mean $\pm$ 1.96-times the standard deviation of the difference between the two systems.

4.3. Results

In Figure 4, a randomly selected trial for the estimated ankle, knee and hip joint flexion and extension angles during a squat exercise is shown. Figure 4 shows the effect of different damping factors on the joint angle estimation. We compared with the pseudo-inverse method ($\kappa = 0$) and DLS method. It is shown clearly that the pseudo-inverse method performs poorly because of instability near singularities (σ_i approaches zero). However, the DLS method tends to act similarly to the pseudo-inverse method away from singularities and effectively smooths out the performance of the pseudo-inverse method in the neighborhood of singularities. All of the joint angles estimated by the DLS method show a great correspondence with the reference system.

For the clarity of the Bland–Altman plots, we randomly selected 4 trials to draw these figures, as shown in Figure 5. The data plotted in Figure 5I show that all points lie along the line of equality, which means a high degree of agreement between the measurements. The PCC values between the two system are also calculated for the data in Figure 5I. For all of these joint angles, the correlations are all equal to or larger than 0.99 ($p < 0.001$), where the p-value is the probability that the measurements by the two methods are not linearly related. As the p-value is extremely small, we can safely conclude that the measurements of joint angles by these two methods are significant. Figure 5II shows the Bland–Altman plot for the selected data by plotting the difference between the two systems against the mean values of ankle, knee and hip joint angles. This plot is used to gauge the limits of agreement between the two methods. The 95% confidence intervals were determined to be $(-6.48°, 6.56°)$ for ankle angles, $(-5.34°, -0.32°)$ for knee angles and $(-1.42°, 4.7°)$ for hip angles.

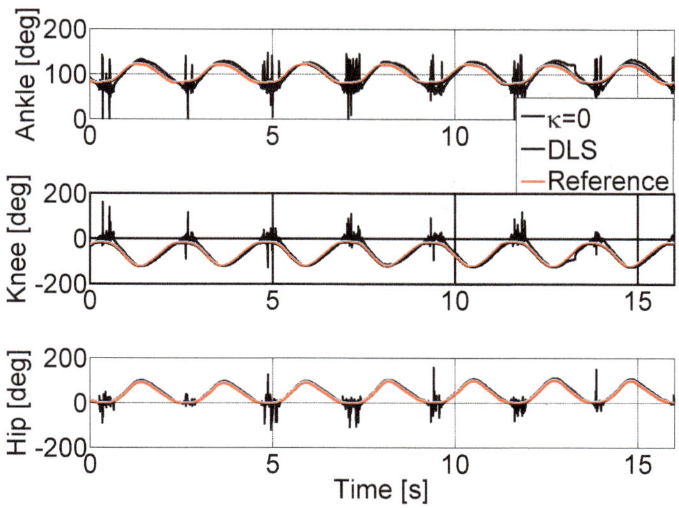

Figure 4. The joint angles obtained from both ultrasonic and optical tracking system during a 15-s consecutive planar squat exercise. The black line makes use of the pseudo-inverse method, that is $\kappa = 0$. The blue line is the damping least-squares method. The red line is a reference result.

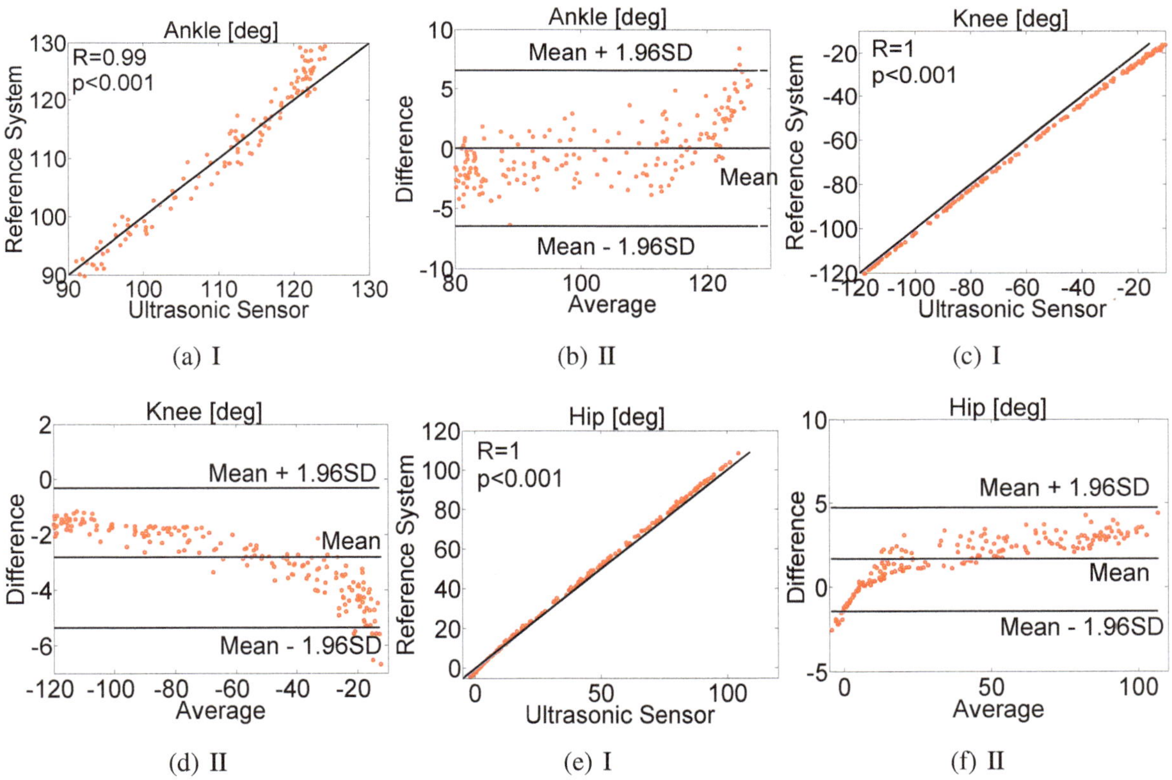

Figure 5. (I) Joint angles estimated by the proposed ultrasonic and reference camera-based measurement system, plotted with the line of equality. (**a**) Ankle joint angle; (**c**) knee joint angle and (**e**) hip joint angle. (II) Bland–Altman plot with the mean and difference between two values estimated by the proposed ultrasonic and reference camera-based measurement system, plotted with the limits of agreement, which are the average difference $\pm$ 1.96-times the standard deviation of the difference. (**b**) Ankle joint angle; (**d**) knee joint angle and (**f**) hip joint angle.

All of the collected data from 8 subjects are used to compute the mean and standard deviation of the difference, RMSE, PCC values and the limits of agreement, which are reported in Table 3. Horizontal displacement was obtained with a difference of 1.08 ± 6.07 mm, an RMSE of 6.50 ± 2.08 mm and a high PCC value of 0.94 ± 0.04. The mean difference and standard deviation of the vertical displacement was 0.75 ± 9.65 mm, which had a high similarity of 0.94 ± 0.06 with the results of the reference system. For all of the collected data, the joint angles of interest were estimated with a mean difference from the reference data of 2.85° (RMSE) and with high similarities greater than 0.99, where all p-values are smaller than (0.001). For all three estimated joint angles, the hip joint has the smallest error, and consequently, tighter lower and upper limits of agreement, closer to zero than the ankle and hip joint. In summary, the great agreement of all joint angles estimated by the two systems is enough for us to be confident that the proposed measurement system can be used for further clinical applications.

Table 3. Comparison of errors in measurements for the proposed and reference system. The mean difference and relevant standard deviation (std) together with Pearson's correlation coefficient (PCC) values are shown. Bland–Altman limits of agreement are also reported. HD: Horizontal Displacement; VD: Vertical Displacement.

	Difference		**RMSE**		**PCC**		**Limits of Agreement**	
	Mean	**Std**	**Mean**	**Std**	**Mean**	**Std**	**Lower**	**Upper**
HD (mm)	1.08	6.07	6.50	2.08	0.94	0.04	−10.82	12.97
VD (mm)	0.75	9.65	9.90	4.79	0.94	0.06	−18.17	19.67
ankle (deg)	0.57	3.10	3.10	1.10	0.99	0.01	−6.07	6.07
knee (deg)	−2.46	1.82	3.28	0.39	1.00	0.00	−6.03	1.11
hip (deg)	1.49	1.56	2.18	0.21	1.00	0.00	−1.55	4.54

5. Discussion and Conclusions

This study has demonstrated that one ultrasound transmitter (attached to the human body) and four receivers can be used to obtain high accuracy ankle, knee, hip joint flexion/extension angles estimates. The extended Kalman filter is applied to estimate the displacements in the vertical and horizontal direction of the ultrasonic sensor, and then, the recorded displacements together with known joint constraints are used to estimate the joint angles of the trunk using the damped least-squares-based technique for the singularity avoidance problem of redundant systems.

The proposed measurement system was compared to a camera-based system, and the results were averaged over eight subjects. These metrics (difference, RMSE, PCC values and limits of agreement) are shown in Table 3 to illustrate the differences between these two systems. The experimental results showed that the horizontal and vertical displacement of the ultrasonic sensor and the joint angles (hip, knee and ankle) can be estimated with relatively small differences. Bland–Altman analysis indicated a great agreement between these two systems, with the PCC value higher than 0.99. The average RMSD of the estimate of the lower extremity joint angles is $2.85^\circ \pm 0.57^\circ$, which is similar to what has been obtained in other studies using inertial sensors to estimate these joint angles [1,13].

Although the positive results showed the feasibility of applying such a system for in-home monitoring, there is an issue to be addressed in further research, that is how to deal with the multipath propagation. All of the experiments in this study are conducted under the line-of-sight condition, where the ultrasonic transmitter faces all of the receivers without any obstacles between them. Therefore, for 3D displacements, according to the spherical positioning technique, a minimum of four anchors with known positions is required. The method used in our experiment to mitigate the multipath propagation is by setting an inhibition time, *i.e.*, the ultrasound detector will be disabled within the inhibition time to detect an ultrasound signal. Then, it will be enabled after this inhibit duration. Another possible solution is that we can use more receivers, which can not only account for multipath propagation, but also increase the measurement volume and accuracy of the proposed system [19].

A key issue in the system is the determination of the damping factor in Equation (16), since it plays an important role in finding a solution to the inverse kinematic model. Larger κ makes the solution for joint velocities behave well in the neighborhood of a singularity, but results in low tracking accuracy and a low convergence rate. Small values of the damping factor give accurate solutions, but not are robust to the singular or near-singular region. Hence, it is essential to select suitable values for the damping factor.

In summary, this paper presents an unrestrained measurement system based on a wearable wireless ultrasonic sensor network to track the lower limb joint and trunk kinematics during a squat exercise with only one ultrasonic sensor attached to the trunk. The camera-based tracking reference system was used to benchmark the performance of the proposed ultrasonic motion analysis system. It demonstrates acceptable accuracy for further applications. Additionally, the proposed system is easy to wear, to use and much less expensive than current camera systems. It does not restrict the movement of patients or subjects with bulky cables.

Acknowledgments

We thank Qu Xingda from School of Mechanical & Aerospace Engineering for his technical support and all of the subjects who participated in this project.

Author Contributions

The work presented here was carried out in collaboration between all authors. Cheong Boon Soh, Erry Gunawan, and Kay-Soon Low proposed the research theme with regular feedbacks of suggestions/ideas during the weekly research meetings . Yongbin Qi and Cheong Boon Soh co-designed the methods and implementation. Yongbin Qi and Rijil Thomas co-worked on associated data collection and carried out the laboratory experiments. All authors have contributed, reviewed and improved the manuscript.

Conflicts of Interest

The authors declare no conflict of interest.

References

1. Bonnet, V.; Mazza, C.; Fraisse, P.; Cappozzo, A. Real-Time Estimate of Body Kinematics during a Planar Squat Task Using a Single Inertial Measurement Unit. *IEEE Trans. Biomed. Eng.* **2013**, *60*, 1920–1926.
2. Fry, A.C.; Smith, J.C.; Schilling, B.K. Effect of knee position on hip and knee torques during the barbell squat. *J. Strength Cond. Res.* **2003**, *17*, 629–633.
3. Bonnet, V.; Mazza, C.; Fraisse, P.; Cappozzo, A. A least-squares identification algorithm for estimating squat exercise mechanics using a single inertial measurement unit. *J. Biomech.* **2012**, *45*, 1472–1477.
4. Cappozzo, A.; Della Croce, U.; Leardini, A.; Chiari, L. Human movement analysis using stereophotogrammetry: Part 1: Theoretical background. *Gait Posture* **2005**, *21*, 186–196.
5. Vlasic, D.; Adelsberger, R.; Vannucci, G.; Barnwell, J.; Gross, M.; Matusik, W.; Popović, J. Practical Motion Capture in Everyday Surroundings; In *ACM Transactions on Graphics (TOG)—Proceedings of ACM SIGGRAPH 2007*; ACM: New York, NY, USA, 2007; Volume 26.
6. Lange, B.; Chang, C.Y.; Suma, E.; Newman, B.; Rizzo, A.S.; Bolas, M. Development and evaluation of low cost game-based balance rehabilitation tool using the Microsoft Kinect sensor. *IEEE Eng. Med. Biol. Soc. Conf. Proc.* **2011**, *2011*, 1831–1834.
7. Clark, R.A.; Pua, Y.H.; Fortin, K.; Ritchie, C.; Webster, K.E.; Denehy, L.; Bryant, A.L. Validity of the Microsoft Kinect for assessment of postural control. *Gait Posture* **2012**, *36*, 372–377.
8. Lee, G.X.; Low, K.S. A Factorized Quaternion Approach to Determine the Arm Motions Using Triaxial Accelerometers with Anatomical and Sensor Constraints. *IEEE Trans. Instrum. Meas.* **2012**, *61*, 1793–1802.
9. Brennan, A.; Zhang, J.; Deluzio, K.; Li, Q. Quantification of inertial sensor-based 3D joint angle measurement accuracy using an instrumented gimbal. *Gait Posture* **2011**, *34*, 320–323.
10. Picerno, P.; Cereatti, A.; Cappozzo, A. Joint kinematics estimate using wearable inertial and magnetic sensing modules. *Gait Posture* **2008**, *28*, 588–595.
11. Brandes, M.; Zijlstra, W.; Heikens, S.; van Lummel, R.; Rosenbaum, D. Accelerometry based assessment of gait parameters in children. *Gait Posture* **2006**, *24*, 482–486.
12. Lai, D.; Begg, R.; Charry, E.; Palaniswami, M.; Hill, K. Measuring toe clearance using a wireless inertial sensing device. In Proceedings of the International Conference on Intelligent Sensors, Sensor Networks and Information Processing, 2008 (ISSNIP 2008), Sydney, NSW, Austrilia, 15–18 December 2008; pp. 375–380.
13. McCamley, J.; Donati, M.; Grimpampi, E.; Mazzà, C. An enhanced estimate of initial contact and final contact instants of time using lower trunk inertial sensor data. *Gait Posture* **2012**, *36*, 316–318.
14. Petruska, A.J.; Meek, S.G. Non-drifting limb angle measurement relative to the gravitational vector during dynamic motions using accelerometers and rate gyros. In Proceedings of the 2011 IEEE/RSJ International Conference on Intellignet Robots and Systems, San Francisco, CA, USA, 25–30 September 2011; pp. 3632–3637.

15. Panahandeh, G.; Mohammadiha, N.; Leijon, A.; Händel, P. Continuous hidden Markov model for pedestrian activity classification and gait analysis. *IEEE Trans. Instrum. Meas.* **2013**, *62*, 1073–1083.
16. Ruiz, A.R.J.; Granja, F.S.; Prieto Honorato, J.C.; Rosas, J.I.G. Accurate pedestrian indoor navigation by tightly coupling foot-mounted IMU and RFID measurements. *IEEE Trans. Instrum. Meas.* **2012**, *61*, 178–189.
17. Corrales, J.A.; Candelas, F.; Torres, F. Hybrid tracking of human operators using IMU/UWB data fusion by a Kalman filter. In Proceedings of the 2008 3rd ACM/IEEE International Conference on Human-Robot Interaction (HRI 2008), Amsterdam, The Netherlands, 12–15 March 2008; pp. 193–200.
18. Qi, Y.; Soh, C.B.; Gunawan, E.; Low, K.S.; Maskooki, A. Using wearable UWB radios to measure foot clearance during walking. In Proceedings of the 2013 35th Annual International Conference of the IEEE Engineering in Medicine and Biology Society (EMBC), Osaka, Japan, 3–7 July 2013; pp. 5199–5202.
19. Qi, Y.; Soh, C.B.; Gunawan, E.; Low, K.S.; Maskooki, A. An Accurate 3D UWB Hyperbolic Localization in Indoor Multipath Environment Using Iterative Taylor-Series Estimation. In Proceedings of the 2012 IEEE Vehicular Technology Conference (VTC Fall), Quebec City, QC, Canada, 3–6 September 2012.
20. Shaban, H.; El-Nasr, M.; Buehrer, R. Toward a highly accurate ambulatory system for clinical gait analysis via UWB radios. *IEEE Trans. Inf. Technol. Biomed.* **2010**, *14*, 284–291.
21. Qi, Y.; Soh, C.B.; Gunawan, E.; Low, K.S.; Maskooki, A. Measurement of knee flexion/extension angle using wearable UWB radios. In Proceedings of the 2013 35th Annual International Conference of the IEEE Engineering in Medicine and Biology Society (EMBC), Osaka, Japan, 3–7 July 2013; pp. 7213–7216.
22. Qi, Y.; Soh, C.B.; Gunawan, E.; Low, K.S.; Maskooki, A. A Novel Approach to Joint Flexion/Extension Angles Measurement Based on Wearable UWB Radios. *IEEE J. Biomed. Health Inform.* **2013**, *18*, 300–308.
23. Carbonelli, C.; Mengali, U. Synchronization algorithms for UWB signals. *IEEE Trans. Commun.* **2006**, *54*, 329–338.
24. Zhou, Y.; Law, C.; Xia, J. Ultra low-power UWB-RFID system for precise location-aware applications. In Proceedings of the 2012 IEEE Wireless Communications and Networking Conference Workshops (WCNCW), Paris, France, 1 April 2012; pp. 154–158.
25. Qi, Y.; Soh, C.B.; Gunawan, E.; Low, K.S. Ambulatory measurement of foot kinematics using wearable ultrasonic sensors. In Proceedings of the 2014 36th Annual International Conference of the IEEE Engineering in Medicine and Biology Society (EMBC), Chicago, IL, USA, 26–30 August 2014.
26. Wahab, Y.; Zayegh, A.; Bakar, N. Foot-to-ground Clearance Measurement: Analysis and Modelling of MEMS Applicability. *Int. J. Res. Rev. Signal Acquis. Process. (IJRRSAP)* **2012**, *2*, 154–158.
27. Huitema, R.B.; Hof, A.L.; Postema, K. Ultrasonic motion analysis system—Measurement of temporal and spatial gait parameters. *J. Biomech.* **2002**, *35*, 837–842.

28. Zhou, Y.; Law, C.; Guan, Y.; Chin, F. Indoor elliptical localization based on asynchronous UWB range measurement. *IEEE Trans. Instrum. Meas.* **2011**, *60*, 248–257.
29. Qi, Y.; Soh, C.B.; Gunawan, E.; Low, K.S. Ambulatory Measurement of 3-Dimensional Foot Displacement during Treadmill Walking Using Wearable Wireless Ultrasonic Sensor Network. *IEEE J. Accept. publ. Biomed. Health Inform.* **2015**, *19*, 446–452.
30. Qi, Y.; Soh, C.B.; Gunawan, E.; Low, K.S.; Thomas, R. Estimation of Spatial-Temporal Gait Parameters Using a Low-Cost Ultrasonic Motion Analysis System. *Sensors* **2014**, *14*, 15434–15457.
31. El-Gohary, M.; McNames, J. Shoulder and Elbow Joint Angle Tracking With Inertial Sensors. *IEEE Trans. Biomed. Eng.* **2012**, *59*, 2635–2641.
32. Denavit, J. A kinematic notation for lower-pair mechanisms based on matrices. *J. Applied Mech. Trans. ASME* **1955**, *22*, 215–221.
33. Fong, D.T.P.; Chan, Y.Y. The use of wearable inertial motion sensors in human lower limb biomechanics studies: A systematic review. *Sensors* **2010**, *10*, 11556–11565.
34. Smith, A.; Balakrishnan, H.; Goraczko, M.; Priyantha, N. Tracking moving devices with the cricket location system. In Proceedings of the 2nd International Conference on Mobile Systems, Applications, and Services, Boston, MA, USA, 6–9 June 2004; ACM: New York, NY, USA, 2004; pp. 190–202.
35. Shareef, A.; Zhu, Y. Localization using extended Kalman filters in wireless sensor networks. In *Kalman Filter: Recent Advances and Applications*; InTechOpen: Rijeka, Croatia, 2009.
36. Hartikainen, J.; Solin, A.; Särkkä, S. *Optimal Filtering with Kalman Filters and Smoothers—A Manual for MATLAB Toolbox EKF/UKF*; Aalto University School of Science: Espoo, Finland, 2011.
37. Liegeois, A. Automatic supervisory control of the configuration and behavior of multibody mechanisms. *IEEE Trans. Syst. Man Cybern.* **1977**, *7*, 868–871.
38. Nenchev, D.N. Redundancy resolution through local optimization: A review. *J. Robot. Syst.* **1989**, *6*, 769–798.
39. Sciavicco, L. *Modelling and Control of Robot Manipulators*; Springer: Berlin, Germany, 2000.
40. Buss, S.R.; Kim, J.S. Selectively damped least squares for inverse kinematics. *J. Graph. Gpu Game Tools* **2005**, *10*, 37–49.
41. Chiaverini, S.; Siciliano, B.; Egeland, O. Review of the damped least-squares inverse kinematics with experiments on an industrial robot manipulator. *IEEE Trans. Control Syst. Technol.* **1994**, *2*, 123–134.
42. Martin Bland, J.; Altman, D. Statistical methods for assessing agreement between two methods of clinical measurement. *Lancet* **1986**, *327*, 307–310.

2

Tidal Volume Estimation Using the Blanket Fractal Dimension of the Tracheal Sounds Acquired by Smartphone

Natasa Reljin *, Bersain A. Reyes and Ki H. Chon

Department of Biomedical Engineering, University of Connecticut, 260 Glenbrook Road, Storrs, CT 06269, USA; E-Mails: bersain.reyes@uconn.edu (B.A.R.); kchon@engr.uconn.edu (K.H.C.)

* Author to whom correspondence should be addressed; E-Mail: reljin@engr.uconn.edu

Academic Editor: Vittorio M.N. Passaro

Abstract: In this paper, we propose the use of blanket fractal dimension (BFD) to estimate the tidal volume from smartphone-acquired tracheal sounds. We collected tracheal sounds with a Samsung Galaxy S4 smartphone, from five ($N = 5$) healthy volunteers. Each volunteer performed the experiment six times; first to obtain linear and exponential fitting models, and then to fit new data onto the existing models. Thus, the total number of recordings was 30. The estimated volumes were compared to the true values, obtained with a Respitrace system, which was considered as a reference. Since Shannon entropy (SE) is frequently used as a feature in tracheal sound analyses, we estimated the tidal volume from the same sounds by using SE as well. The evaluation of the performed estimation, using BFD and SE methods, was quantified by the normalized root-mean-squared error (NRMSE). The results show that the BFD outperformed the SE (at least twice smaller NRMSE was obtained). The smallest NRMSE error of 15.877% ± 9.246% (mean ± standard deviation) was obtained with the BFD and exponential model. In addition, it was shown that the fitting curves calculated during the first day of experiments could be successfully used for at least the five following days.

Keywords: blanket fractal dimension; tidal volume; tracheal sounds; smartphone

---

1. Introduction

Tracheal sounds are defined as those that are detected or heard over the extrathoracic part of the trachea [1]. Tracheal sounds are strong, and cover a wide frequency range [2]. As part of respiratory sounds, they play an important role in monitoring respiratory activity, as well as in detection of pulmonary diseases [1–3].

Respiratory activity is one of the vital signs, and as such requires an adequate attention. Tidal volume is one of the parameters for monitoring respiratory activity [4]. It plays an important role for both healthy people and people with respiratory diseases, hence measuring and checking volume's values can be helpful, especially in assessing risky situations involving respiratory failure [4–6]. Tidal volume is defined as the volume of air exchanged in one breath, and is commonly measured at the mouth [1,2,7]. The average value is about 500 mL per breath at rest [2,7]. Various methods exist for measuring the tidal volume, such as spirometry, whole-body plethysmography, inductance plethysmography, and electrocardiography [2,8–10]. However, these methods require the use of specialized equipment, and cannot be easily applied in nonclinical settings. Therefore, there is a need for a miniature monitoring device that can be used in everyday situations and not only in clinical and/or research settings [11]. In addition, with an extensive growth of electronic devices and their computational capabilities, the development of portable tidal volume estimation systems is now possible [12].

Several efforts have been made in the research oriented towards the estimation of tidal volume. In [13], the authors estimated volume by optically tracking reflective markers in three dimensions. Petrovic *et al.* proposed a technique for measuring tidal volumes by using a single fiber-grating sensor [14], while in [15] the authors estimated the tidal volume using Doppler radar signals. Chen *et al.* estimated tidal volume from the energy of the tracheal sounds [6]. To the best of our knowledge, there are no studies exploring the possibility to estimate tidal volume directly from smartphone-acquired tracheal sounds.

Smartphones are widely used nowadays. They have fast microprocessors, large storage capacities and a lot of media capabilities. In addition, the mobility of the smartphones is making them more popular for usage outside the clinics or research facilities, when they can be used for measuring vital signs and health monitoring, as shown in some of the previous works of our research group [16–18].

In this paper, we propose the use of blanket fractal dimension (BFD) for estimating the tidal volume from tracheal sounds acquired by a commercially available Android smartphone. Tracheal sounds, as part of respiratory sounds, are non-stationary and stochastic signals [2,19]. Due to this fact, some past studies investigated and showed successful applications of fractal analysis on tracheal and lung sounds [20–24]. None of these efforts was concerned with the tidal volume estimation using fractal analysis. In this study, we explore the possibility to estimate tidal volume using BFD, which, to the best of our knowledge, was not used for respiratory sound analysis. The estimated volumes were compared to peak-to-peak volumes obtained from a Respitrace signal, which was considered as a reference. In addition, we estimated volumes by obtaining Shannon entropy (SE) from the same tracheal sounds, and compared them to reference volumes. For testing the proposed method and comparing it with SE method, we collected signals from healthy and non-smoker volunteers for six days, for a total of 30 recordings. As a figure of merit, the normalized root-mean-squared errors (NRMSEs) were calculated in both cases. Repeated experiments were performed to investigate if the models for fitting data obtained during the first day of collecting signals could be successfully used on the data from the remaining days.

2. Materials and Methods

2.1. Subjects

Five healthy non-smoker volunteers (four males and one female), with the mean age and standard deviation of 27 ± 7.5 years, weight of 63.5 ± 5 kg, and height 173.2 ± 8.4 cm, were asked to participate in this study. Individuals with previous pneumothorax, chronic respiratory illnesses, and common cold were excluded from the study. This group of participants consisted of students and staff members from the University of Connecticut (UConn, Storrs, CT, USA). All participants signed a consent form approved by the Institutional Review Board of UConn.

2.2. Equipment and Acquisition of the Signals

In this study, two signals were acquired simultaneously: tracheal sounds and Respitrace signal. The tracheal sounds were collected using an acoustical sensor, which contained a subminiature electret microphone BT-21759-000 (Knowles Electronics, Itasca, IL, USA) placed in a plastic bell, which consisted of a conical coupler chamber [25], in accordance to previous findings [26]. The importance of this shape is that it provides an efficient transducer of air pressure fluctuations from the skin over the trachea to the microphone [27]. The acoustic sensor used in this study was developed by our colleagues at the Metropolitan Autonomous University at Mexico City, Mexico, and have been successfully applied for respiratory sound acquisitions [18,25,28]. The acoustic sensor was connected to the audio jack of the Samsung Galaxy S4 smartphone (Samsung Electronics Co., Seoul, Korea). The tracheal sounds were recorded using the built-in audio recorder application (Voice Recorder), with 16-bit per sample and 44.1 kHz sampling rate, and saved in the .wav format. Afterwards, the recorded files were transferred to a personal computer and processed offline using Matlab (R2012a, The Mathworks, Inc., Natick, MA, USA).

The Respitrace (nowadays known as Inductotrace) signal was obtained simultaneously with the tracheal sounds, from two Respibands (Ambulatory Monitoring, Inc., Ardsley, NY, USA), placed over the rib cage and abdomen. Respibands' signals were digitized using 16-bit A/D converter (PowerLab/4SP, ADInstruments, Inc., Dunedin, New Zealand) at 10 kHz sampling rate, using the manufacturer's software (LabChart 7, ADInstruments, Inc.). Prior to every participant's recording, the Respibands were calibrated using a spirometer system (FE141 Spirometer, ADInstruments, Inc.) following the manufacturer's manual, and the corresponding signal was considered as the reference for volume estimation. Calibration errors between Respibands and spirometer were obtained for every recording, and were less than 10%, which is in accordance to the manufacturer's manual.

Experiments were performed in a regular dry lab which was held quiet. Respibands were placed over the participant's rib cage and abdomen, while the acoustical sensor was fixed at the suprasternal notch using a double-sided adhesive ring (BIOPAC Systems, Goleta, CA, USA). The experiment consisted of three stages, and all were performed in standing posture:

1. Participants were asked to breathe through an 800 mL Spirobag (Ambulatory Monitoring, Inc., Ardsley, NY, USA) for about six respiratory cycles;

2. Participants were asked to follow a maneuver that consisted of increasing tidal volumes and then decreasing with each breath, ranging from participant's comfortable lowest to highest volume, while breathing through a paper tube (tube's length: 20 cm, internal diameter: 1.5 cm, external diameter: 2 cm), for approximately 2 min;
3. Participants were asked to repeat the same maneuver as in the second stage while breathing without the tube.

In everyday situations people do not have access to spirometers or Respibands, and the lack of portable and easily accessible device with possibility to control and limit the tidal volume is needed. Thus, in this research, we use a Spirobag, since it is easy to find and carry, and has an almost fixed volume (800 mL). The exact volume of the bag changes at each volunteers' breathe. Hence, we used the Respitrace system as reference in order to know this volume, since the use of spirometer with a bag was practically prohibited in the experimental setup.

Since breathing through a tube adds some resistance to the respiratory tract and changes the natural way of breathing, one of the objectives was to investigate if this apparatus influences the estimation results. This was the reason for recording the third stage of the experiment.

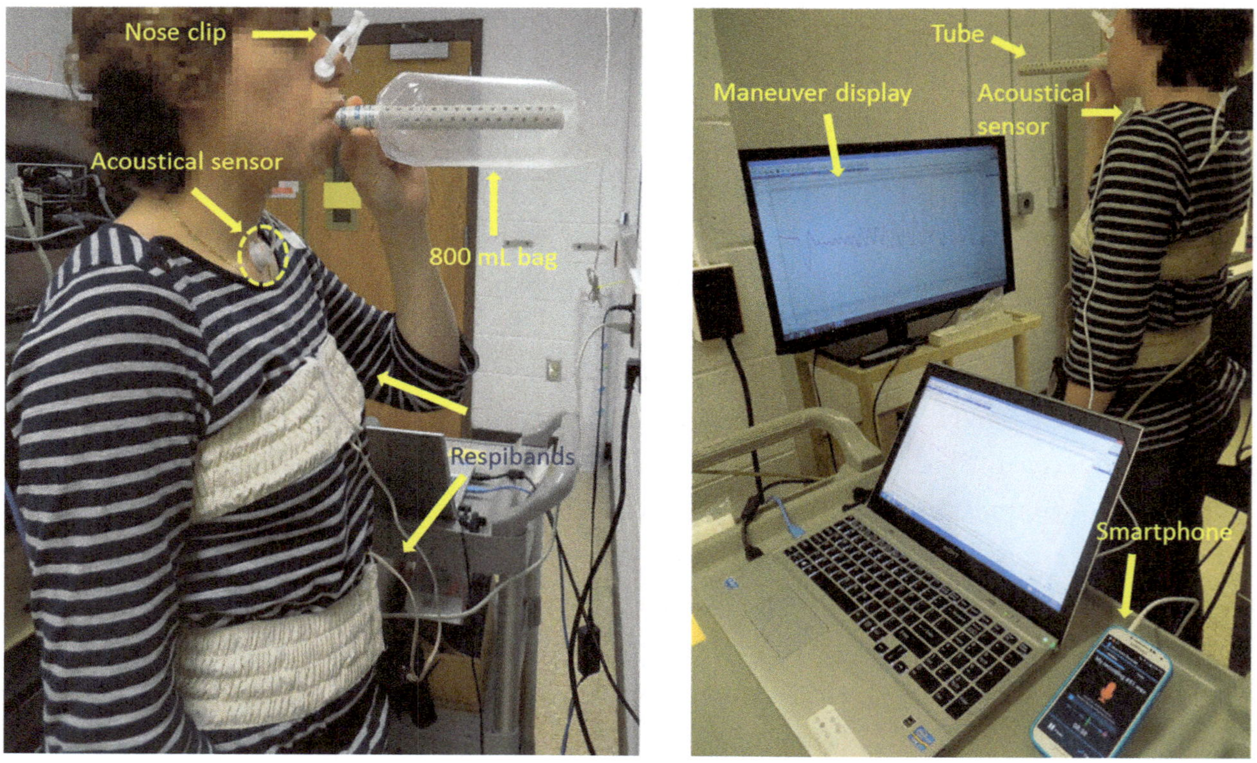

(a) **(b)**

Figure 1. Simultaneous recordings of the tracheal sound (using a smartphone) and the volume signal (using Respibands). (**a**) The participant is breathing through 800 mL bag; (**b**) The participant is breathing through a tube while performing the respiratory maneuver.

In all three stages, initial and final apnea phases of approximately 5 s were acquired for automatic alignment purposes between the two recordings, as well as for recording the ambient noise levels. In the last two stages, after the initial apnea, participants were instructed to take a forced respiration cycle before performing the maneuver. In order to provide the visual feedback during the second and the third stage, the volume signal was displayed on a 40” monitor, placed in front of the participant. During the experiment, nose clips (MLA1008, ADInstruments, Inc.) were used to clamp the nostrils. An example of the set-up of the experiment is shown in Figure 1. Figure 1a depicts the first stage of the experiment, when the 800 mL bag was used, while Figure 1b shows the breathing maneuver through a tube (the second stage of the experiment).

2.3. Data Processing

Figure 2 shows the flowchart of the data processing steps. The acquired tracheal sounds were first downsampled from 44.1 kHz to 6.3 kHz, and then digitally filtered with a 4th order bandpass Butterworth filter with cutoff frequencies 100 and 3000 Hz to minimize the effects of heart sounds and muscle interferences [27,29]. The volume signal was first downsampled from 10 kHz to 5 kHz, and then interpolated to 6.3 kHz in order to achieve the same sampling frequency as the tracheal sounds. Lastly, the volume signal was lowpass filtered at 2 Hz with a 4th order Butterworth filter.

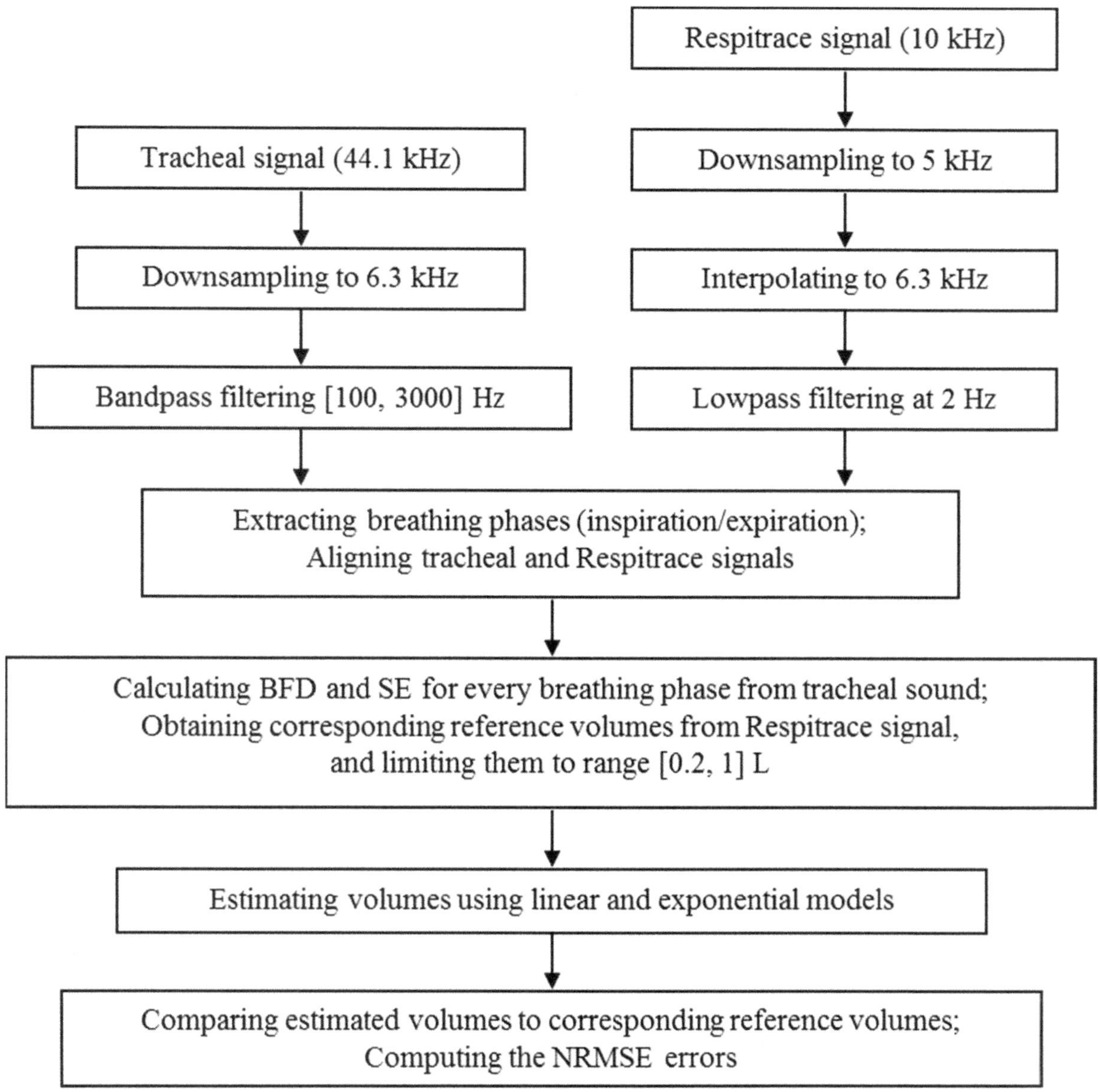

Figure 2. The flowchart showing the steps for tracheal sounds’ and Respitrace signal’s processing.

The automatic extraction of the breathing phases (inspiration/expiration) was performed from the volume signal, by finding its corresponding local maxima and minima during the respiratory maneuver and computing the slope of the volume at each phase [18]. The tracheal sounds and the volume signal were recorded simultaneously, however, due to the different times of pressing the start buttons, the two signals were aligned manually. Figure 3 depicts an example of the filtered, detrended and aligned tracheal sounds and volume signal during the respiratory maneuver.

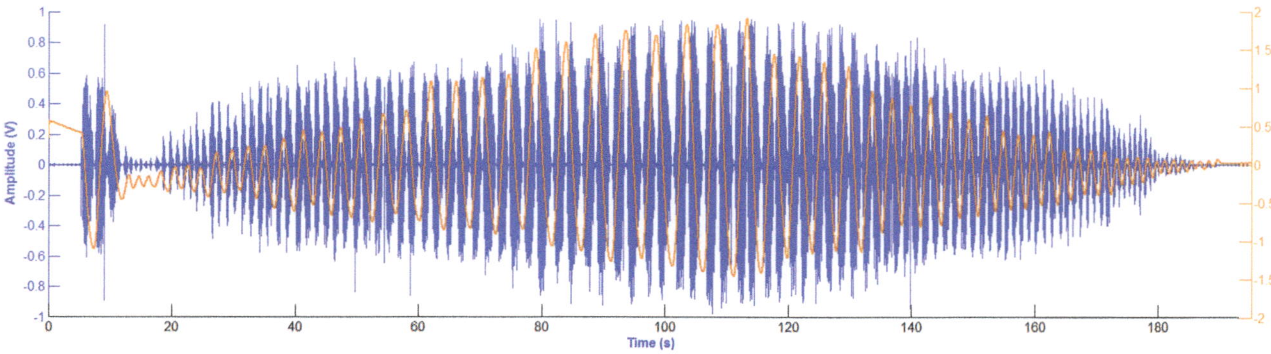

Figure 3. Filtered, detrended and aligned tracheal sounds and volume signal during the respiratory maneuver. Tracheal sound (in volts) is represented in blue, while volume signal (in liters) is in orange.

The volume signal, acquired with the Respibands, was assumed as the reference. For every breathing phase, the absolute volume difference between two consecutive extrema from the volume signal was calculated, and was considered as the true tidal volume value, V_T. Two features were used for estimating the tidal volume from the tracheal sounds acquired by smartphone: blanket fractal dimension (BFD) and the integral of the Shannon entropy (SE). Every breathing phase (inspiration/expiration) from the tracheal sound was represented with one BFD and one SE value. In order to estimate the volume from these features, linear and exponential fitting curves were used. The estimated volumes are defined with the following:

$$V_{est\_l} = a \cdot F + b$$
$$V_{est\_e} = c \cdot e^{d \cdot F} \tag{1}$$

where V_{est_l} and V_{est_e} are the estimated volumes with linear and exponential models, respectively, a, b, c and d are coefficients, and F is the value of the BFD or SE feature computed from the tracheal sounds.

The last step in the data processing is the comparison of the estimated volumes to the corresponding reference volume values, and the evaluation of the performed estimation via computation of the normalized root-mean-squared error (NRMSE) defined as follows:

$$NRMSE = \frac{RMSE}{mean(V_T)} \cdot 100\%$$
$$RMSE = \sqrt{\frac{\sum_{i=1}^{N} (V_T(i) - V_{est}(i))^2}{P}} \tag{2}$$

where V_T is the volume obtained from Respitrace, V_{est} denotes the estimated volume, *i.e.*, V_{est_I} or V_{est_e}, and P is the number of breathing phases during the maneuver.

Shannon entropy is a measure of uncertainty or irregularity of a process [30]. It is one of the features frequently used for analysis of respiratory sounds, and has been successfully applied to airflow estimation in the field of tracheal sound analysis [31]. For a random signal with a probability density function (pdf), p, SE is defined as:

$$SE(p) = -\sum_{i=1}^{M} p_i \cdot \log p_i \tag{3}$$

where M is the number of outcomes of the random variable with pdf p. In this study, pdf is estimated using the method of Parzen's windows with a Gaussian kernel [32,33]. More details on this method can be found in [18,31]. In this study we were concerned with the tidal volume estimation rather than respiratory airflow, and based on the relationship between these two variables over time, the integral of the SE over each corresponding breathing phase was used as feature for tidal volume estimation.

2.4. Blanket Fractal Dimension

Fractals are defined as 'a set having the fractal dimension strictly greater than its integer dimension', and are used to describe non-regular and non-stationary structures [34–36]. There are two types of fractals: natural and deterministic. Natural fractals are structures that could be found in the nature, such as lungs, while deterministic fractals are constructed artificially, by applying predetermined replicating rules (e.g., the Von Koch curve, the Cantor set) [36,37]. Fractal structures may be quantified by fractal dimension, which is a number (usually non-integer) expressing the manner in which the irregular structure replicates itself through different scales [36,37]. Among various fractal dimensions, in this study we used blanket fractal dimension (BFD). The BFD was initially proposed for estimating fractal dimension of digital images (2D signals) [38], and is further extended to 1D signals [39].

In the case of 1D signals, the set of points within maximal distance ε from a curve is considered. Therefore, a strip of width 2ε that surrounds the curve is observed [40]. Blanket method creates the strip around the signal, defined by the upper and lower limiting lines, defined as follows [39]:

$$\begin{aligned} u_\varepsilon(i) &= \max\left\{u_{\varepsilon-1}(i)+1, \max_{|m-i|\le 1} u_{\varepsilon-1}(m)\right\} \\ b_\varepsilon(i) &= \min\left\{b_{\varepsilon-1}(i)-1, \min_{|m-i|\le 1} b_{\varepsilon-1}(m)\right\} \\ u_0(i) &= b_0(i) = x(i) \end{aligned} \tag{4}$$

where $x(i)$ represents the observed 1D signal, $u_\varepsilon(i)$ and $b_\varepsilon(i)$ are the upper and lower lines, respectively, i is the current sample of the signal, m denotes samples within the window around the current sample of the signal, and ε is the predefined maximal distance of upper/lower line from the signal. As can be noted from Equation (4), the upper/lower line is always calculated for the three consecutive samples: $i-1$, i, and $i+1$.

The area of the strip between upper and lower lines is defined as:

$$A_\varepsilon = \sum_i \left\{u_\varepsilon(i) - b_\varepsilon(i)\right\} \tag{5}$$

from which the length of the curve x can be estimated as [39]:

$$L(\varepsilon)=\frac{A_{\varepsilon}-A_{\varepsilon-1}}{2} \tag{6}$$

On the other hand, the length of the curve follows the power law [36]:

$$L(\varepsilon)=C\cdot\varepsilon^{1-D} \tag{7}$$

where C is the constant and D is the blanket fractal dimension (BFD). By combining Equations (6) and (7), and using the least square approximation, blanket fractal dimension is calculated.

3. Results

All five participants performed the experiments described in Section 2.2 six times in six distinct days, thus creating a database of 30 recordings. The data collected on the first day were used for obtaining the linear and exponential models, while the data from the remaining five days were used for testing the previously obtained models. Each breathing phase, inspiration and expiration, was analyzed separately.

The linear and exponential fitting curves were calculated only from the first stage of the experiment performed during the first day, using two and three points, respectively, when the participant was breathing through an 800 mL bag for about six respiratory cycles. BFD and SE features were calculated from the smartphone acquired tracheal sounds, while the reference volume values were obtained from the Respitrace signal. This was performed for every inspiratory and expiratory phase, as well as for the portion of the signal during the initial apnea (denoted as background). For the linear fitting curve, for both BFD and SE features, it was found, experimentally, that two points, A and B, with the following coordinates:

$$\begin{aligned}A&=(x_1,y_1)=(\text{mean}(\text{feature values for 800 mL}),\text{mean}(\text{volumes for 800 mL}))\\B&=(x_2,y_2)=(\text{feature value of background for 800 mL, volume of background for 800 mL})\end{aligned} \tag{8}$$

are sufficient for determining the fitting line.

Similarly, for exponential fitting curves, we found empirically that three points are sufficient, as follows. When using BFD features, the three points (C, D, E) are:

$$\begin{aligned}C&=(x_3,y_3)=(\text{mean}(\text{BFD for 800 mL}),\text{mean}(\text{volumes for 800 mL}))\\D&=(x_4,y_4)=(0.8,0)\\E&=(x_5,y_5)=(2,2)\end{aligned} \tag{9}$$

and with SE features (points F, G, H):

$$\begin{aligned}F&=(x_6,y_6)=(\text{mean}(\text{SE for 800 mL}),\text{mean}(\text{volumes for 800 mL}))\\G&=(x_7,y_7)=(0,0.2)\\H&=(x_8,y_8)=(6,2)\end{aligned} \tag{10}$$

After investigating values of the BFD and SE features from all participants, we noticed that the upper limits were 2 and 6, for BFD and SE respectively. Therefore, we used these asymptotic values as abscissae of points E and H. Figure 4 illustrates the computation of the linear and exponential models.

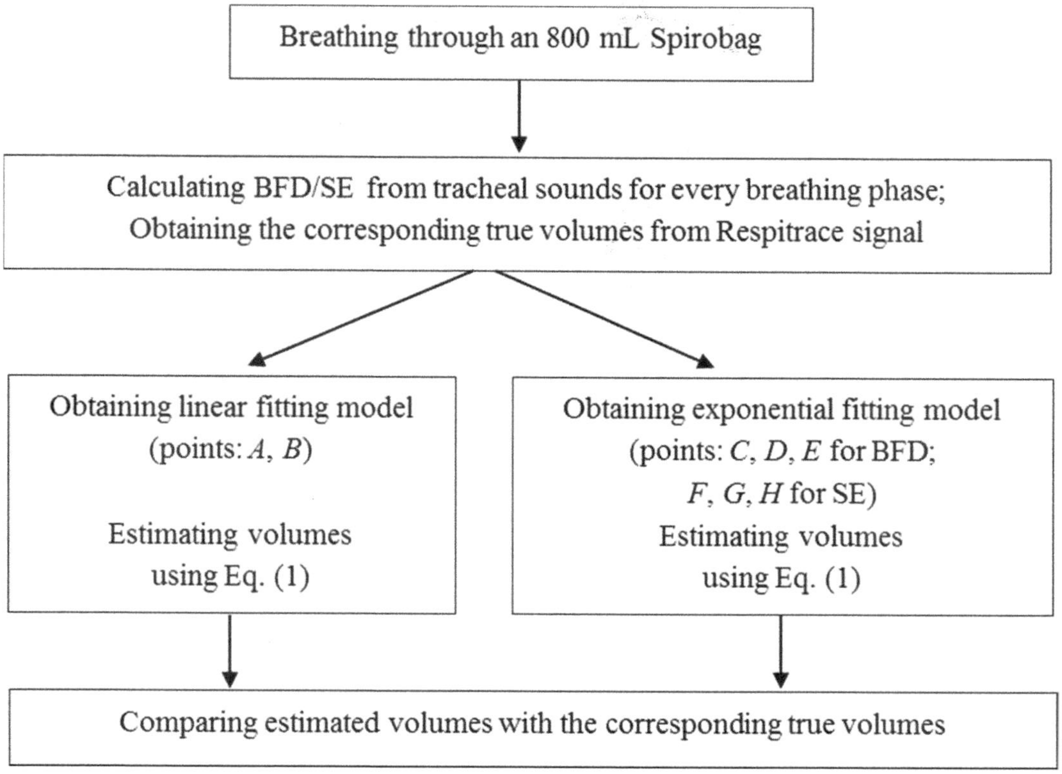

Figure 4. The flowchart showing the computation of the fitting models.

After the linear and exponential curves are calculated, data from the second and the third stages of the experiment (breathing with and without a tube) were used to fit the curves, separately. BFD and SE features were calculated from the smartphone acquired tracheal sounds, and the corresponding volumes were estimated using Equation (1) for the linear and exponential models. Simultaneously, the true volume values were obtained from the reference Respitrace signal. Since the volume range for normal breathing is between 0.2 and 1 L [7], we limited the true volume values to this range, and used only the corresponding portions of tracheal sounds for analysis.

An example of the volume estimation from smartphone acquired tracheal sounds using BFD features and exponential model, for both inspiration and expiration, of one subject is shown in Figure 5. The true tidal volume values (from Respitrace system) and their corresponding BFD values when breathing through 800 mL bag and tube are represented in blue squares and green circles, respectively, while the estimated volumes and their corresponding BFD features are depicted as brown triangles. The three points, shown as black marks in Figure 5 and given with Equation (9), are used for obtaining the exponential fitting curve, which is shown as a solid red curve.

For every inspiration and expiration phase, when a true volume value was between 0.2 and 1 L, the estimated volumes were compared to their corresponding true volumes, and NRMSEs were calculated using Equation (2). In Figure 6 are shown the estimated and reference volumes, as well as the corresponding NRMSE errors for every inspiratory and expiratory phase for the same example as in Figure 5.

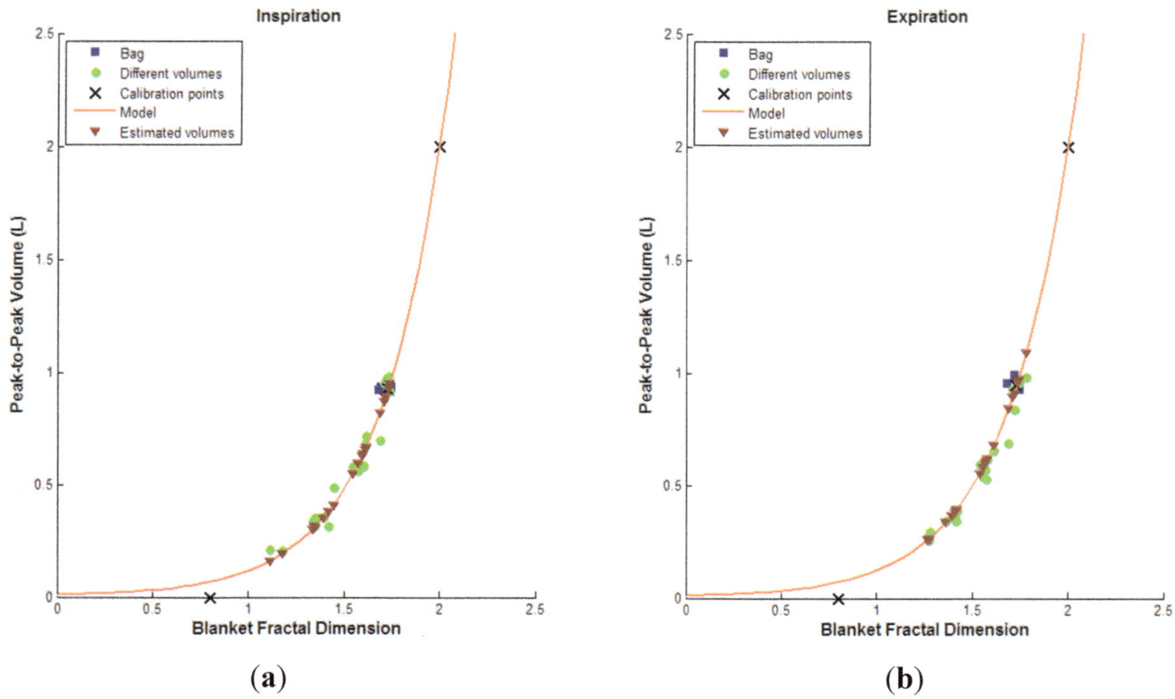

Figure 5. An example of the volume estimation from smartphone acquired tracheal sounds using BFD features and exponential model of one subject. The true volumes while breathing through a tube (green circles) are limited to a range from 0.2 to 1 L. (**a**) The inspiration phase; (**b**) The expiration phase.

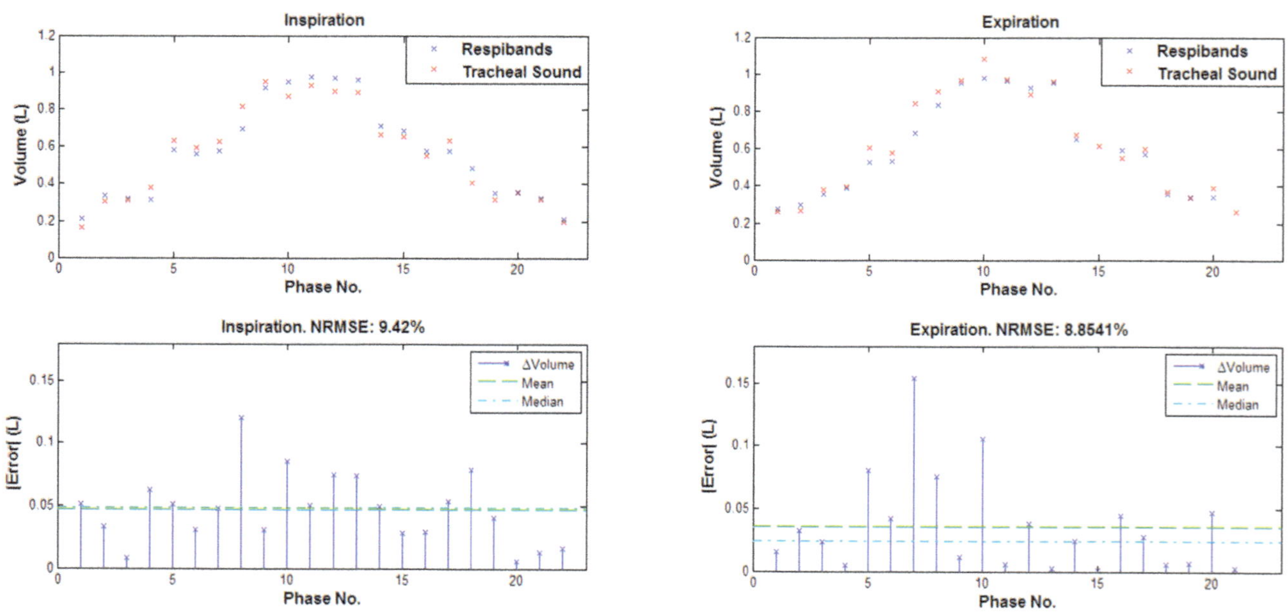

Figure 6. Top: Reference and estimated volumes for the same example as in Figure 5. **Bottom:** The corresponding NRMSE errors.

As can be noted from Figure 6, values of the volumes estimated from a smartphone acquired tracheal sounds using the BFD features are very similar to the volume values obtained from a Respitrace (reference) signal; and the NRMSE errors in both inspiration and expiration phases are low (less than 10%).

After the first day of experiments (later denoted as training), the participants repeated breathing maneuvers with and without a tube for five days (denoted as tests 1–5). The BFD and SE features were calculated from the tracheal sounds, and the volumes were estimated using the first day's fitting curves. Simultaneously, the true volume values were obtained from the Respitrace signal. Again, the estimated volumes were compared to the true volumes, and NRMSEs were calculated.

In this study, we compared the volume estimation results when the proposed blanket fractal dimension is used as feature, with results obtained with Shannon entropy. Conditions of comparisons included: the type of the model (exponential, linear), the type of the apparatus (tube, no tube), and the breathing phase (inspiration, expiration). All combinations of conditions were made, and the corresponding ones were tested statistically, using the two-tailed paired *t*-tests (SPSS Statistics 20, IBM Corporation, Armonk, NY, USA). Table 1 contains the list of combinations and their corresponding *p*-values when statistically significant differences occurred ($p < 0.05$).

Table 1. Combinations of conditions when statistically significant differences were obtained, and their corresponding *p*-values. Results are grouped into 4 groups, based on the type of comparisons performed, *i.e.*, BFD *vs.* SE; inspiration *vs.* expiration; no tube *vs.* tube; exponential *vs.* linear model.

Type	Day	Conditions	*p*-value
BFD *vs.* SE	Test 4	Exponential, tube, inspiration	0.049
		Exponential, tube, expiration	0.015
		Exponential, no tube, expiration	0.011
		Linear, tube, inspiration	0.037
		Linear, tube, expiration	0.013
		Linear, no tube, expiration	0.002
	Test 5	Exponential, tube, expiration	0.017
		Linear, tube, expiration	0.006
		Linear, no tube, expiration	0.007
Inspiration *vs.* Expiration	Test 1	BFD, linear, tube	0.033
	Test 4	SE, linear, tube	0.025
	Test 5	BFD, linear, tube	0.022
		SE, exponential, tube	0.029
		SE, linear, tube	0.031
No tube *vs.* Tube	Training	SE, exponential, inspiration	0.016
	Test 4	BFD, linear, inspiration	0.042
	Test 5	BFD, linear, inspiration	0.033
Exponential *vs.* Linear	Training	BFD, tube, expiration	0.008
		BFD, no tube, expiration	0.038
	Test 4	BFD, tube, expiration	0.028
		SE, tube, expiration	0.018
	Test 5	SE, tube, expiration	0.028

In addition, for each combination, the comparisons between results (NRMSE errors) of the training day and the five test days were performed, and tested statistically using the repeated measures ANOVA with Bonferroni *post-hoc* tests (SPSS Statistics 20). The NRMSE errors are grouped into four parts, based on the apparatus and breathing phase, so that comparisons between features and models can be performed, and are depicted in Figure 7. These graphs show the changes in NRMSE errors throughout six days of experiments for all combinations of features and models simultaneously.

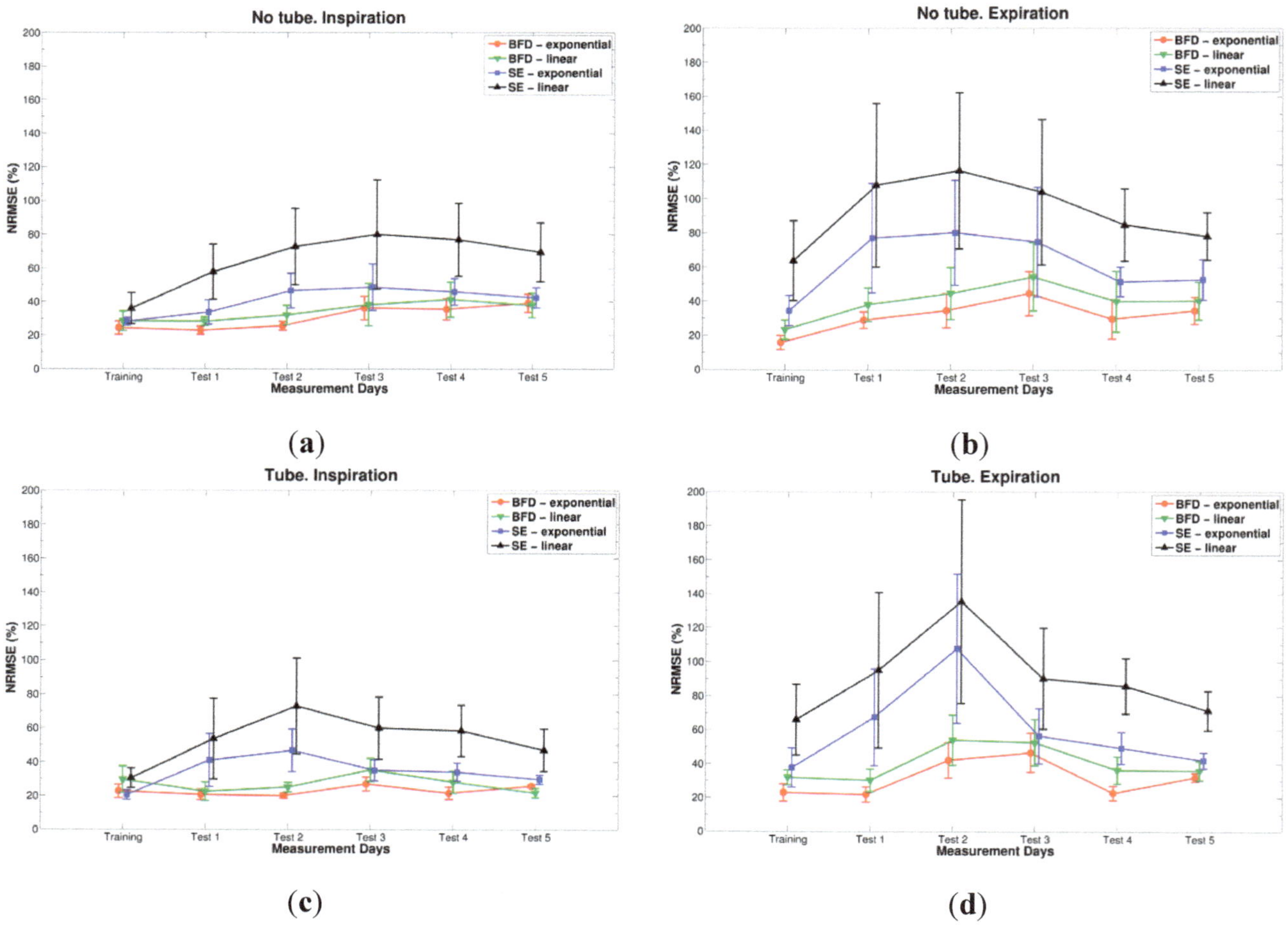

Figure 7. NRMSE errors (represented with its mean and standard error of the mean) when: BFD and exponential model (red circles), BFD and linear model (green downward triangles), SE and exponential model (blue squares), and SE and linear model (black triangles) are used. (**a**) No tube and inspiration; (**b**) No tube and expiration; (**c**) Tube and inspiration; (**d**) Tube and expiration.

As can be concluded from the graphs in Figure 7, when blanket fractal dimension was used for volume estimation (red and green lines), the errors were lower at least two times than when Shannon entropy was used (blue and black lines), especially with the exponential model (red circles). Moreover, note that standard errors are also smaller when BFD is used. Statistically significant differences between the two features appeared during the fourth test day (for: exponential and linear models, with tube and both inspiration and expiration phases; and for both models, without tube and expiration) and the fifth test day (for: both models, with tube and expiration phase; and linear model, without a tube and expiration), as shown in Table 1.

The smallest NRMSE error, with mean and standard deviation of 15.877% ± 9.246%, was obtained during the first day of experiments (training), when BFD feature with the exponential model was used, for expiratory phase, while the participants were breathing without a tube, Figure 7b. The Bland-Altman analysis showed a bias and standard deviation of 0.0226 ± 0.0918 L, and the corresponding results are presented in Figure 8.

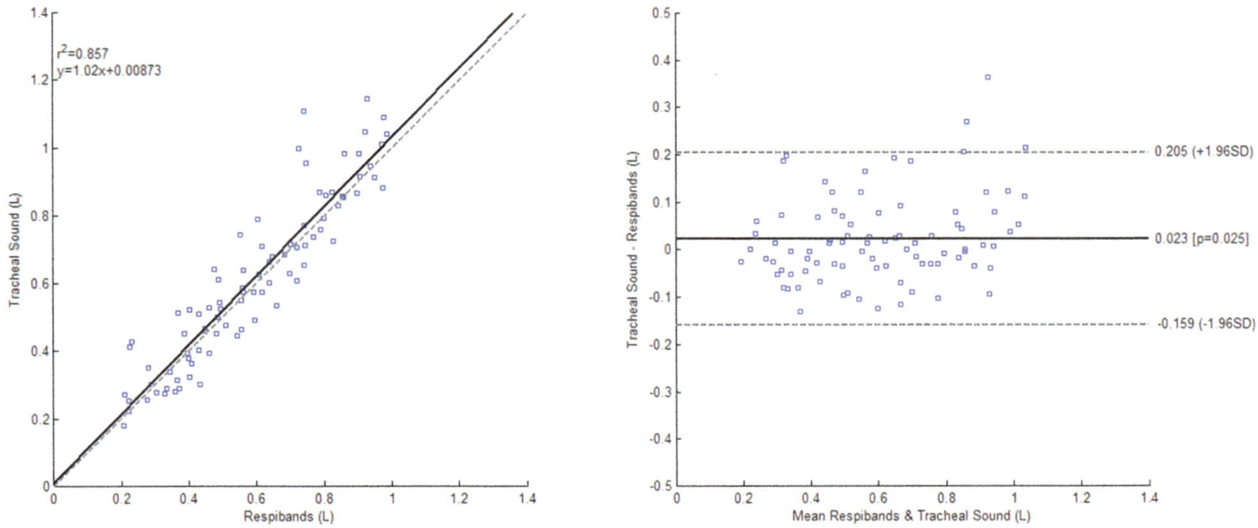

(a) (b)

Figure 8. Bland-Altman plot for BFD feature with the exponential model, for expiratory phase, while the participants (N = 5) were breathing without a tube during the first day of experiments. (**a**) The regression plot: The unitary line is shown as gray dashed line, while the regression line is represented as black solid line; (**b**) Bland-Altman plot: The bias is represented as a solid black line and the 95% limits of agreement as gray dashed lines.

By looking at the NRMSEs calculated for the remaining 5 days (test days), one can conclude that the smallest was always obtained with the BFD feature, exponential model and inspiration while breathing through a tube (errors ranging from 20% to 27%), Figure 7c, except for the fifth day, when linear model provided better estimation (error around 21%). No statistically significant differences were found between BFD exponential model from inspiratory and expiratory phases, as deduced from Table 1.

As was mentioned above, when BFD feature was used the errors were always smaller than with SE. In addition, one can conclude that the fitting curves obtained during the first day of experiments (training) can be successfully used for the following test days. This way, the participants do not need to perform all three stages of the experiments, and the fitting curves do not need to be calculated every day, as the previously determined could be used. In order to statistically compare errors throughout all six days of experiments, repeated measures ANOVA with Bonferroni *post-hoc* tests were performed, and was determined that there were no statistically significant differences between the days of experiments when BFD or SE was used as feature. According to Table 1, for the BFD using exponential model, no statistically significant differences were found between breathing through the tube or not.

4. Discussions and Conclusions

The goal of this study was to estimate tidal volume from the smartphone acquired tracheal sounds. The main challenge was to find a suitable feature to describe these sounds, such that the volume could be estimated directly from the sounds as accurate as possible. Respiratory sounds, and hence tracheal sounds, are non-stationary and stochastic signals [2], and as such they are suitable for fractal analysis [36]. We tested several ways for estimating fractal dimension, and decided to use the blanket fractal dimension because it was more suitable for describing and following the dynamics of the tracheal sounds, which was evident after exploring the results. Possible explanation could be the definition of the blanket fractal dimension itself. Blanket method creates a strip around the tracheal signal, closely following the changes in the signal. As the signal changes faster, the value of blanket fractal dimension becomes higher. In some past studies fractal analysis and fractal dimensions were used for analyzing tracheal and lung sounds [20–24]. Moreover, blanket fractal dimension was not used in respiratory sound analysis yet, and especially not for estimating the tidal volume, which are some of the novelties of this manuscript. In addition, to the best of our knowledge, none of the studies on tidal volume estimation has reported results based on tracheal sounds acquired by a smartphone.

In addition to BFD features, we used Shannon entropy (SE), as it is one of the features frequently used for analysis of respiratory sounds. In [41], the authors proposed a method to estimate airflow from tracheal sounds using SE. In [42], the authors proposed tidal volume estimation method by integrating airflow derived from tracheal sounds, which takes advantage of airflow/sound intensity relationship. As can be noted, the straightforward comparison between our method and method used in [42] is difficult to perform, since the conditions are not exactly the same. We estimated the tidal volume directly from tracheal sounds, using BFD as a feature, while Que *et al.* [42] obtained first the relationship between sounds' amplitude and airflow, and then the volume by integrating the flow. Consequently, according to the provided results, the range of volume values in [42] was roughly between 0.3 and 0.8 L, while we limited volumes to a broader range [0.2, 1] L. That being said, the Bland-Altman analysis results of [42] were 0.009 ± 0.046 L (bias ± SD), while we found a bias and standard deviation of 0.0226 ± 0.0918 L. Chen *et al.* estimated tidal volume from the energy of the tracheal sounds [6]. The comparison of our results with those reported in [6] is not easy to perform since they are reported separately for each individual participant. If we compute the average results from the provided individually-based values reported in Table 1 [6]), we can conclude that the results are comparable. The volumes ranged from 0.15 to 0.5 L in [6], which is notably smaller range than the one used in this study. Note that in contrast to these two studies, the only external information needed to compute the calibration model with our proposed method was obtained with a simple bag at a known fixed value and not from a spirometer-like device.

After volumes were estimated from the smartphone acquired tracheal sounds, they were compared to the true volume values, obtained from Respitrace signal, which was considered as a reference in this study. The Respitrace signal was calibrated against the spirometer signal prior every recording and the obtained calibration errors were less than 10%, which is in accordance to the manufacturer's manual. These reference volumes were limited to a range from 0.2 to 1 L, as it is the normal breathing range [7]. Inspiratory and expiratory phases were analyzed separately. Two fitting models, exponential and linear, were used for estimation. Our results indicate that the best estimation was obtained using blanket fractal dimension with exponential model, during expiratory phase, while participants were breathing without a tube, when the NRMSE error was 15.877% ± 9.246% (expressed as mean ± standard deviation). In addition, when the BFD is used as a feature, the NRMSEs were always smaller, at least twice, compared to the SE.

The experiments involved acquisition during six days. Data from the first day of experiments were used to construct estimation models, while the data from the remaining five days were plotted against the obtained models. The results show the possibility to successfully apply previously obtained fitting curves and to monitor tidal volume for at least five days. This way we introduce an easy calibration procedure, where there is no need to calculate fitting curves prior every consecutive experiment. In our future work, we plan to determine for how many days the existing models can be used.

This is a preliminary study, with the objective to estimate tidal volume in healthy participants, and not in patients with pulmonary diseases. Therefore, it was performed on five healthy participants, and for the future work we plan to expand the group. This study was limited to acquisition of tracheal sounds in standing posture without head movements. We expect that the results obtained with the proposed methodology would be in agreement with the study reported in [42], where the effects of body movements and posture changes on tidal volume estimates were investigated. Accordingly, we foresee that head movements without neck extension will not modify the obtained results and we do not anticipate an increase in estimation errors when moving to seated posture, but we do when moving from standing to supine posture, where a new calibration in latter posture would be required. It is worth to mention that all recordings were made in a regular dry lab, that was held quiet, and not in a special soundproof environment, hence making it applicable to real-life situations. Since spirometer is not a portable device, not easily accessed and fixed values of tidal volumes are hard to control, which results in additional turbulences and changes in breathing patterns, we used a Spirobag in order to obtain information at a known volume which in turn was employed in the estimation model. In addition, due to high performance capabilities of smartphones, by connecting an adequate acoustical sensor to a smartphone and using a Spirobag, a portable system for tidal volume estimation can be obtained.

In summary, in this manuscript we proposed a novel technique for estimation of tidal volume directly from the blanket fractal dimension of the tracheal sounds. The proposed method provided promising results and outperformed a method based on the Shannon entropy, which is frequently used in tracheal sounds analysis. Furthermore, we introduced an easy calibration procedure that does not require specialized devices and when combined with the proposed signal processing technique allows reasonable estimation for at least five days, which makes this method easier to use in everyday situations. The employment of smartphone-acquired tracheal sounds was also introduced for all of the above mentioned purposes. We foresee that similar efforts to the one presented here represent a step forward to the development of a mobile breathing monitoring system easily available for the general population.

Acknowledgments

This work is supported in part by the US Army Medical Research and Material Command (US-AMRMC) under grant No. W81XWH-12-1-0541. The authors would like to thank Sonia Charleston-Villalobos, Thomas Aljama-Corrales, and Ramon Gonzales-Camarena for introducing a microphone sensor used in this research work. The authors would also like to thank Milorad Paskas for his help in programs for calculating blanked fractal dimension. We also acknowledge the feedback provided by the reviewers and editor to improve the quality of the paper.

Author Contributions

Natasa Reljin and Bersain Reyes designed and performed the experiments, analyzed the data, and wrote the manuscript. Ki Chon is the Principal Investigator, and oversaw and guided the study and the writing of the manuscript.

Conflicts of Interest

The authors declare no conflict of interest.

References

1. Sovijarvi, A.R.A.; Dalmasso, F.; Vanderschoot, J.; Malmberg, L.P.; Righini, G.; Stoneman, S.A.T. Definition of terms for applications of respiratory sounds. *Eur. Respir. Rev.* **2000**, *10*, 597–610.
2. Moussavi, Z. *Fundamentals of Respiratory Sounds and Analysis*, 1st ed.; Morgan & Claypool Publishers: San Rafael, CA, USA, 2006.
3. Sovijarvi, A.R.A.; Malmberg, L.P.; Charbonneau, G.; Vanderschoot, J.; Dalmasso, F.; Sacco, C.; Rossi, M.; Earis, J.E. Characteristics of breath sounds and adventitious respiratory sounds. *Eur. Respir. Rev.* **2000**, *10*, 591–596.
4. Folke, M.; Cernerud, L.; Ekström, M.; Hök, B. Critical review of non-invasive respiratory monitoring in medical care. *Med. Biol. Eng. Comput.* **2003**, *41*, 377–383.
5. Kuratomi, Y.; Okazaki, N.; Ishihara, T.; Arai, T.; Kira, S. Variability of breath-by-breath tidal volume and its characteristics in normal and diseased subjects. *Jpn. J. Med.* **1985**, *24*, 141–149.
6. Chen, G.; de la Cruz, I.; Rodriguez-Villegas, E. Automatic lung tidal volumes estimation from tracheal sounds. In Proceedings of the 36th Annual International Conference of the IEEE Engineering in Medicine and Biology Society (EMBC), Chicago, IL, USA, 26–30 August 2014; pp. 1497–1500.
7. Sherwood, L. *Fundamentals of Human Physiology*, 4th ed.; Cengage Learning: Boston, MA, USA, 2012.
8. Semmes, B.J.; Tobin, M.J.; Snyder, J.V.; Grenvik, A. Subjective and objective measurement of tidal volume in critically ill patients. *Chest* **1985**, *87*, 577–579.
9. Grossman, P.; Spoerle, M.; Wilhelm, F.H. Reliability of respiratory tidal volume estimation by means of ambulatory inductive plethysmography. *Biomed. Sci. Instrum.* **2006**, *42*, 193–198.
10. Sayadi, O.; Weiss, E.H.; Merchant, F.M.; Puppala, D.; Armoundas, A.A. An optimized method for estimating the tidal volume from intracardiac or body surface electrocardiographic signals: Implications for estimating minute ventilation. *Am. J. Physiol. Heart Circ. Physiol.* **2014**, *307*, H426–H436.
11. Corbishley, P.; Rodriguez-Villegas, E. Breathing detection: Towards a miniaturized, wearable, battery-operated monitoring system. *IEEE Trans. Biomed. Eng.* **2008**, *55*, 196–204.
12. Earis, J.E.; Cheetham, B.M.G. Future perspectives for respiratory sound research. *Eur. Respir. Rev.* **2000**, *10*, 641–646.

13. Cala, S.J.; Kenyon, C.M.; Ferrigno, G.; Carnevali, P.; Aliverti, A.; Pedotti, A.; Macklem, P.T.; Rochester, D.F. Chest wall and lung volume estimation by optical reflectance motion analysis. *J. Appl. Physiol.* **1996**, *81*, 2680–2689.
14. Petrovic, M.D.; Petrovic, J.; Danicic, A.; Vukcevic, M.; Bojovic, B.; Hadzievski, Lj.; Allsop, T.; Lloyd, G.; Webb, D.J. Non-invasive respiratory monitoring using long-period fiber grating sensors. *Biomed. Opt. Express* **2014**, *5*, 1136–1144.
15. Lee, Y.S.; Pathirana, P.N.; Steinfort, C.L.; Caelli, T. Monitoring and analysis of respiratory patterns using microwave doppler radar. *IEEE J. Transl. Eng. Health Med.* **2014**, *2*, 1–12.
16. Scully, C.G.; Lee, J.; Meyer, J.; Gorbach, A.M.; Granquist-Fraser, D.; Mendelson, Y.; Chon, K.H. Physiological parameter monitoring from optical recordings with a mobile phone. *IEEE Trans. Biomed. Eng.* **2012**, *59*, 303–306.
17. Lee, J.; Reyes, B.A.; McManus, D.D.; Mathias, O.; Chon, K.H. Atrial fibrillation detection using an iPhone 4S. *IEEE Trans. Biomed. Eng.* **2013**, *60*, 203–206.
18. Reyes, B.A.; Reljin, N.; Chon, K.H. Tracheal sounds acquisition using smartphones. *Sensors* **2014**, *14*, 13830–13850.
19. Ahlstrom, C.; Johansson, A.; Hult, P.; Ask, P. Chaotic dynamics of respiratory sounds. *Chaos Solitons Fractals* **2006**, *29*, 1054–1062.
20. Yap, Y.L.; Moussavi, Z. Respiratory onset detection using variance fractal dimension. In Proceedings of 23rd Annual International Conference of the IEEE Engineering in Medicine and Biology Society, Istanbul, Turkey, 25–28 October 2001; pp. 1554–1556.
21. Gnitecki, J.; Moussavi, Z. Variance fractal dimension trajectory as a tool for heart sound localization in lung sounds recordings. In Proceedings of 25th Annual International Conference of the IEEE Engineering in Medicine and Biology Society, Cancun, Mexico, 17–21 September 2003; pp. 2420–2423.
22. Hadjileontiadis, L.J.; Rekanos, I.T. Detection of explosive lung and bowel sounds by means of fractal dimension. *IEEE Signal Process. Lett.* **2003**, *10*, 311–314.
23. Gnitecki, J.; Moussavi, Z. The fractality of lung sounds: A comparison of three waveform fractal dimension algorithms. *Chaos Solitons Fractals* **2005**, *26*, 1065–1072.
24. Hadjileontiadis, L.J. A novel technique for denoising explosive lung sounds: Empirical mode decomposition and fractal dimension filter. *IEEE Eng. Med. Biol. Mag.* **2007**, *26*, 30–39.
25. Charleston-Villalobos, S.; Albuerne-Sanchez, L.; Gonzalez-Camarena, R.; Mejia-Avila, M.; Carrillo-Rodriguez, G.; Aljama-Corrales, T. Linear and nonlinear analysis of base lung sound in extrinsic allergic alveolitis patients in comparison to healthy subjects. *Methods Inf. Med.* **2013**, *52*, 266–276.
26. Druzgalski, C.K.; Donnerberg, R.L.; Campbell, R.M. Techniques of recording respiratory sounds. *J. Clin. Eng.* **1980**, *5*, 321–330.
27. Pasterkamp, H.; Kraman, S.S.; Wodicka, G.R. Respiratory sounds: Advances beyond the stethoscope. *Am. J. Respir. Crit. Care Med.* **1997**, *156*, 974–987.
28. Charleston-Villalobos, S.; Martinez-Hernandez, G.; Gonzalez-Camarena, R.; Chi-Lem, G.; Carrillo, J.G.; Aljama-Corrales, T. Assessment of multichannel lung sounds parameterization for two-class classification in interstitial lung disease patients. *Comput. Biol. Med.* **2011**, *41*, 473–482.

29. Reichert, S.; Gass, R.; Brandt, C.; Andres, E. Analysis of respiratory sounds: State of the art. *Clin. Med. Insights Circ. Respir. Pulm. Med.* **2008**, *2*, 45–58.
30. Papoulis, A.; Pillai, S.U. *Probability, Random Variables, and Stochastic Processes*; McGraw-Hill: New York, NY, USA, 2002.
31. Yadollahi, A.; Moussavi, Z.M.K. A robust method for heart sounds localization using lung sounds entropy. *IEEE Trans. Biomed. Eng.* **2006**, *53*, 497–502.
32. Parzen, E. On estimation of a probability density function and mode. *Ann. Math. Stat.* **1962**, *33*, 1065–1076.
33. Duda, R.O.; Hart, P.E.; Stork, D.G. *Pattern Classification*, 2nd ed.; Wiley-Interscience: New York, NY, USA, 2000.
34. Mandelbrot, B.B. How long is the coast of Britain? Statistical self-similarity and fractional dimension. *Science* **1967**, *156*, 636–638.
35. Mandelbrot, B.B. Stochastic models for the Earth's relief, the shape and the fractal dimension of the coastlines, and the number-area rule for islands. *Proc. Natl. Acad. Sci.* **1975**, *72*, 3825–3828.
36. Mandelbrot, B.B. *The Fractal Geometry of Nature*; W. H. Freeman and Company: New York, NY, USA, 1982.
37. Peitgen, H-O.; Juergens, H.; Saupe, D. *Chaos and Fractals*, 2nd ed.; Springer: New York, NY, USA, 2004.
38. Peleg, S.; Naor, J.; Hartley, R.; Avnir, D. Multiple resolution texture analysis and classification. *IEEE Trans. Pattern Anal. Mach. Intell.* **1984**, *PAMI-6*, 518–523.
39. Paskas, M.P.; Gavrovska, A.M.; Reljin, N.B. Identification of fundamental heart sounds from PCG using blanket fractal dimension. In Proceedings of the 8th Conference of the European Study Group on Cardiovascular Oscillations, Trento, Italy, 25–28 May 2014; pp. 123–124.
40. Turner, M.J.; Blackledge, J.M. Analysis of the limitations of fractal dimension texture segmentation for image characterisation. In *Fractal Geometry: Mathematical Methods, Algorithms, Applications*; Blackledge, J.M., Evans, A.K., Turner, M.J., Eds.; Horwood Publishing Limited: Chichester, England, 2002; pp. 114–137.
41. Yadollahi, A.; Moussavi, Z.M.K. A robust method for estimating respiratory flow using tracheal sounds entropy. *IEEE Trans. Biomed. Eng.* **2006**, *53*, 662–668.
42. Que, C.-L.; Kolmaga, C.; Durand, L.-G.; Kelly, S.M.; Macklem, P.T. Phonospirometry for noninvasive measurement of ventilation: Methodology and preliminary results. *J. Appl. Physiol.* **2002**, *93*, 1515–1526.

3

Optimization and Application of Reflective LSPR Optical Fiber Biosensors Based on Silver Nanoparticles

Jiangping Chen [1], Se Shi [1], Rongxin Su [1,2,*], Wei Qi [1,2,*], Renliang Huang [3], Mengfan Wang [1], Libing Wang [1] and Zhimin He [1]

[1] State Key Laboratory of Chemical Engineering, Tianjin Key Laboratory of Membrane Science and Desalination Technology, School of Chemical Engineering and Technology, Tianjin University, Tianjin 300072, China; E-Mails: cjpjm@126.com (J.C.); shise@tju.edu.cn (S.S.); mwang@tju.edu.cn (M.W.); wanglb1@163.com (L.W.); enzyme@tju.edu.cn (Z.H.)

[2] Collaborative Innovation Center of Chemical Science and Engineering (Tianjin), Tianjin 300072, China

[3] School of Environmental Science and Engineering, Tianjin University, Tianjin 300072, China; E-Mail: tjuhrl@tju.edu.cn

* Authors to whom correspondence should be addressed; E-Mails: surx@tju.edu.cn (R.S.); qiwei@tju.edu.cn (W.Q.)

Academic Editor: W. Rudolf Seitz

Abstract: In this study, we developed a reflective localized surface plasmon resonance (LSPR) optical fiber sensor, based on silver nanoparticles (Ag NPs). To enhance the sensitivity of the LSPR optical sensor, two key parameters were optimized, the length of the sensing area and the coating time of the Ag NPs. A sensing length of 1.5 cm and a 1-h coating time proved to be suitable conditions to produce highly sensitive sensors for biosensing. The optimized sensor has a high refractive index sensitivity of 387 nm/RIU, which is much higher than that of other reported individual silver nanoparticles in solutions. Moreover, the sensor was further modified with antigen to act as a biosensor. Distinctive wavelength shifts were found after each surface modification step. In addition, the reflective LSPR optical fiber sensor has high reproducibility and stability.

Keywords: localized surface plasmon resonance; optical fiber; silver nanoparticles; biosensors

---

1. Introduction

Localized surface plasmon resonance (LSPR) is a nanoscale phenomenon based on a localized electromagnetic field around noble metal surfaces, which are sensitive to the surrounding refractive indices [1]. LSPR has many advantages, including label-free and real-time detection and tunable resonant wavelengths, which are controlled by careful adjustment of the size, shape and material of nanoparticles [2–8]. LSPR optical fiber sensors are based on the plasmon resonance of metal nanoparticles, coated on optical fiber surfaces, that are sensitive to changes in the surrounding medium. Accordingly, they may be used to detect label-free biomolecule interactions and to monitor the adsorption and desorption kinetics by continuous optical measurements [9]. Hence, LSPR optical fiber sensors have been widely used for biotin-streptavidin [10] and antigen-antibody interaction monitoring [11], metal ion detection, and other applications [12–17].

To date, many researchers have developed optical fiber LSPR sensors with various structures, e.g., the in-line transmission type [18–20] and terminated reflection type [21–28], a U-bend probe [29–31]. Compared to other structures, the terminated reflection type has shown greater merits. It possesses a much more efficient sensing area, and it does not require the use of flow-cells [32]. For this structure, a micro-fabricated silver mirror is usually coated at the end of the sensor probe to allow the incident light to travel twice through the sensing area without an increase in the sensing length. During the detection, the sensor probe is immersed into the testing solution without a need for flow-cells, which makes the optical setup much cheaper and simpler. Therefore, we selected the reflection type for further study, which includes fabrication and optimization of LSPR optical fiber sensor probes.

There are few studies concerning the preparation of LSPR optical fiber sensors based on silver nanoparticles (Ag NPs), although silver nanofilms have been proven to be much more sensitive to surrounding medium changes than other metal films [33]. Recently, Dutta *et al.* [34,35] produced a U-bend LSPR optical fiber sensor based on Ag NPs, but they did not further modify their LSPR sensor as a biosensor for detecting biomolecular interactions.

Therefore, the objectives of this study were to (1) fabricate a stable and sensitive reflective LSPR optical fiber sensor based on Ag NPs, (2) optimize the fabrication process, including two key parameters (the sensing length and the coating time), and (3) modify the surfaces of Ag NPs on the optical fiber, monitor the antigen-antibody binding process by continuous optical measurements, and observe the wavelength shifts during each modification step.

2. Experimental Section

2.1. Materials

Silver nitrate ($AgNO_3$), 3-aminopropyl trimethoxysilane 97% (APTMS), 11-mercaptoundecanoic acid (MUA), 2-morpholinoethanesulfonic acid, *N*-(3-dimethylaminopropyl)-*N'*-ethylcarbodiimide hydrochloride (EDC), *N*-hydroxysuccinimide (NHS), and bovine serum albumin (BSA) were purchased from Sigma-Aldrich. All other chemicals were of analytical reagent grade. Human IgG (purified immunoglobulin, reagent grade), anti-human IgG (whole molecule, produced in rabbit) were purchased from the Wuhan Boster Biotechnology Company, China. Ultrapure water from a Milli-Q ultrapure water purification system (Millipore, Billerica, MA, USA) was used throughout the experiments. Multimode optical fibers of diameter 400 μm and NA = 0.22 were purchased from Ocean Optics.

2.2. Preparation of Silver Nanoparticles

The silver nanoparticles with a mean diameter of 30 nm chosen for this experiment were prepared as previously described [36]. Typically, an Ag NPs solution was synthesized by dissolving 0.026 g of sodium citrate in 50 mL of ethylene glycol. After the solution was stirred and heated to 90 °C, 0.012 g of $AgNO_3$ was rapidly added, resulting in a color change from colorless to dark yellow after a few seconds. The mixture was continuously stirred and heated for approximately 10 min. The solution was diluted with 100 mL of deionized water and stored at 4 °C for future use.

2.3. Preparation of Ag NP-Based LSPR Sensor Probes

First, a polyimide-coating silica multimode optical fiber was cut into several 10-cm-long sections for use as sensor substrates. Both end surfaces of each fiber section were then carefully polished with polishing papers (using types of 12 μm, 3 μm, and 1 μm roughness, in sequential order, Ocean Optics). Each fiber section was left with a 2-cm unclad portion from the end to be used as a sensing area. The 2-cm portion was treated with 98% H_2SO_4 and 40% HF, respectively. The sensing area was then cleaned and hydrolyzed with Piranha solution (with a volume ratio of 7:3 of H_2SO_4:H_2O_2) for 30 min at 85 °C (it should be noted that this solution is extremely dangerous and needed to be handled with extreme care). Following that, the optical fiber was rinsed with deionized water, blow-dried with N_2 and placed in an oven for 20 min at 110 °C.

For the Ag NP coating, the clean fiber was first immersed in a 10% solution of 3-aminopropyl trimethoxysilane (APTMS) in methanol for 1 h at 40 °C. The functionalized optical fiber was rinsed sequentially with ethanol and deionized water to remove unbound APTMS. After being blow-dried with N_2, the optical fiber was placed in the oven for 20 min at 110 °C. Finally, the optical fiber was dipped into a Ag NPs solution for several hours. The incubated condition should be controlled at 4 °C to avoid random aggregation of Ag NPs on the fiber core surface. It has been shown that the aggregation of nanoparticles would significantly reduce the activity of nanoparticles, resulting in inferior performance of optical sensor probes [35].

To produce a reflective optical fiber sensor probe, a thin silver film was coated onto the end surface of the fiber probe using a chemical technique with Tollen's reagent [37]. The preparation of Tollen's reagents started with adding 2% ammonia solution drop-by-drop into 2 mL of 0.1 M $AgNO_3$ solution. After the brown precipitate was totally dissolved, 1.4 mL of 0.8 M KOH was added to form a black precipitate in the solution. Subsequently, ammonia solution was added drop wise into the solution. Once the black precipitate was dissolved, Tollen's reagent was acquired. Before silver mirror coating, the end surface of the fiber probe was dipped into a 0.2% tin (II) chloride dehydrate solution to sensitize the coating area. After rinsing thoroughly with deionized water and blow-drying with N_2, the coating area was dipped into Tollen's reagent. Subsequently, 0.4 mL of 0.25 M glucose was rapidly added; the fiber was then kept in the reagent for a period of 5 min to form the silver mirror. Afterwards, the fiber was removed from the reagent and rinsed with deionized water and blow-dried with N_2. The sensor probe, now in its final form, was stored in deionized water at 4 °C for future use.

2.4. Functionalization of Sensor Probes

As shown in Figure 1, in order to create a Ag NP-based LSPR optical fiber biosensor for detecting the anti-human IgG, the as-prepared sensor probe was further functionalized (the human IgG was immobilized onto the sensor probe). The functionalization was carried out using a two-step procedure as follows. In the first step, we incorporated NHS-activated carboxyl groups onto the sensor probe. The sensor probe was immersed in a 10 mM MUA ethanol solution for 12 h to create a layer of carboxyl groups on the surface of the Ag NPs. The sensor probe was rinsed with ethanol and 0.01 M PBS (pH = 7.4) to remove unbound MUA. The sensor probe was then immersed into a mixed solution of 2 mM EDC and 5 mM NHS in 0.1 M MES buffer (pH = 6.0) with a volume ratio of 1:1 for 10 min to activate the carboxyl group of the MUA. After activation, the sensor probe was rinsed with PBS buffer. In the second step, we immobilized the human IgG onto the sensor probe surface and deactivated the residual NHS-activated carboxyl groups. The sensor probe was immersed in a 100 $\mu g \cdot mL^{-1}$ human IgG solution for 2 h to immobilize human IgG on the surface of the Ag NPs. After the immobilization, the sensor probe was immersed in 2 $mg \cdot mL^{-1}$ bovine serum albumin (BSA) solution for 1 h to suppress nonspecific binding. After that, the sensor probe was rinsed again with PBS to remove unbound BSA. Finally, the biosensor probe was created and stored in the PBS buffer at 4 °C for future use.

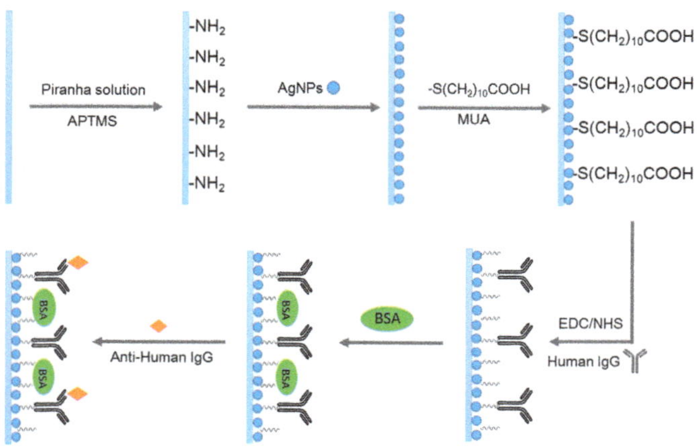

Figure 1. Schematic of bio-sensing approaches using the Ag NPs-based LSPR sensor probe developed in this work.

2.5. Monitoring of the Antigen-Antibody Interaction

After the functionalized procedure, the biosensor probe was ready to detect the specific anti-human IgG because the interaction between the anti-human IgG and the human IgG immobilized on the biosensor probe surface yields a change of a resonance wavelength in the reflection spectrum. To detect the anti-human IgG, the biosensor probe was immersed into the rabbit anti-human IgG solution to start the antibody-antigen reaction, and the reflection spectrum was monitored every five minutes.

2.6. Optical Setup and Data Processing

The optical measurement set-up used in this experiment is shown in Figure 2A. In this optical system, a 1 × 2 fiber coupler was used to connect the sensor probe, the light source and the optical spectrometer. As illustrated in Figure 2A, the 1 × 2 fiber coupler has three light delivery paths. Part a and part c were connected to a tungsten halogen light source (HL-2000, Ocean Optics) and a spectrometer (HR2000+, Ocean Optics), respectively, and part b was connected to a homemade LSPR sensor probe by a splice bushing (21-02 SMA, Ocean Optics). Light emitted by the light source was coupled into the 1 × 2 fiber and travelled to the sensor probe. Due to the silver mirror at the end surface of the sensor probe, light was reflected back along the fiber and fed into the spectrometer. Thus, a computer could monitor, display, and record the optical signals measured by the spectrometer. The original data were analyzed using a homemade MATLAB code to identify the resonance wavelength from the obtained reflectance spectra over time.

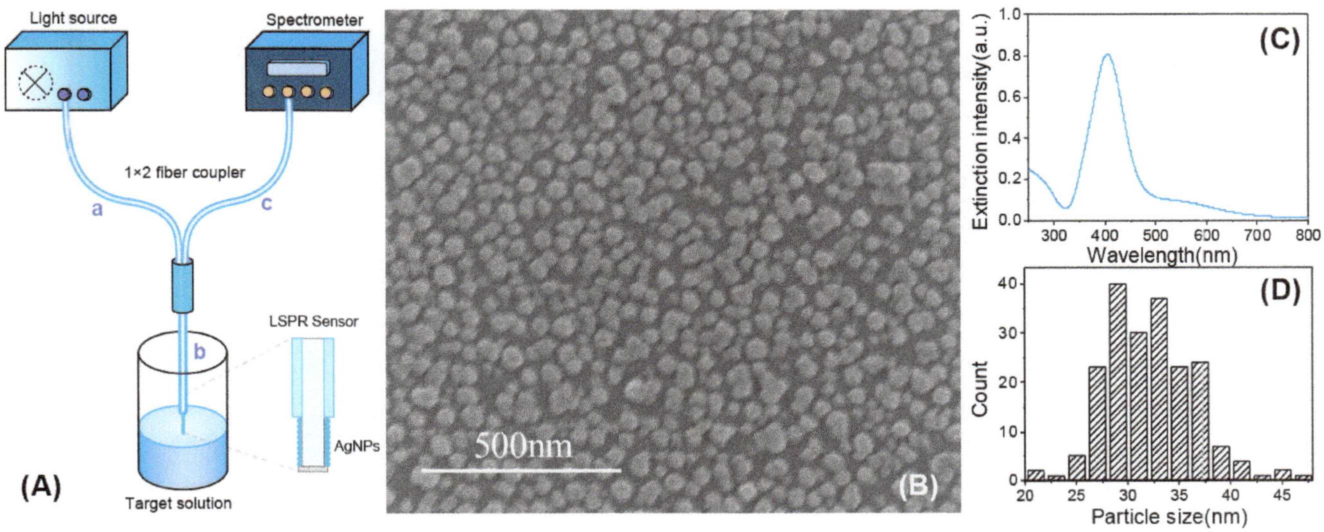

Figure 2. (**A**) Schematic of the experimental set-up used in this study; (**B**) An SEM image of AgNPs after being immobilized on optical fiber; (**C**) Absorption spectra of AgNPs solutions; (**D**) Histogram showing the corresponding particle size distribution of AgNPs.

3. Results and Discussion

3.1. Characterization of the Sensor Probe

The as-prepared Ag NPs were measured by UV-Vis absorption spectroscopy (TU-1810, Persee) over the wavelength range from 300 nm to 800 nm, as shown in Figure 2C. The peak located near 405 nm is typically observed for Ag NPs. As shown in Figure 2B, the Ag NPs immobilized on the surface of sensor probe were characterized by FE-SEM (S4800, Hitachi, Japan). According to the image, the particle diameter of Ag NPs was, on average, 30 nm (the counting histogram was shown in Figure 2D), and most of them were nearly spherical. In addition, most of the Ag NPs immobilized on the sensing surface were separated from each other. According to a previous report, the aggregation was a source of LSPR band broadening and wavelength red-shift [35]. Therefore, careful control of temperature of the self-assembly process to avoid the aggregation of nanoparticles was necessary in this experiment.

To determine the sensitivity of the as-prepared sensor probe to the surrounding refractive index, the sensor probe was immersed in a series of testing solutions of varying refractive indices, ranging from 1.33 to 1.40. The testing solutions were a series of different mass concentrations of sucrose solutions. The refractive indices of the sucrose solutions were measured by an Abbe refractometer. It is noticed that once the sensor probe was immersed in the testing solution, the spectrum changed immediately and stabilized quickly at the specific position corresponding to the refractive index of the testing solution.

The spectra acquired from the sensitivity test are illustrated in Figure 3A. With an increase in the refractive index, the relative reflectivity decreased accordingly. The LSPR wavelength showed a red-shift with an increase in the refractive index. This observation verified the sensitivity of the as-prepared sensor probe to refractive index changes to the surrounding medium.

The refractive index sensitivity was defined as follows [9,38]:

$$S = \frac{d\lambda_{min}}{dn} \quad (1)$$

$$S = \frac{dI_{min}}{dn} \quad (2)$$

where S represents sensitivity, λ_{min} is the peak value in the reflectivity spectrum, I_{min} is the optical intensity in the reflectivity spectrum and n is the refractive index value. The reflective peak wavelength and optical intensity observed from the sensor probe corresponding to each sensitivity test was recorded and calculated by a homemade MATLAB routine based on Equations (1) and (2). Figure 3B shows the shift in the LSPR wavelengths of the various sucrose solutions with different reflective indices. The sensitivity of the sensor probe to reflective indices was 387 nm/RIU, and the total peak wavelength shift was approximately 30 nm. Both the sensitivity and the total peak wavelength of the as-prepared Ag NP-based sensor probe were equivalent to those of another reported Ag NP-based fiber sensor probe with a different structure [34] and were much higher than those of other reported individual silver nanoparticles in solutions. Figure 3C shows the shift in the reflectivity of the sensor probe corresponding to the various sucrose solutions with different reflective indices. The sensitivity of the sensor probe was 160%/RIU in absolute value and the total optical intensity shift was approximately 12%.

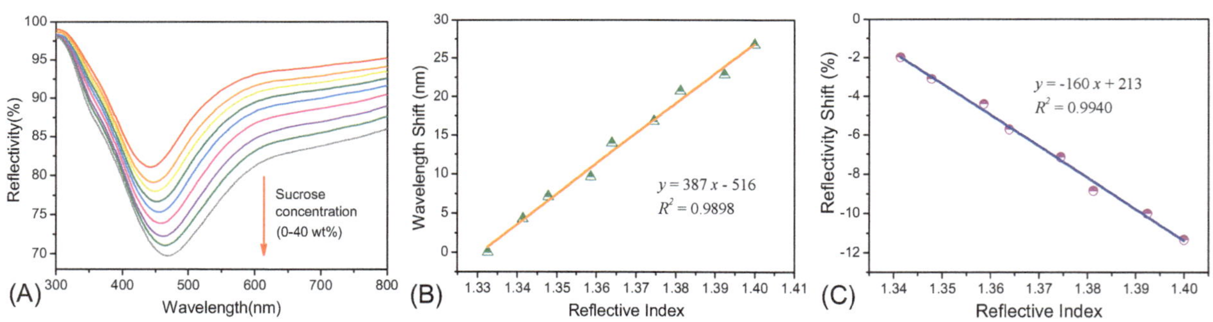

Figure 3. (**A**) The reflective spectra of the Ag NP-based sensors of sucrose solutions with different concentrations; (**B**) Changes in the LSPR wavelength shift to the refractive indices; (**C**) Changes in the LSPR reflectivity shift to the refractive indices.

According to a previous report [39], the resolution was also a key performance characteristic of an LSPR sensor and could be defined as the smallest change in the bulk refractive index that produced a detectable change in the sensor output. That is, the resolution of an LSPR sensor was related to the level of the standard deviation of noise of the sensor output. The formula was shown as follows:

$$r_{RI} = \frac{\sigma_{SO}}{S_{RI}} \quad (3)$$

where r_{RI} represented resolution, σ_{SO} was the standard deviation of noise of the sensor output, S_{RI} was the refractive index value. As we used the HR2000+ spectrometer (Ocean Optics), the number of σ_{SO} of wavelength and reflectivity was 0.35 nm and 0.75%, respectively. Take the data from Figure 3 for example.

As shown in Table 1, the r_{RI} based on wavelength shift was smaller than the r_{RI} based on reflectivity change, which was consistent with the compare result of refractive index sensitivity. With this method, we could compare the refractive index sensitivity of the two approaches: wavelength shift *versus* reflectivity change. Moreover, Cao *et al.* [12] had mentioned that wavelength-based technique was much more insensitivity to the interference of the surrounding environment. Therefore, we preferred to use the sensitivity based on wavelength peak shift rather than reflectivity in our following discussions.

Table 1. Compare of the resolution of the as-prepared LSPR sensor based on wavelength shift and optical intensity change respectively.

	σ_{SO}	S_{RI}	r_{RI}
Wavelength shift	0.35 nm	387 nm/RIU	9.04×10^{-4} RIU
Reflectivity change	0.75%	160%/RIU	4.68×10^{-3} RIU

3.2. Optimization of Sensor Performance

Prior to the biosensor applications of the sensor probe, two parameters, including the sensing length and the coating time, were optimized during the fabrication process.

To study the influence of the sensing length of the sensor probe, we prepared and tested a series of sensor probes with sensing lengths of 0.5 cm, 1 cm and 1.5 cm. The refractive index sensitivities are demonstrated in Figure 4, and the detailed reflective spectra can be seen in Figure S1 (Supplementary materials). As illustrated in Figure 4A, with an increase in the sensing length of the sensor probe, the refractive index sensitivity changed from 275 nm/RIU to 379 nm/RIU. The increase in the sensitivity may be due to the increase in the sensing area. A longer sensing length meant a larger sensing surface area; hence, more silver nanoparticles were immobilized on the sensing surface. Therefore, the interactions between the incident light and the nanoparticles would increase accordingly. In addition, according to Figure 4B, the optical intensity shift of the sensor probe increased with an increase in the sensing length, with a change from 5% to 11%, and the sensitivity increased from 86%/RIU to 160%/RIU. The optical intensity change is clearly evident and the sensitivity is high, indicating that both the peak wavelength and the optical intensity changed with the environmental refractive indices. What is more, the peak wavelength shift turned out to be much more sensitive.

A similar experiment was carried out to investigate the influence of the Ag NP coating time on the sensitivity of the sensor probe. We produced a series of sensor probes with different coating times of 0.5 h, 1 h and 3 h. The results obtained are shown in Figure 5. As observed, with an increase in coating time, the sensitivity initially increased from 173 nm/RIU to 461 nm/RIU (Figure 5A). However, as the coating time further increased, reaching 3 h, the sensitivity decreased to 355 nm/RIU and a new plasmon resonance band was observed in the higher wavelength region from 600 nm to 800 nm in the reflectivity spectrum, as shown in Figure S2 (Supplementary materials). This observation appears to be different from a finding reported previously [40]. As predicted, there was an increase in the refractive index sensitivity relative to the decrease in the particle pair partner's distances. As for this contradiction, we believed that the reason may be the aggregation of the absorbed nanoparticles. The appearance of a new plasmon resonance band further proves our assumption, which is inconsistent with that of other reported literature we mentioned above [35]. Hence, the total 1 h time was sufficient for the coating process using our approach. With longer coating times, a higher level of aggregation of the nanoparticles may occur, making it difficult to obtain a sensitive optical sensor probe. Moreover, we also studied the relationship between the optical intensity shift and the coating times; the results are illustrated in Figure 5B. With an increase in the coating time, the reflectivity decreased, and the sensitivity also decreased from 195%/RIU to 57%/RIU. The aggregation of the Ag NPs did not impart changes to the optical intensity.

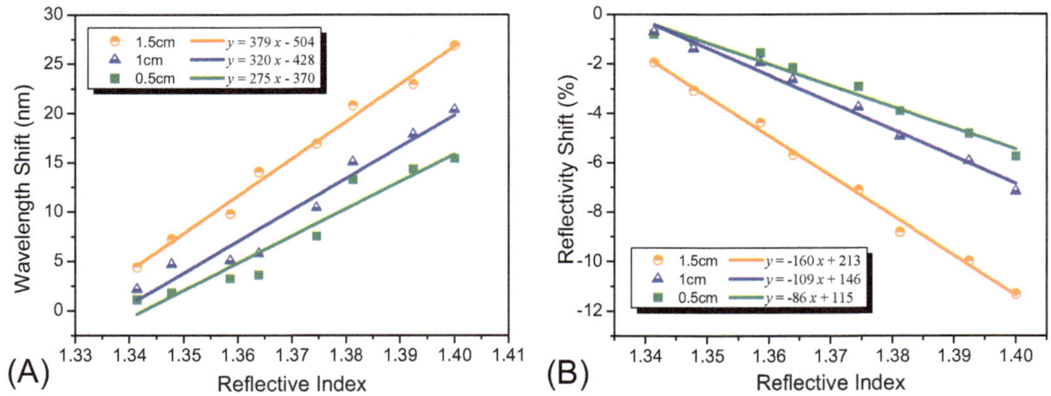

Figure 4. Refractive index sensitivities of the Ag NP-based sensors with different sensing lengths of 0.5 cm, 1 cm, 1.5 cm. (**A**) illustrates the wavelength shifts of the Ag NPs-based sensors with three different sensing lengths; (**B**) illustrates the reflectivity shift of the Ag NP-based sensors with three different sensing lengths.

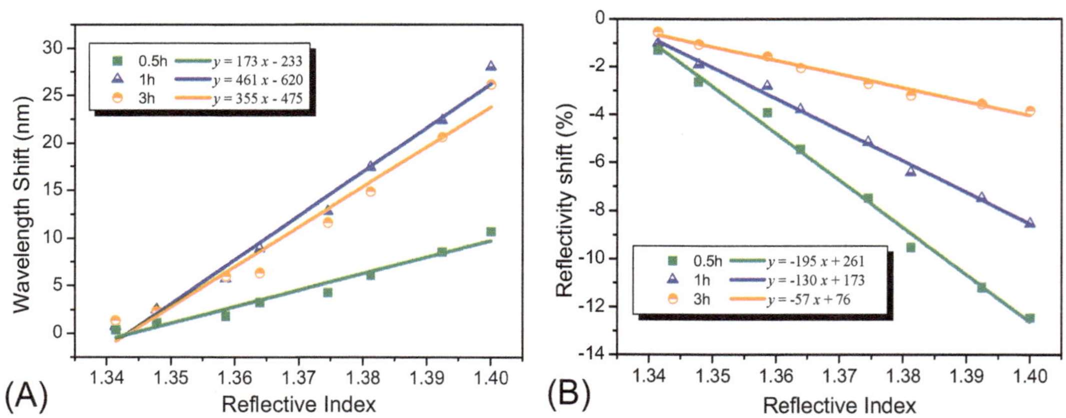

Figure 5. Refractive index sensitivities of the Ag NP-based sensors with different coating time of 0.5 h, 1 h, and 3 h. (**A**) illustrates the wavelength shifts comparison of the Ag NP-based sensors with three different coating times; (**B**) illustrates the optical intensity shifts comparison of the Ag NP-based sensors with three different coating time.

3.3. Stability of the Sensor Probe

To evaluate the stability of the sensor probes, a specific experiment was repeated every three days over a period of 18 days. For each test, the resonance wavelength of one freshly made Ag NP-based

LSPR sensor probe was recorded by immersing the sensor probe into deionized water and then into ethanol. After each test, the sensor probe was stored at 4 °C in deionized water. As shown in Figure 6, only a small fluctuation in resonance wavelength for each testing solution was observed, indicating that storing the sensor probe in the deionized water at 4 °C was effective, as the Ag NP-based LSPR sensor probe persevered in this way had a good stability over the test period of at least 18 days.

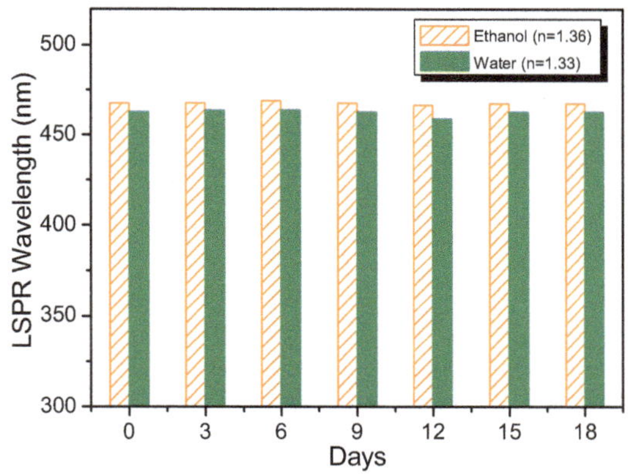

Figure 6. LSPR wavelength monitoring of the Ag NP-based sensor subjected in solutions with different refractive indices for a period of eighteen days.

3.4. Biosensor Applications (the Functionalization and the Bimolecular Interaction Monitoring of the Sensor Probe)

After optimization of the sensor performances, we chose a Ag NP-based sensor probe with a sensing length of 1.5 cm and a coating time of 1 h for further biosensing applications, as illustrated in Figure 1. In this experiment, we used the interaction of human IgG and rabbit anti-human IgG as an analytical model for testing the application of LSPR sensors on the study of antigen-antibody interaction. Before the study of the antigen-antibody interaction, the Ag NP-based sensor probe was treated by a conventional EDC/NHS method to bind human IgG. The sensor probe was immersed in a PBS buffer (pH = 7.4) to obtain and record a stable spectrum, which was used as a baseline for further study of the peak wavelength shift of each modification stage. Afterwards, the sensor probe was immersed into a solution containing 200 $\mu g \cdot mL^{-1}$ human-IgG to begin the binding process. As shown in Figure 7A, the wavelength shift of each modification step was monitored (the error bars were the sample-to-sample deviation, N = 3). Each wavelength shift was based on the baseline obtained from the sensor probe immersed in the PBS buffer. After the sensor probe was functionalized with human-IgG, the wavelength shift was approximately 7 nm. To decrease the specific binding, the biosensor activated by EDC/NHS was then immersed in a BSA solution to cap the active sites. After the BSA blocking, a further wavelength shift of 3 nm was observed. The wavelength shift was not obvious, indicating that there were few bare NHS-activated carboxyl groups at the surface of sensor probe. Afterwards, the biosensor was immersed into 100 $\mu g \cdot mL^{-1}$ anti-human IgG to begin the antigen-antibody binding process. A further wavelength shift of approximately 8 nm was detected. The peak wavelength shift for each modification step was high enough to be observed indicating that the as-prepared optical fiber biosensor has potential application for monitoring antigen-antibody binding processes.

The functionalization process was monitored by recording the spectrum every 5 min. As shown in Figure 7B, a gradual red shift of LSPR peak wavelength was observed with an increase in the functionalized time. The gradual red shift was caused by the refractive index changing at the sensing surface caused by the immobilization of anti-human IgG on the Ag NP-based sensor surface. The shift of the spectrum eventually leveled out indicating that the antibody immobilization of the LSPR biosensor was finished.

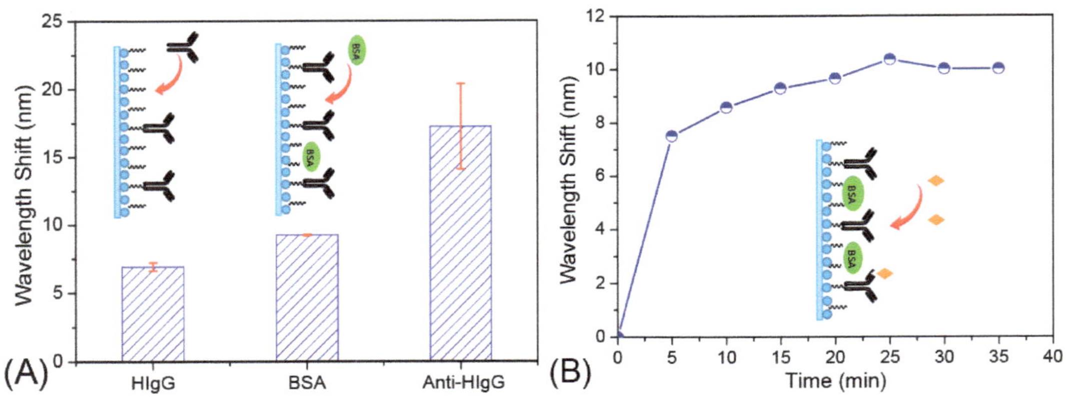

Figure 7. (**A**) Shifts of LSPR spectra of the Ag NP-based sensor after different stages of surface modification relative to the unmodified Ag NP sensor probe; (**B**) LSPR peak wavelength changes during the process of anti-human IgG immobilization on the Ag NP-based sensor surface over a period of 35 min.

4. Conclusions

In this work, we produced a Ag NP-based LSPR optical fiber biosensor and evaluated its refractive index sensitivity based on wavelength shift and reflectivity shift detection. We optimized the probe manufacturing process based on the sensing length of the sensor probe and the coating time of the Ag NPs. After the optimization process, the results showed that the refractive index sensitivity of the Ag NP-based LSPR sensor probe was 387 nm/RIU (wavelength shift) and 160%/RIU (reflectivity shift). Furthermore, we functionalized the sensor probe to demonstrate its use in bio-sensing applications. The fabricated biosensor probe could be used to monitor the interaction of an antigen and antibody. Compared to previously reported LSPR optical fiber sensors, we not only fabricated a stable and sensitive Ag NP-based reflective optical fiber sensor probe by a fast and simple method and optimized the fabrication process, we also used it to monitor the antigen-antibody binding process. With these systematic works, we demonstrated that the Ag NP-based reflective optical fiber sensor was able to be used as a biological sensor.

Acknowledgments

This work was supported by the Ministry of Science and Technology of China (Nos. 2012YQ090194, 2012AA06A303, and 2012BAD29B05), the Natural Science Foundation of China (Nos. 51473115, 21276192, 51173128), and the Ministry of Education (No. B06006 and NCET-11-0372).

Author Contributions

Jiangping Chen, Se Shi, Rongxin Su, Wei Qi, Mengfan Wang, Renliang Huang, Libing Wang and Zhimin He conceived and designed the experiments; Jiangping Chen performed the experiments; Jiangping Chen, Se Shi, Rongxin Su analyzed the data; Jiangping Chen, Se Shi, Rongxin Su and Wei Qi wrote the paper.

Conflicts of Interest

The authors declare no conflict of interest.

References

1. Chung, T.; Lee, S.Y.; Song, E.Y.; Chun, H.; Lee, B. Plasmonic nanostructures for nano-scale bio-sensing. *Sensors* **2011**, *11*, 10907–10929.
2. Zhang, X.-Y.; Hu, A.; Zhang, T.; Lei, W.; Xue, X.-J.; Zhou, Y.; Duley, W.W. Self-assembly of large-scale and ultrathin silver nanoplate films with tunable plasmon resonance properties. *ACS Nano* **2011**, *5*, 9082–9092.
3. Mock, J.J.; Smith, D.R.; Schultz, S. Local refractive index dependence of plasmon resonance spectra from individual nanoparticles. *Nano Lett.* **2003**, *3*, 485–491.
4. Shao, Y.; Xu, S.; Zheng, X.; Wang, Y.; Xu, W. Optical fiber LSPR biosensor prepared by gold nanoparticle assembly on polyelectrolyte multilayer. *Sensors* **2010**, *10*, 3585–3596.
5. Cennamo, N.; D'Agostino, G.; Dona, A.; Dacarro, G.; Pallavicini, P.; Pesavento, M.; Zeni, L. Localized surface plasmon resonance with five-branched gold nanostars in a plastic optical fiber for bio-chemical sensor implementation. *Sensors* **2013**, *13*, 14676–14686.
6. Ma, W.; Yang, H.; Wang, W.; Gao, P.; Yao, J. Ethanol vapor sensing properties of triangular silver nanostructures based on localized surface plasmon resonance. *Sensors* **2011**, *11*, 8643–8653.
7. Zhang, T.; Song, Y.J.; Zhang, X.Y.; Wu, J.Y. Synthesis of silver nanostructures by multistep methods. *Sensors* **2014**, *14*, 5860–5889.
8. Haes, A.J.; Zou, S.; Schatz, G.C.; Duyne, R.P.V. A nanoscale optical biosensor: The long range distance dependence of the localized surface plasmon resonance of noble metal nanoparticles. *J. Phys. Chem. B* **2003**, *108*, 109–116.
9. Mayer, K.M.; Hafner, J.H. Localized surface plasmon resonance sensors. *Chem. Rev.* **2011**, *111*, 3828–3857.
10. Wan, M.; Luo, P.; Jin, J.; Xing, J.; Wang, Z.; Wong, S.T. Fabrication of localized surface plasmon resonance fiber probes using ionic self-assembled gold nanoparticles. *Sensors* **2010**, *10*, 6477–6487.
11. Kim, H.M.; Jin, S.M.; Lee, S.K.; Kim, M.G.; Shin, Y.B. Detection of biomolecular binding through enhancement of localized surface plasmon resonance (LSPR) by gold nanoparticles. *Sensors* **2009**, *9*, 2334–2344.
12. Cao, J.; Tu, M.H.; Sun, T.; Grattan, K.T.V. Wavelength-based localized surface plasmon resonance optical fiber biosensor. *Sens. Actuators B Chem.* **2013**, *181*, 611–619.
13. Mishra, S.K.; Gupta, B.D. Surface plasmon resonance based fiber optic sensor for the detection of CrO_4^{2-} using Ag/ITO/hydrogel layers. *Anal. Methods* **2014**, *6*, 5191–5197.

14. Fan, M.; Thompson, M.; Andrade, M.L.; Brolo, A.G. Silver nanoparticles on a plastic platform for localized surface plasmon resonance biosensing. *Anal. Chem.* **2010**, *82*, 6350–6352.
15. Yang, S.; Wu, T.; Zhao, X.; Li, X.; Tan, W. The optical property of core-shell nanosensors and detection of atrazine based on localized surface plasmon resonance (LSPR) sensing. *Sensors* **2014**, *14*, 13273–13284.
16. Chen, I.C.; Lin, S.S.; Lin, T.J.; Du, J.K. Detection of hydrofluoric acid by a SiO_2 sol-gel coating fiber-optic probe based on reflection-based localized surface plasmon resonance. *Sensors* **2011**, *11*, 1907–1923.
17. Lee, T.H.; Lee, S.W.; Jung, J.A.; Ahn, J.; Kim, M.G.; Shin, Y.B. Signal amplification by enzymatic reaction in an immunosensor based on localized surface plasmon resonance (LSPR). *Sensors* **2010**, *10*, 2045–2053.
18. Sharma, A.K.; Gupta, B.D. Fiber optic sensor based on surface plasmon resonance with nanoparticle films. *Photon. Nanostr. Fundam. Appl.* **2005**, *3*, 30–37.
19. Sharma, A.K.; Jha, R.; Gupta, B.D. Fiber-optic sensor based on surface plasmon resonance with Ag-Au alloy nanoparticle films. *Nanotechnology* **2006**, *7*, 124–131.
20. Cheng, S.-F.; Chau, L.-K. Colloidal gold-modified optical fiber for chemical and biochemical sensing. *Anal. Chem.* **2003**, *75*, 16–21.
21. Tu, M.H.; Sun, T.; Grattan, K.T.V. LSPR optical fibre sensors based on hollow gold nanostructures. *Sens. Actuators B Chem.* **2014**, *191*, 37–44.
22. Tu, H.; Sun, T.; Grattan, K.T. SPR-based optical fiber sensors using gold–silver alloy particles as the active sensing material. *IEEE Sens. J.* **2013**, *13*, 2192–2199.
23. Sciacca, B.; Monro, T.M. Dip biosensor based on localized surface plasmon resonance at the tip of an optical fiber. *Langmuir* **2014**, *30*, 946–54.
24. Ortega-Mendoza, J.G.; Padilla-Vivanco, A.; Toxqui-Quitl, C.; Zaca-Moran, P.; Villegas-Hernandez, D.; Chavez, F. Optical fiber sensor based on localized surface plasmon resonance using silver nanoparticles photodeposited on the optical fiber end. *Sensors* **2014**, *14*, 18701–18710.
25. Chau, L.-K.; Lin, Y.-F.; Cheng, S.-F.; Lin, T.-J. Fiber-optic chemical and biochemical probes based on localized surface plasmon resonance. *Sens. Actuators B Chem.* **2006**, *113*, 100–105.
26. Consales, M.; Ricciardi, A.; Crescitelli, A.; Esposito, E.; Cutolo, A.; Cusano, A. Lab-on-fiber technology: Toward multifunctional optical nanoprobes. *ACS Nano* **2012**, *6*, 3163–3170.
27. Lin, Y.; Zou, Y.; Mo, Y.; Guo, J.; Lindquist, R.G. E-beam patterned gold nanodot arrays on optical fiber tips for localized surface plasmon resonance biochemical sensing. *Sensors* **2010**, *10*, 9397–9406.
28. Mitsui, K.; Handa, Y.; Kajikawa, K. Optical fiber affinity biosensor based on localized surface plasmon resonance. *Appl. Phys. Lett.* **2004**, *85*, 4231–4233.
29. Sai, V.V.; Kundu, T.; Mukherji, S. Novel U-bent fiber optic probe for localized surface plasmon resonance based biosensor. *Biosens. Bioelectron.* **2009**, *24*, 2804–2809.
30. Srivastava, S.K.; Arora, V.; Sapra, S.; Gupta, B.D. Localized surface plasmon resonance-based fiber optic U-shaped biosensor for the detection of blood glucose. *Plasmonics* **2011**, *7*, 261–268.
31. Bharadwaj, R.; Mukherji, S. Gold nanoparticle coated U-bend fibre optic probe for localized surface plasmon resonance based detection of explosive vapours. *Sens. Actuators B Chem.* **2014**, *192*, 804–811.

32. Jorgenson, R.C.; Yee, S.S. A fiber-optic chemical sensor based on surface plasmon resonance. *Sens. Actuators B Chem.* **1993**, *12*, 213–220.
33. Sagle, L.B.; Ruvuna, L.K.; Ruemmele, J.A.; Van Duyne, R.P. Advances in localized surface plasmon resonance spectroscopy biosensing. *Nanomedicine* **2011**, *6*, 1447–1462.
34. Dutta, R.; Bharadwaj, R.; Mukherji, S.; Kundu, T. Study of localized surface-plasmon-resonance-based optical fiber sensor. *Appl. Opt.* **2011**, *50*, E138–E144.
35. Dutta, R.; Singh, B.P.; Kundu, T. Plasmonic Coupling Effect on spectral response of silver nanoparticles immobilized on an optical fiber sensor. *J. Phys. Chem. C* **2013**, *117*, 17167–17176.
36. Chen, Y.-Q.; Lu, C.-J. Surface modification on silver nanoparticles for enhancing vapor selectivity of localized surface plasmon resonance sensors. *Sens. Actuators B Chem.* **2009**, *135*, 492–498.
37. Cao, J.; Galbraith, E.K.; Sun, T.; Grattan, K.T. Cross-comparison of surface plasmon resonance-based optical fiber sensors with different coating structures. *IEEE Sens. J.* **2012**, *12*, 2355–2361.
38. Lin, T.J.; Chung, M.F. Detection of cadmium by a fiber-optic biosensor based on localized surface plasmon resonance. *Biosens. Bioelectron.* **2009**, *24*, 1213–1218.
39. Homola, J. Surface plasmon resonance sensors for detection of chemical and biological species. *Chem. Rev.* **2008**, *108*, 462–493.
40. Jain, P.K.; El-Sayed, M.A. Noble metal nanoparticle pairs: Effect of medium for enhanced nanosensing. *Nano Lett.* **2008**, *8*, 4347–4352.

4

Embedded Ultrasonic Transducers for Active and Passive Concrete Monitoring

Ernst Niederleithinger *, Julia Wolf †, Frank Mielentz †, Herbert Wiggenhauser † and Stephan Pirskawetz †

BAM Federal Institute for Materials Research and Testing, Berlin 12200, Germany;
E-Mails: julia.wolf@bam.de (J.W.); frank.mielentz@bam.de (F.M.);
herbert.wiggenhauser@bam.de (H.W.); stephan.pirskawetz@bam.de (S.P.)

† These authors contributed equally to this work.

* Author to whom correspondence should be addressed; E-Mail: ernst.niederleithinger@bam.de

Academic Editor: Thomas Schumacher

Abstract: Recently developed new transducers for ultrasonic transmission, which can be embedded right into concrete, are now used for non-destructive permanent monitoring of concrete. They can be installed during construction or thereafter. Large volumes of concrete can be monitored for changes of material properties by a limited number of transducers. The transducer design, the main properties as well as installation procedures are presented. It is shown that compressional waves with a central frequency of 62 kHz are mainly generated around the transducer's axis. The transducer can be used as a transmitter or receiver. Application examples demonstrate that the transducers can be used to monitor concrete conditions parameters (stress, temperature, …) as well as damages in an early state or the detection of acoustic events (e.g., crack opening). Besides application in civil engineering our setups can also be used for model studies in geosciences.

Keywords: ultrasound; transmission; concrete; damages; cracks; stress

---

1. Introduction

Concrete is a complex, multi-phase material. It is made of hydraulic cement, water and aggregates in lots of variations. The first two ingredients start to hydrate and crystallize when in contact with each other. This process is fast in the first hours and days, but can continue for months or years. The aggregates, gravel or crushed stone of various types and sizes (µm-cm range), are used as a filler to save cost and energy. Concrete contains pores, either filled with unbound water or air.

Concrete is the material most produced by mankind. It is considered to be strong, resistive and durable. Some early concrete structures as the cupola of the Roman Pantheon are standing tall after almost 2000 years. However, under certain conditions (hostile environments, adverse load conditions) concrete constructions require attention. For example the apparently ever increasing traffic load (number and individual load of trucks) on bridges may lead to deterioration much earlier than expected at the time of design. Currently, inspections are still mainly based on visual methods, but sophisticated non-destructive methods are used more and more often. The use of bridge instrumentation increases rapidly [1]. However, monitoring is so far limited to local sensors (e.g., strain gauges), which are probing just their close vicinity, or global methods as modal analysis, which looks to the structure as a whole. There is a gap in between. A method which would look at a certain critical volume of concrete with a very limited number of sensors would be of great value.

Ultrasonic transducers with frequencies from 25 kHz to 400 kHz have been used for concrete since decades. They are used in the lab on samples (and sometimes on site) in transmission mode to measure elastic properties and to assess degradation. New point contact transducers have revolutionized the use of echo techniques for thickness measurements and structural imaging. Even the detection of voids in tendon ducts seems to be possible [2]. All transducers used in practice today are for surface mounting. For monitoring this approach shows three strong disadvantages. First, the need for constant coupling, which is hard to realize on the surface in practice. Second, the high influence of surface and external effects (temperature and others) leads to unwanted effects on the measurements. Third, in practical field applications the transducers are prone for accidents or vandalism. Therefore we started to develop a novel transducer, which can be permanently embedded in concrete.

This is not the first or only attempt to embed ultrasonic transducers in concrete. Similar ideas have been proposed e.g., by [3,4], but only in experimental setups for lab applications. For practical applications a much more robust approach would be required. An experiment conducted more than 35 years ago to monitor the hardening of concrete at a massive water dam in Saxony, Germany, by embedded ultrasonic transmitters and receivers recently gave us the opportunity to prove that this kind of sensors might survive in concrete for decades [5].

2. A novel Ultrasonic Transducer to Be Embedded in Concrete

2.1. Transducer Design and Description

New ultrasonic transducers ("SO807") have been designed by Acoustic Control Systems, Ltd. (ACS, Moscow, Russia) in cooperation with and exclusively for BAM (Figure 1). The main part is a hollow piezoceramic cylinder of 20 mm diameter and 35 mm length. The electric connections are on the inside. On both ends metallic pieces are clamped to the piezoceramic part. The outer diameter of

15 mm allows stacking of several transducers along a line using standard PVC tubes. Having all cables inside ensures good coupling of the piezo to the concrete and protects the electrical connections during installation. The total length of the transducer is 75 mm.

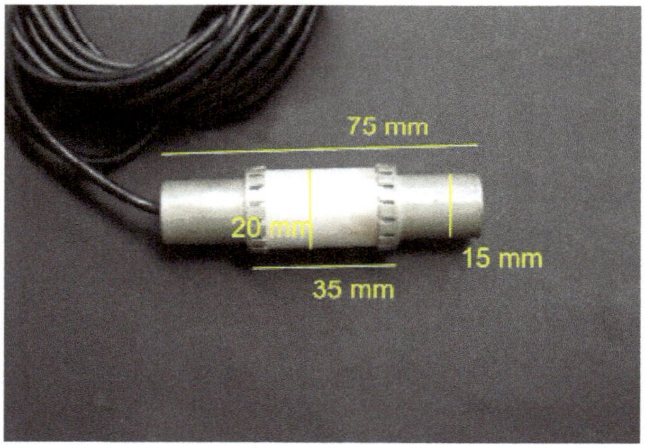

Figure 1. Photograph of the novel ultrasonic transducer for embedment in concrete (Manufacturer: Acsys Ltd., Moscow, Russia. Photo: BAM).

2.2. Installation

The transducers can easily be installed at the time of construction. To ensure that they keep their position during casting and vibration, they have to be mounted either direct to the reinforcement or using some kind of stabilizing construction. In our latest bridge installation we have used L-shaped pieces of rebar welded to the reinforcement to hold the transducers at the specified position (Figure 2). Another possibility would be to mount stacked series of transducers using PVC tube segments (similar to the method used for installation in existing structures described below).

Figure 2. Photograph of an ultrasonic transducer mounted at a bridge construction site before casting (Photo: Neostrain S.A.).

For existing structures we have developed a method to install one or several transducers at the required depth(s) in a drill hole and ensuring sufficient coupling to the concrete. The hole is drilled with a slightly larger diameter than the transducers and slightly deeper than the installation depth. The transducer(s) and the tube segments are connected with a sealing cap to the structures' surface

(Figure 3). The cap contains an inlet, which is connected to a (liquid) grout reservoir and to the space between the transducers/tubes and the concrete. By connecting a suction pump via an outlet in the cap to the inner hollow space of the transducers and the tubes the grout is sucked into the drill hole and via the crown of the tubes into the inner space (Figure 3c). When the grout appears at the outlet we can be sure the entire space is (at least almost) filled and the transducers are coupled to the concrete. We are using a fast hardening, slightly expanding type of grout.

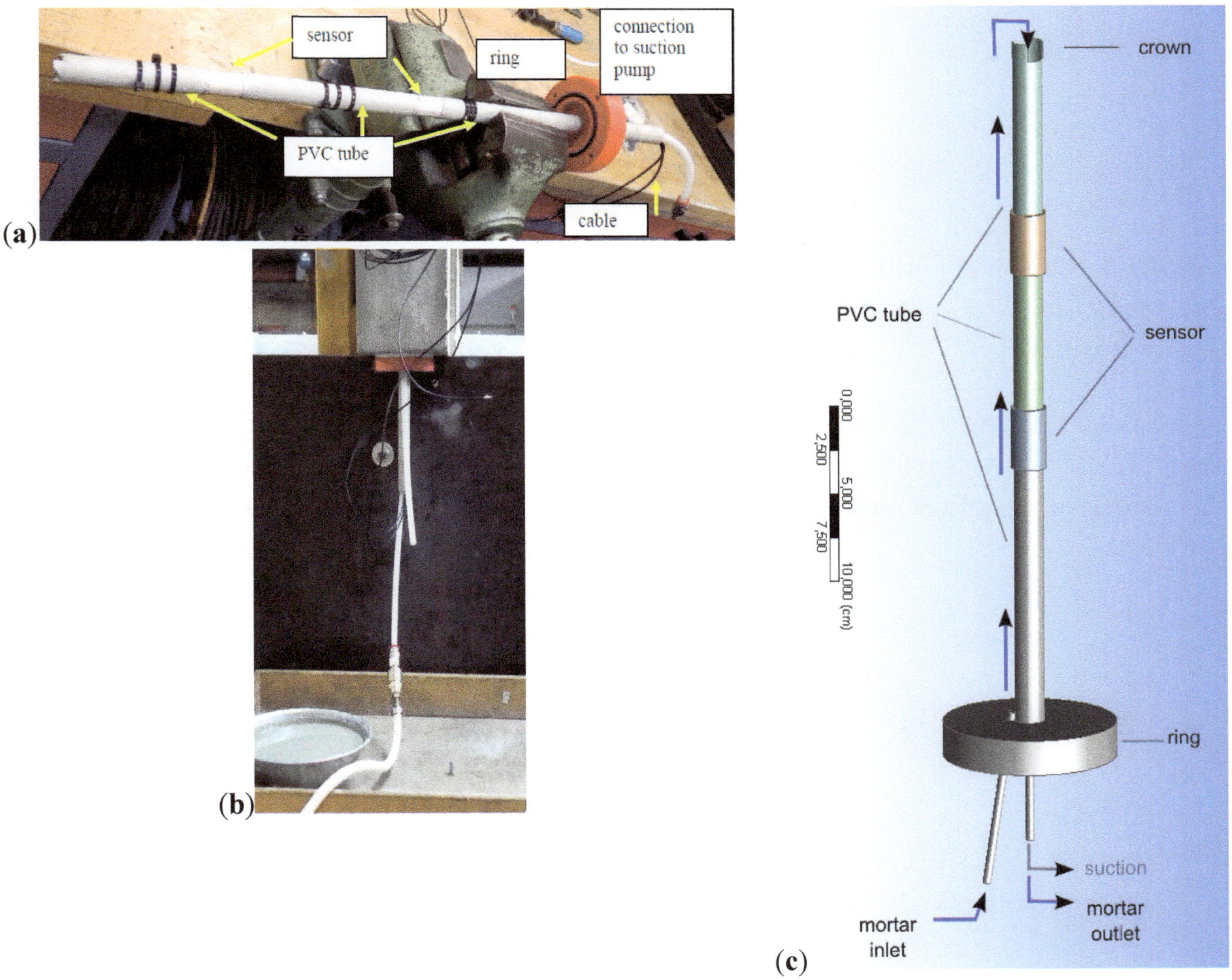

Figure 3. (**a**): Two transducers connected by PVC tube segments and equipped with sealing cap for post-concreting installation; (**b**): Installation into a concrete specimen; (**c**): Sketch of grout flow during installation.

2.3. Characterization

The transducers have been characterized by some basic experiments. First, two transducers of the same type have been installed back to back directly without any medium in between to evaluate the frequency spectrum. A short impulse (t = 2 μs) was used as input. The signal recorded by the transducer is displayed in Figure 4. The signal recorded shows a wavelet with about 20 periods and a total duration of ca. 0.3 ms. The reverberations show a lack of damping of the piezo material. As we don't intend to use the transducers for imaging applications, where a sharp response (broadband in frequency domain) would be more beneficial, this behavior is fully satisfactory.

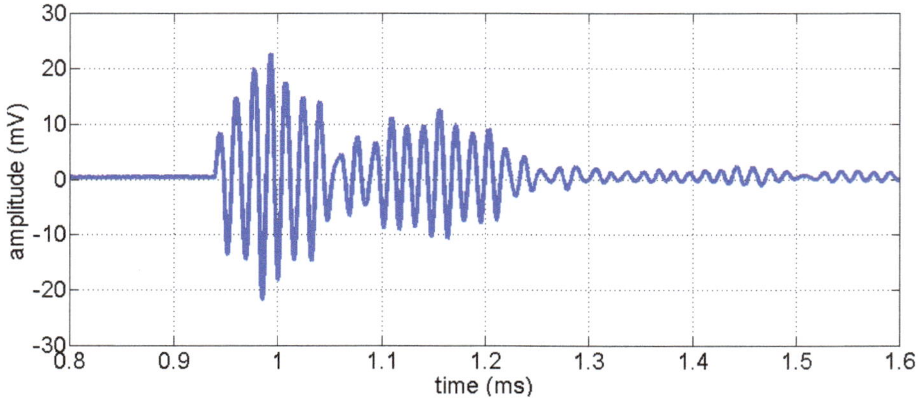

Figure 4. Response of the novel transducer to a short electrical impulse (2 μs), recorded by a second, identical one mounted back to back. Time of onset due to recording hardware settings.

The amplitude spectrum of the data of Figure 4 is shown in Figure 5. There is a prominent frequency peak at 62 kHz and a significant second one at 65 kHz. Smaller peaks appear around 50 and 85 kHz. There is no significant energy with frequencies lower than 40 or higher than 90 kHz.

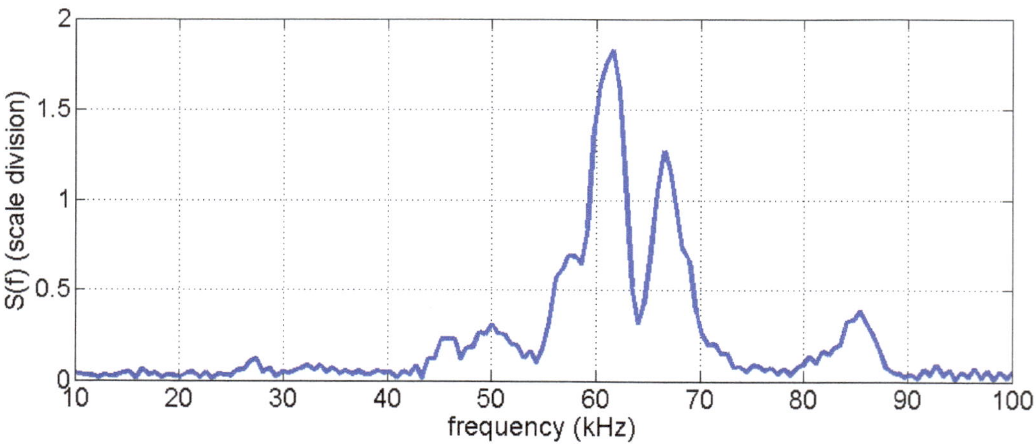

Figure 5. Amplitude spectrum of data shown in Figure 4.

A frequency of 62 kHz relates to a wavelength in concrete (compressional wave speed of about 4000 m/s) of ca. 65 mm. This is at least double the size of most aggregates (max aggregate size 32 mm in many types of concrete).

To evaluate the directivity pattern of the new transducer, we have conducted two experiments, the first in a water pool, the second using transducers embedded horizontally or vertically, respectively, in two separate cylindrical concrete blocks (Figure 6). In the first case (water pool) an identical transducer (vertically oriented) was used as receiver. In the second experiment a laser vibrometer was used to record the surface vibrations of the concrete. The laser vibrometer's position was fixed while the block was rotated in 10° intervals before each excitation. In both cases the measurements were taken using (a) vertical and (b) horizontal orientation of the transmitting transducers.

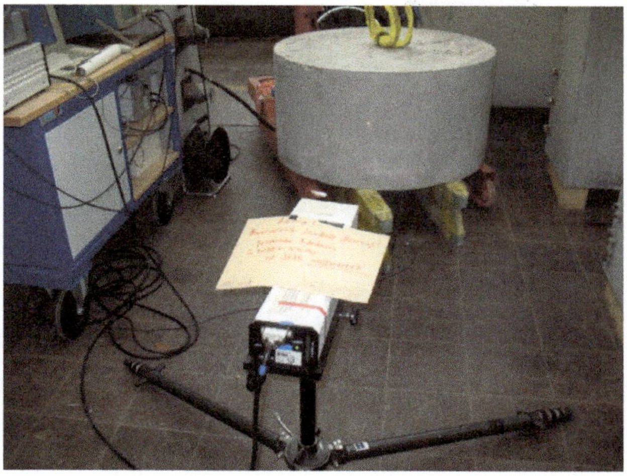

Figure 6. Experimental setup for measurement of directivity patterns. Transmitting transducers are embedded in the center of the concrete block. Surface movements are recorded by a laser vibrometer (in front).

The results of the directivity measurements are shown in Figure 7. To produce these plots, the amplitude of the direct arrivals (maximum of the first cycle) have been taken in 10° steps and normalized to the maximum value of all of them. For a vertical transmitter (right in Figure 7) we have recorded an almost perfect circle in water. This had to be expected as the transmitter is rotational symmetric round its vertical axis. In concrete the circular radiation pattern is still recognizable, but far from being perfect, probably due to the inherent inhomogeneities in concrete.

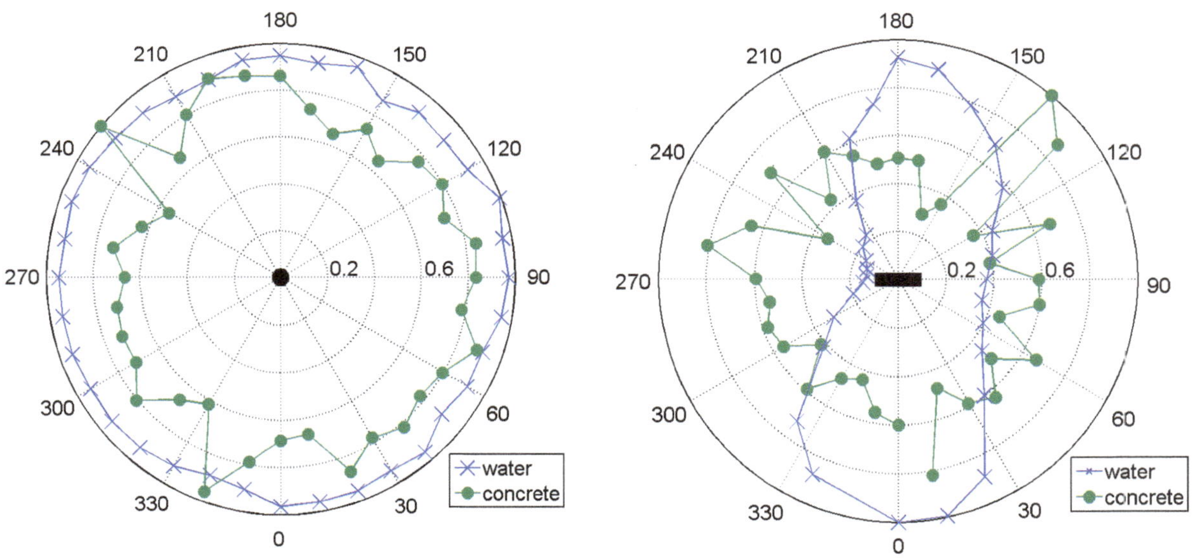

Figure 7. Directivity patterns for the new transducer, measured in water (blue crosses) and concrete bodies (green dots). (**Left**) vertical orientation; (**Right**) horizontal orientation. Axis from center to perimeter: relative amplitude. Seen from above.

For horizontal transmitter orientation (left in Figure 7) the directivity pattern in water resembles something close to a number eight. Apparently (and not unexpectedly as applying a voltage to the cylindrical piezoceramic leads mainly to changes in diameter, not in length) the amplitudes emitted in the direction of the transducer's rotational axis are much smaller than in the perpendicular direction.

The experiment with the same configuration in concrete showed that the amplitude is similar in all directions (but has significant variation). Our interpretation is that due to scattering at the concrete's aggregates, pores and cracks, the directivity pattern is somewhat equalized.

Additional experiments have been performed to get an idea on the range of our new transducers. This was done to assess the maximum possible distance between transmitters and receivers in future sensor networks in massive concrete structures. For this purpose a set of transducers was installed in two concrete blocks partly with, partly without reinforcement. Some transducers were installed before casting, some afterwards (see Section 2.2). One block had a maximum aggregate size of 16 mm, the other 32 mm. For all scenarios (aggregate size, reinforcement, and installation type) we have used the same measurement setup. One of the transducers, excited with a 100 V square pulse, was used as transmitter. At two others (distance r = 0.25 and 0.75 m, respectively, from the transmitter) the amplitudes A (maximum of first cycle) of the first arrivals were recorded. The amplitude ratios are shown in Figure 8. It shows, that the attenuation is mainly dependent on the installation type, but not on aggregate size or reinforcement (the latter not shown here). The applicable range of our transducers was estimated from these data to be around three meters [6,7]. For this we calculated the material and frequency specific damping constant α from the amplitudes (A_1, A_2) of the first arrival after r_1 = 0.25 m and r_2 = 0.75 m using:

$$-\alpha = \left[ln\left(\frac{A_2}{A_1} \cdot \frac{r_2}{r_1} \right) \right] \frac{1}{(r_2 - r_1)} \tag{1}$$

As an estimate for the maximum applicable range r_E we have replaced r_2, A_2 in this equation by r_E, A_E and resolved for r_E .Under the assumption for the amplitude A_E at r_E to be at least twice the noise level, which might be different for every measurement environment, we have calculated values between 3 m and 5.5 m for our experimental setups [6,7]. The amplitudes for transducers installed after concreting are larger, probably because the special expanding grout used guarantees perfect coupling.

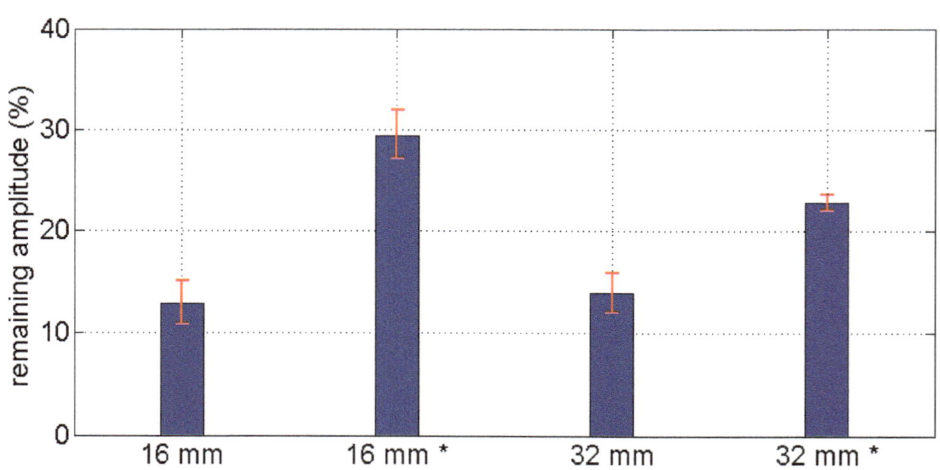

Figure 8. Amplitude ratios between near and far receivers in the attenuation experiment for the two specimens (16 and 32 mm max. aggregate size). The * marks values for transducers installed after concreting.

3. Short Notes on Ultrasonic Transmission Experiments

3.1. Wave Propagation in Concrete

Piezotransducers induce elastic waves into a concrete body. The propagation of these wave fields depends on the transducer, signal type and frequency as well as the body's material, and geometry. In general, three types of waves are generated: Surface (Rayleigh) waves, which are slightly less important in our case, as we are using embedded transducers, and two types of body waves: compressional and shear. In a homogeneous linear elastic material the velocities of these waves (c_p, c_s) depend just on a few parameters: Young's modulus E, shear modulus G, Poisson's ratio ν, and density ρ:

$$c_p = \sqrt{\frac{E(1-\upsilon)}{\rho(1+\nu)(1-2\nu)}} \tag{2}$$

and:

$$c_s = \sqrt{\frac{G}{\rho}} \tag{3}$$

respectively.

In reality concrete is neither homogeneous nor fully linear elastic. In addition, ultrasonic waves are subject to reflections at internal boundaries or scattering at small (similar or smaller than the wavelength used) objects as aggregates, small cracks or reinforcement rebars. At the same time the amplitudes of the ultrasonic waves are affected by geometrical and intrinsical (material related) attenuation. From literature it is known that the following factors (and potentially also other) have an influence on ultrasonic velocities and the amplitudes:

- Concrete type and compressive strength ([8–10]);
- Stress ([11–15]);
- Temperature ([7,9,10,16,17]);
- Moisture ([18–20]);
- Degradation (microcracking) ([13,21–23]).

Traditional time of flight measurements are considering direct waves only (first arrival at the receiver in most cases). The area of influence is limited to a narrow band ("first Fresnel zone") between transmitter and receiver (Figure 9 left).

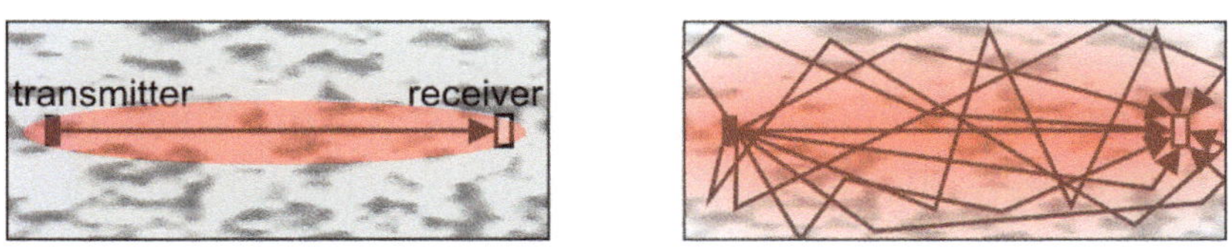

Figure 9. Sketch of ray paths (black lines)and area of influence (red) for the direct wave (**left**) and the full signal including coda (**right**).

Lots of other waves, which arrive much later at the receiver, have undergone reflections and scattering, potentially even wave type conversion. It is quite often difficult to evaluate these arrivals separately. However, they contain useful information as they are covering larger areas of the concrete body (Figure 9 right) and are more sensible to velocity changes due to the longer travel paths.

3.2. Data Evaluation

The traditional way to interpret transmission data ("time of flight") is described in the corresponding standards on ultrasonic pulse velocity measurements (see e.g., [10,24]). Picking of the first arrival travel times is often done manually. A lot of different automatic picking algorithms have been proposed from simple threshold pickers to more sophisticated ones based on statistical criteria. Based on our experience we prefer the Aikake Information Criterion (AIC) picker as described in [25]. This picker has been successfully used in many of our applications (e.g., [5]). Main disadvantage of the time of flight method is that changes in the medium under investigation may result in very small changes of the arrival times which are hard to detect. In addition, we would need a dense network of transducers in a construction to cover the entire volume by the relatively narrow Fresnel zones.

Recently methods have been introduced to ultrasound investigations in concrete, which are using the entire signal and not just the first arrival. Online monitoring systems can benefit from the use of correlation calculations (of a time series measured during/after load/damage against a reference one). Some more details and preliminary results have been published in [26]. For quantitative evaluation a novel method called coda wave interferometry may be used, which is able to calculate velocity changes from ultrasonic data with a sensitivity of about 2×10^{-5} [27]. The method can be expanded to tomographic imaging applications. So far, only a few applications to concrete have been reported [26–29].

4. Application Examples

4.1. Monitoring of Load Changes

A $1.5 \times 1.5 \times 1.5$ m^3 concrete block ("GK32") has been cast in the BAM labs for various tests of the embedded ultrasonic transducers (Figure 10). The lower half contains a certain amount of reinforcement, the upper one is unreinforced. A total number of 18 ultrasonic transducers have been embedded, partly during partly after casting the block. Just ten of those have been used for a load monitoring experiment due to limitation of the available data acquisition equipment. The experiment has already been described in more detail in [26]. A two channel multiplexer had connected the transducers to an ultrasonic transmitter (rectangle, 50 kHz) or data recording system, respectively. All 90 transducer combinations could be interrogated within seconds or minutes, depending on the number of repetitions. In the upper half of the concrete block a hole was drilled to insert a thread bolt. Some of the transducers have been just a few cm away from the center of the load, some almost 1 m. Nuts, 10×10 cm^2 load distributing plates and a piezo load cell provided a way to introduce localized compressional stress in a controlled, repeatable way. Direction of the main compressional load is perpendicular to the front surface shown in Figure 10. However, stresses parallel to the front face are generated as well. Load steps of 5 or 10 kN were applied in various cycles up to a maximum load between 20 and 100 kN, more than one order of magnitude below the compressive strength of the concrete.

Figure 10. Concrete specimen GK32 with embedded transducers and load application system. From [26].

The applied loads, even very small ones, had a clear influence on the ultrasonic signals. A simple but valuable tool to provide a measure for the change is calculating the correlation coefficient between a reference measurement (here: zero load) and all consecutive measurement under various load conditions. Figure 11 shows the development of the correlations coefficient for transmission data between two embedded transducers close to the loading point. Both transducers have the same embedment depth (seen from the front face). Thus, direct waves are traveling perpendicular to the main load direction, but later parts of the signals (reflections, scattering) contain also information from different propagation directions.

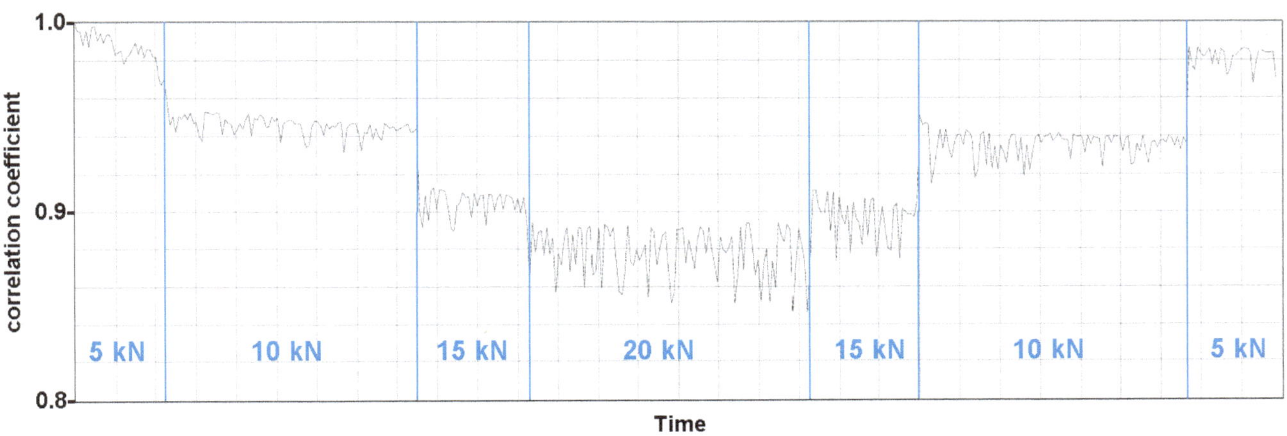

Figure 11. Correlation coefficient of 5 ms time series (reference: measurement at zero load) of ultrasonic signals measured by embedded transducers in a concrete block under small local compressional load. Line represents very dense consecutive measurements. From [26].

Even small load changes of 5 kN can clearly be seen, even in the presence of noise. Tomographic coda wave evaluation of the entire embedded transducer data set showed that the area of significant influence of the load is limited to about 0.3 m away from the loading point [26].

4.2. Acoustic Emission

The acoustic emission (AE) technique monitors acoustic waves produced by newly developing micro cracks, opening and closure of existing cracks, friction, *etc*., all of which are caused by internal stress variations [30]. Based on techniques used in seismology the source of the acoustic emission can be localized. Traditionally the sensors used to detect AE events are arranged on the surface of the monitored element. The distance between them is restricted by the attenuation of the waves within the material. The frequency spectrum of the acoustic events in concrete is below 200 kHz [31]. If a set of embedded transducers would be able to perform passive (AE, localization of active cracks) and active measurements (determination of changes of velocity/attenuation/material parameters) at the same time, this would be a great step forward for structural health monitoring of concrete structures.

Laboratory tests on the use of the embedded transducers for AE were conducted using a similar specimen as discussed in Section 4.1 ("GK16"). All twelve embedded transducers were used with an AMSY6 recording system (Vallen Systeme GmbH, Icking, Germany) to detect acoustic emission events within the specimen and on its surface. All following measurements have been repeated three times. In a first step, the embedded transducers were used one at a time as artificial acoustic emission sources by sending a voltage pulse to each transducer successively. The average velocity of the compressional waves, measured during this automatic sensor test, was around 4500 m/s. Figure 12 shows the corresponding event localizations (green dots) calculated by the location processor option "Planar, plane" of the software Vallen VisualAE. They are coincident with the transducer locations.

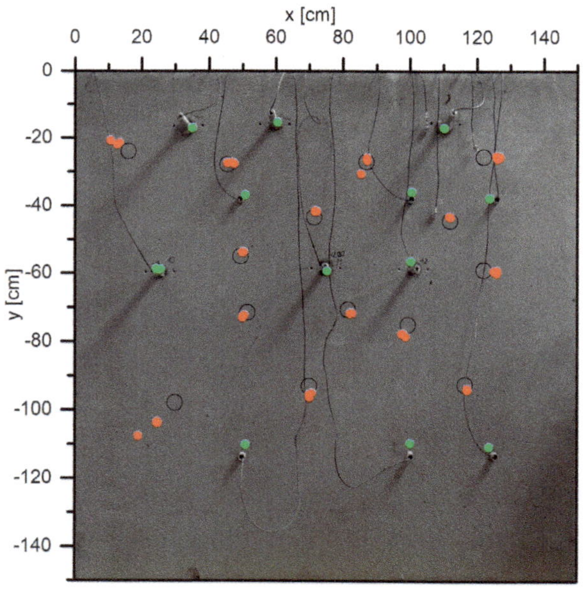

Figure 12. Location of acoustic emissions monitored with embedded ultrasonic transducers. Green marks: events localized by AE within the specimen (sources: one of the embedded transducers). Red marks: event localized by AE on the specimen's surface (source: pencil breaks in black circles).

In a second set of experiments pencil breaks on the specimen's surface (a commonly used test for AE systems according to ASTM E976) have been used. Again, the experiments have been repeated three times at points marked by circles in Figure 12 and the events have been recorded and evaluated using the AMSY6 system. The localization (red dots in Figure 12) in areas with good transducer coverage lies on the mark ±3 cm. Those events outside the transducer array were not located accurately.

The experiment shows that the joint use of the embedded transducers in active and passive measurements is possible. The source needs to be within the sensor network to allow for an accurate localization. Future research will focus on the appropriate distance between the receivers for detecting and localizing acoustic emissions of different frequency and energy.

4.3. Time Reversal Experiment

This application example illustrates the use of the embedded transducers for model experiments in geophysical research. In geophysics, the validation of new approaches for measurement techniques or data evaluation is quite often limited to simulations as the real subsurface (especially in exploration when talking about several kilometers depth) can never fully be explored. So full scale experiments often lack of ground truth. Scaled model experiments might be of help.

Time reversal is a technique to backproject elastic waves, which have been generated by events (e.g., cracks) inside a medium and registered by some sensors inside the medium as well or at the surface, numerically or physically back to their source point. Very few sensors are required for scattering media. In fact, the signals at the sensor, which have started as a spike at the source but then have been reflected, scattered, attenuated and dispersed on the way, can be focused back to the source point (as a perfect spike under ideal conditions). Recently a method to improve the results using deconvolution was proposed [32]. The transducers described in this paper have been used to verify the benefits of this method in case of multiple events overlapping in time [33]. For this purpose three transducers were embedded in a small concrete body (Figure 13) and served as ultrasonic sources (mimicking real acoustic emissions) in the first step. The data recorded at a single external receiver were then processed by the technology proposed in [32] and re-emitted by the external transducer into the model. Our embedded transducers now served as receivers. According to theory energy should focus at the position of a specific transducer at the time corresponding to the original event. This results in distinct sharp peaks in the recorded data (Figure 14). They correspond to the original events in time and shape. It could be shown, that the processed data focused much better in time and space at the transducers positions compared to unprocessed data. This experiment using our new transducers has definitively helped to prove the concept of using deconvolution in time reversal experiments, which would have been difficult under field conditions (events in some km depth) or doubted by practitioners if just done numerically. Details of the methodology, the experiment and the implications of the results are discussed in [33].

4.4. Other Applications

The embedded transducers have been applied in two other fields of applications so far. In several experiments we have used them to instrument concrete lab samples (typical size 15 × 15 × 40 cm^3). These samples have been put into climate chambers or CDF test devices to evaluate the influence of temperature [16] and moisture on ultrasonic data. This evaluation is required in real world monitoring application as the influence of loads or deterioration is often covered by environmental influences. Instrumented lab samples may also be used in CDF test for freeze thaw resistance evaluations. The measurement of ultrasonic pulse velocity is already part of these evaluations, but the samples have to be removed from the CDF test chambers so far. Embedded transducers would make these tests much easier.

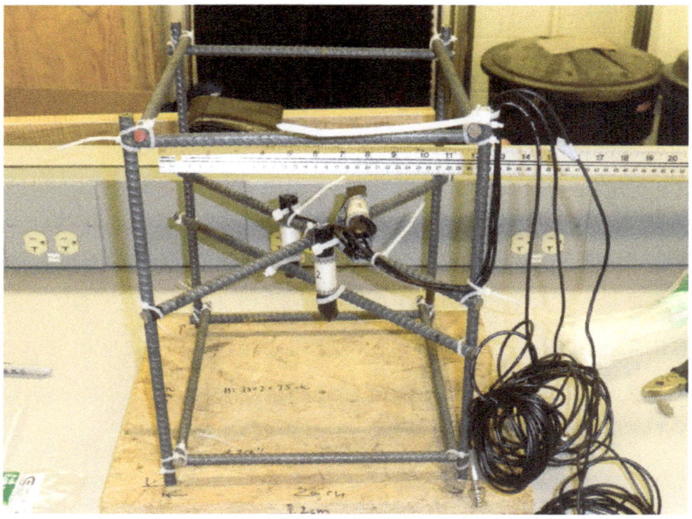

Figure 13. Setup of the three embedded transducers on the time reversal model experiment before concreting of the model, with permission from [33].

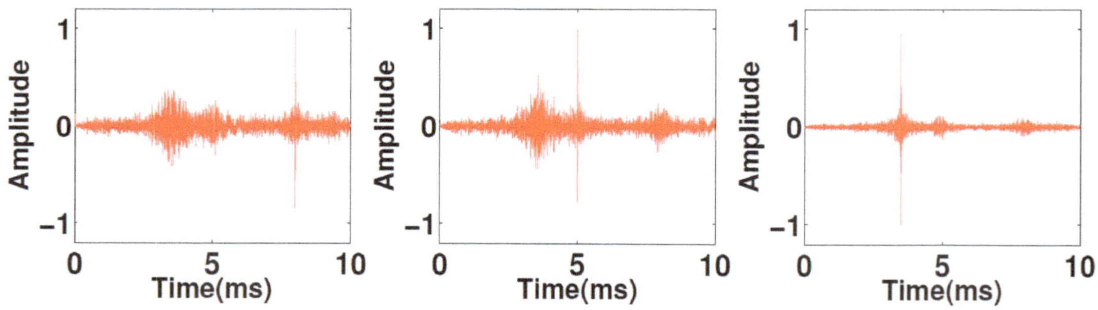

Figure 14. Backpropagated signals recorded at the three embedded transducers. Note the sharp peaks at specific times for each transducers reflecting perfectly the original time and place of the acoustic events, with permission from [33].

We have started to use the embedded transducers in full scale monitoring applications as well. An existing, degraded bridge in Turkey is monitored since more than two years by a set of eight transducers. A new bridge in Poland has been equipped with eight transducers before concreting. The hardening of the concrete could be monitored during the first 28 days. Load experiments (a requirement in Poland) will follow.

5. Conclusions

The ultrasonic transducers developed for embedment in concrete have shown to be valuable tools for various tasks in structural monitoring. They met our expectations in frequency (around 60 kHz), directivity (almost circular around the main axis) and range (at least three meters). The transducers proved to be very robust. We have developed deployment systems for existing and newly built structures. Early versions are now embedded and used in lab samples and real structures for a few years and are still fully operational.

We have shown that the transducers are useful in a lot of applications. They can be used for active transmission experiments as well as for collecting passive acoustic emission data. They proved to be useful in lab samples, scale experiments and real world monitoring systems. Load changes can be detected and localized as well as environmental influences (temperature, moisture) and various degradation mechanisms. In connection with novel interpretation tools as coda wave interferometry we have very sensible methods for the detection of changes in concrete at hand.

The embedment in concrete has various advantages: The coupling to the concrete is constant, transducers at lab samples have not to be removed before putting them into climate chambers, chemical baths or similar and installations are more vandal proof and less accident prone on real constructions.

Acknowledgments

The authors wish to acknowledge the help of various colleagues at BAM, the excellent cooperation with the manufacturer of the transducers (Acsys Ltd., Moscow, Russia) as well as the fruitful discussions on coda waves with Sens-Schönfelder (Helmholtz Geoforschungszentrum Potsdam) and Roel Snieder (Colorado School of Mines).

Author Contributions

Ernst Niederleithinger wrote the main part of paper and worked on the load experiment as well as the time reversal application example.

Julia Wolf performed the experiments on transducer characterization as well as the AE application examples. She designed most of the figures and contributed in writing.

Frank Mielentz is the supervisor of Julia Wolf and contributed greatly to all ultrasonic experiments by designing and evaluating the electronic setups and some of the experiments for transducer characterization.

Herbert Wiggenhauser gave the original idea for the design of the transducers and worked with the manufacturer on the implementation. He also contributed to the experimental design and to data evaluation.

Stephan Pirskawetz performed the experiments and evaluated the data for the AE application example.

Conflicts of Interest

The authors declare no conflict of interest.

References

1. Karbhari, V.M.; Ansari, M. *Structural Health Monitoring of Civil Infrastructure Systems*; Woodhead Publishing: Cambridge, UK, 2009, ISBN 978-1-84569-392-3.
2. Krause, M.; Mayer, K.; Friese, M.; Milmann, B.; Mielentz, F.; Ballier, G. Progress in ultrasonic tendon duct imaging. *Eur. J. Environ. Civil Eng.* **2011**, *15*, 461–485.
3. Song, G.; Gu, H.; Mo, Y.-L. Smart aggregates: Multi-functional sensors for concrete structures—A tutorial and a review. *Smart Mater. Struct.* **2008**, *17*, doi:10.1088/0964-1726/17/3/033001.
4. Kee, S.H.; Zhu, J. Using piezoelectric sensors for ultrasonic pulse velocity measurements in concrete. *Smart Mater. Struct.* **2013**, *22*, doi:10.1088/0964-1726/22/11/115016.
5. Niederleithinger, E.; Krompholz, R.; Müller, S.; Lautenschläger, R.; Kittler, J. 36 Jahre Talsperre Eibenstock—36 Jahre Überwachung des Eibenstock—36 Years Monitoring the concrete condition by Ultrasound, in German). In Proceedings of 38. Dresdner Wasserbaukolloquium, Germany, 2015. Available online: http://www.researchgate.net/publication/274708685_36_Jahre_Talsperre_Eibenstock__36_Jahre_berwachung_des_Betonzustands_durch_Ultraschall (accessed on 24 April 2015).
6. Wolf, J.; Mielentz, F.; Milmann, B.; Helmerich, R.; Köpp, C.; Wiggenhauser, H. Ultrasound based monitoring system for concrete monolithic objects. In Procedings of the 6th International Conference on Structural Health Monitoring of Intelligent Infrastructure (SHMII-6), Hong Kong, China, 9 December 2013.
7. Wolf, J.; Niederleithinger, E.; Mielentz, F.; Grothe, S; Wiggenhauser, H. Überwachung von Betonkonstruktionen mit eingebetteten Ultraschallsensoren. *Bautechnik* **2014**, *91*, doi:10.1002/bate.201400073.
8. Jones, R.; Facaoaru, I. Recommendations for testing concrete by the ultrasonic pulse method. *Mater. Constr.* **1969**, *2*, 275–284.
9. Crawford, G.I. Guide to Nondestructive Testing of Concrete. *Technical Report FHWA-SA-97-105*; Federal Highway Administration: Washington, DC, USA, 1997.
10. British Standards Institute (BSI). Testing Concrete in Structures. Determination of Ultrasonic Pulse Velocity. BS EN 12504-4:2004 (equivalent German version: DIN EN 12504–4:2004). Available online: http://shop.bsigroup.com/ProductDetail/?pid=000000000030102823 (accessed on 24 April 2015).
11. Sayers, C.M. Stress-induced ultrasonic wave velocity anisotropy in fractured rock. *Ultrasonics* **1988**, *26*, 311–317.
12. Shokouhi, P.; Zoega, A. Surface Wave Velocity-Stress Relationship in Uniaxially Loaded Concrete. *ACI Mater. J.* **2012**, *109*, 141–148.
13. Suaris, W.; Fernando, V. Ultrasonic Pulse Attenuation as a Measure of Damage Growth during Cyclic Loading of Concrete. *ACI Mater. J.* **1987**, *84*, 185–193.
14. Wu, T.T. The Stress Effect on the Ultrasonic Velocity Variations of Concrete under Repeated Loading. *ACI Mater. J.* **1998**, *95*, 519–524.

15. Zhang, Y.; Abraham, O.; Grondin, F.; Loukili, A.; Tournat, V.; Duff, A.-L.; Lascoup, B.; Durand, O. Study of stress-induced velocity variation in concrete under direct tensile force and monitoring of the damage level by using thermally-compensated Coda Wave Interferometry. *Ultrasonics* **2012**, *52*, 1038–1045.
16. Niederleithinger, E.; Wunderlich, C. Influence of small temperature variations on the ultrasonic velocity in concrete. *Rev. Prog. Quant. Nondestruct. Eval.* **2013**, *1511*, 390–397.
17. Zhang, Y.; Abraham, O.; Tournat, V.; Duff, A.L.; Lascoup, B.; Loukili, A.; Grondin, F.; Durand, O. Validation of a thermal bias control technique for Coda Wave Interferometry (CWI). *Ultrasonics* **2013**, *53*, 658–664.
18. Ohdaira, E.; Masuzawa, N. Water content and its effect on ultrasound propagation in concrete—The possibility of NDE. *Ultrasonics* **2000**, *38*, 546–552.
19. Hedenblad, G. *Moisture Permeability of Some Porous Materials*; Report TVBM (Intern 7000-Rapport); Byggnadstysik, DTH, Lyngby, September 1993. Available online: http://lup.lub.lu.se/luur/download?func=downloadFile&recordOId=1653173&fileOId=1653174 (accessed on 24 April 2015).
20. Lencis, U. Moisture Effect on the Ultrasonic Pulse Velocity in Concrete Cured under Normal Conditions and at Elevated Temperature. *Constr. Sci.* **2013**, *14*, 71–78.
21. Payan, C.; Quiviger, A.; Garnier, J.F. Applying diffuse ultrasound under dynamic loading to improve closed crack characterization in concrete. *J. Acoust. Soc. Am.* **2013**, *134*, doi:10.1121/1.4813847.
22. Ramamoorthy, S.K.; Kane, Y.; Turner, J.A. Ultrasound diffusion for crack depth determination in concrete. *J. Acoust. Soc. Am.* **2004**, *115*, 523–529.
23. Payan, C.; Garnier, V. Determination of third order elastic constants in a complex solid applying coda wave interferometry. *Appl. Phys. Lett.* **2009**, *94*, doi:10.1063/1.3064129.
24. ASTM. Standard Test Method for Pulse Velocity through Concrete. ASTM C597-09. Available online: http://www.astm.org/Standards/C597.htm (accessed on 24 April 2015).
25. Tronicke, J. The Influence of High Frequency Uncorrelated Noise on First-Break Arrival Times and Crosshole Traveltime Tomography. *J. Environ. Eng. Geophys.* **2007**, *12*, 173–184.
26. Niederleithinger, E.; Sens-Schönfelder, C.; Grothe, S.; Wiggenhauser, H. Coda Wave Interferometry Used to Localize Compressional Load Effects on a Concrete Specimen. In Proceedings of the 7 th European Workshop on Structiral Health Monitoring (EWSHM 2014), Nantes, France, 8–11 July 2014.
27. Larose, E.; Hall, S. Monitoring stress related velocity variation in concrete with a 2×10^{-5} relative resolution using diffuse ultrasound. *J. Acoust. Soc. Am.* **2009**, *125*, 1853–1856.
28. Planes, T.; Larose, E.; Margerin, L.; Rossetto, L.; Sens-Schönfelder, C. Decorrelation and phaseshift of coda waves induced by local changes: Multiple scattering approach and numerical validation. *Waves Random Complex Media* **2014**, *24*, 99–125.
29. Kanu, C. Time-Lapse Monitoring of Localized Changes within Heterogeneous Media with Scattered Waves. Ph.D. Thesis, Colorado School of Mines, Golden, CO, USA, 5 December 2014.

30. Huang, M.; Jiang, L.; Liaw, P.K.; Brooks, C.R.; Seelev, R.; Klarstrom, D.L. Using Acoustic Emission in Fatigue and Fracture Materials Research. *Memb. J. Miner. Metals Mater. Soc.* **1998**, *50*. Available online: http://www.tms.org/pubs/journals/JOM/9811/Huang/Huang-9811.html (accessed on 24 April 2015).
31. Große, C.U.; Schumacher, T. Anwendungen der Schallemissionsanalyse an Betonbauwerken. *Bautechnik* **2013**, *90*, 721–731.
32. Anderson, B.E.; Douma, J.; Ulrich, T.J.; Snieder, R. Improving Spatio-Temporal Focusing and Source Reconstruction through Deconvolution. *Wave Motion* **2015**, *52*, 151–159.
33. Douma, J.; Niederleithinger, E.; Snieder, R. Locating Events Using Time Reversal and Deconvolution: Experimental Application and Analysis. *J. Nondestruct. Eval.* **2015**, in press.

5

Ultrasonic Sensor Signals and Optimum Path Forest Classifier for the Microstructural Characterization of Thermally-Aged Inconel 625 Alloy

Victor Hugo C. de Albuquerque [1,*], Cleisson V. Barbosa [1], Cleiton C. Silva [2], Elineudo P. Moura [2], Pedro P. Rebouças Filho [3], João P. Papa [4] and João Manuel R. S. Tavares [5]

[1] Programa de Pós-Graduação em Informática Aplicada, Universidade de Fortaleza, Fortaleza, Ceará 60811-905, Brazil; E-Mail: cleissonvb@gmail.com

[2] Departamento de Engenharia Metalúrgica e de Materiais, Universidade Federal do Ceará, Fortaleza, Ceará 60455-900, Brazil; E-Mails: cleiton@ufc.br (C.C.S.); elineudo@ufc.br (E.P.M.)

[3] Programa de Pós-Graduação em Ciências da Computação, Instituto Federal de Educação, Ciência e Tecnologia do Ceará, Fortaleza, Ceará 61939-140, Brazil; E-Mail: pedrosarf@ifce.edu.br

[4] Departamento de Ciência da Computação, Universidade Estadual Paulista, Bauru, São Paulo 17033-360, Brazil; E-Mail: papa@fc.unesp.br

[5] Instituto de Ciência e Inovação em Engenharia Mecânica e Engenharia Industrial, Departamento de Engenharia Mecânica, Faculdade de Engenharia, Universidade do Porto, Porto 4200-465, Portugal; E-Mail: tavares@fe.up.pt

* Author to whom correspondence should be addressed; E-Mail: victor.albuquerque@unifor.br

Academic Editor: Vittorio M.N. Passaro

Abstract: Secondary phases, such as laves and carbides, are formed during the final solidification stages of nickel-based superalloy coatings deposited during the gas tungsten arc welding cold wire process. However, when aged at high temperatures, other phases can precipitate in the microstructure, like the γ'' and δ phases. This work presents an evaluation of the powerful optimum path forest (OPF) classifier configured with six distance functions to classify background echo and backscattered ultrasonic signals from samples of the inconel 625 superalloy thermally aged at 650 and 950 °C for 10, 100 and 200 h. The background echo and backscattered ultrasonic signals were acquired using transducers with frequencies of 4 and 5 MHz. The potentiality of ultrasonic sensor signals combined with the OPF to characterize the microstructures of an inconel 625 thermally aged and in the as-welded condition were confirmed by the results. The experimental results revealed that the OPF classifier is sufficiently fast (classification total time of 0.316 ms) and accurate (accuracy of 88.75% and harmonic mean of 89.52) for the application proposed.

Keywords: ultrasonic sensor; metric function; optimum path forest; signal classification; microstructural characterization

---

1. Introduction

Nb-bearing nickel-based superalloys, in particular inconel 625, has greater applicability, especially in highly corrosive environments, such as the ones in the oil and gas industry, than many other nickel (Ni)-based alloys. Nowadays, this alloy is used widely in the weld overlay of the inner surface of carbon steel pipes and other equipment of the offshore industry. However, further studies about this alloy, such as the one presented in this paper, are necessary to increase the overall knowledge of its properties.

During the welding of an inconel 625 alloy, there is an intensive microsegregation of some elements, such as niobium (Nb) and molybdenum (Mo), within the interdendritic regions, causing the supersaturation of the liquid metal with these chemical elements in its final stage of solidification, which results in the precipitation of the Nb-rich laves phase and MCprimary carbides of type NbC [1,2]. This segregation and precipitation of the secondary phases can change the mechanical properties of the alloy and decrease its resistance to corrosion [3]. In addition, the Nb-rich laves phase has a low melting point that causes an increase in the temperature solidification range, making the alloy susceptible to solidification cracking [4].

Nondestructive testing based on ultrasonic signals has been commonly used to study this kind of material. For example, in the evaluation of the embrittlement kinetics and elastic constants of the SAF2205 duplex stainless steel for different aging times at 425 and 475 °C [5], spinodal decomposition mechanism study on the UNSS31803 duplex stainless steel [6], evaluation of grain refiners' influence on the mechanical properties in a CuAlBe shape memory alloy [7], sigma phase detection on a UNS S31803 duplex stainless steel [8], characterization of welding defects [9], characterization of cast iron microstructure [10], pattern classification in nondestructive materials inspection [11], nondestructive characterization of microstructures and determination of elastic properties in plain carbon steel [12] and in the phase transformations evaluation on a UNS S31803 duplex stainless steel based on nondestructive testing [13].

In this sense, the main goal of this work was to evaluate the influence of six distance functions, mainly the Euclidean, chi-square, Manhattan, Canberra, squared chi-squared and Bray–Curtis distances, in the performance of the recent and powerful optimum path forest classifier to detect/identify, based on ultrasonic signals, the kinetics of the phase transformation of a Ni-based alloy thermally aged at 650 and 950 °C for 10, 100 and 200 h, as well as in the as-welded state. Raw data ultrasonic background echo and backscattered signals acquired with two types of transducers (4 and 5 MHz) were used. For a further assessment of the distance functions' performance, the results obtained were very satisfactory in terms of accuracy rate, train and test times, confusion matrix and harmonic mean between specificity and sensitivity, which makes the results presented and discussed of noteworthy value.

The OPF has been evaluated in different applications as, for example, EEG signal classification for epilepsy diagnosis [14], ECG arrhythmia classification [15], automatic characterization of graphite particles in metallographic images [16], intrusion detection in computer networks [17], aquatic weed automatic classification [18] and spoken emotion recognition [19], among others.

2. Materials and Methods

This section describes the experimental work done for the temperatures of 650 and 950 °C for 10, 100 and 200 h, as well as for the as-welded state. First, the ultrasonic sensor signals acquired and the related fundamentals are introduced. Afterwards, the optimum path forest classifier used to classify the ultrasonic signals is presented. Finally, the metrics used in the classifier evaluation are described.

2.1. Ultrasonic Sensor Signals

After the welding and preparation of the samples, described in detail in [20,21], the background echo and backscattered signals were acquired to evaluate the effect of aging on the inconel 625 alloy samples.

The pulse echo technique and the direct contact method were used to collect the background echo and backscattered ultrasonic signals [8]. All of the signals were obtained using commercial nondestructive testing (NDT) ultrasonic transducers: one of 4 MHz (Krautkramer, Model MB4S, Lewistown, PA, USA) and another one of 5 MHz (Krautkramer, Germain, Model MSW-QCG). The choice of these transducers was based on the authors previous experience in this kind of NDT and knowledge concerning the material under study [22–25]. In fact, Albuquerque *et al.*, in [21], showed that these frequencies revealed were to be the most adequate to analyze the material under study, as a transducer with a frequency of 10 MHz completely attenuated the ultrasonic signal; and one with a frequency of 2.25 MHz led to an adjacent echo that overlapped the signal extensively, seriously compromising the accuracy of the results.

As a coupling material, the SAE 15W40 lube oil was used for the longitudinal measurements. A Krautkramer ultrasonic device (GE Inspection Technologies, Lewistown, PA, USA, model USD15B) was used connected to a 100-MHz digital oscilloscope (Tektronix, Portland, OR, USA, model TDS3012B), which transmitted the ultrasonic signals to a computer for processing, according to a sampling rate of 1 GS/s.

The microstructural characterization was carried out using the OPF classifier configured with the Euclidean, chi-square, Manhattan, Canberra, squared chi-squared and Bray–Curtis distances on the original background echo and backscattered signals. In order to assure statistical significance, 40 signals were acquired for each sample, and each background echo signal had 10,000 points; *i.e.*, a total of 400,000 points was attained, and each backscattered signal had 500 points, resulting in a total of 20,000 points for this study.

Albuquerque *et al.*, in [21], did not consider echo signals without preprocessing, claiming that the large number of points made their use impracticable. However, this problem has been overcome, because the classifier used here is faster and more powerful, which is one of the important contributions attained with this work. Nunes *et al.* [20] compared the OPF, configured only with the Euclidean distance, with the support vector machine and Bayesian classifiers and showed its superiority in terms of the processing time and accuracy rate. Thus, another contribution of this work was to analyze the influence of six distance functions on the OPF's performance to detect/identify microstructural changes from the ultrasonic signals due to aging.

2.2. Optimum Path Forest Classifier

The OPF classifier models the problem of pattern recognition as a graph partition in a given feature space. The nodes are represented by the ultrasonic signal feature vectors, and all pairs are connect by edges, defining a complete graph. This kind of representation is straightforward, given that the graph does not need to be explicitly represented, and has low memory requirements. The partition of the graph is carried out by a competition process between some key samples, known as prototypes, which offer optimum paths to the remaining nodes of the graph. Each prototype sample defines its optimum path tree (OPT), and the collection of all OPTs defines the optimum path forest, which gives the name to the classifier [26].

The OPF can be seen as a generalization of the well-known Dijkstra algorithm to compute optimum paths from a source node to the remaining ones [27]. The main difference relies on the fact that OPF uses a set of source nodes, *i.e.*, the prototypes, with any path-cost function. In the case of Dijkstra's algorithm, a function that summed the arc-weights along a path was applied. For OPF, a function that gives the maximum arc-weight along a path is used [26].

Let $Z = Z_1 \cup Z_2$ be a dataset labeled with a function λ, in which Z_1 and Z_2 are, respectively, training and test sets, and let $S \subseteq Z_1$ be a set of prototype patterns (ultrasonic signal feature vectors). Essentially, the OPF classifier builds a discrete optimal partition of the feature space, such that any sample $s \in Z_2$ can be classified according to this partition. This partition is an optimum path forest (OPF) computed in $\Re^n$ by the image foresting transform (IFT) algorithm [28].

The OPF algorithm may be used with any smooth path-cost function that can group ultrasonic signal features with similar properties [28]. This work used the path-cost function f_{max}, which is computed as:

$$\begin{aligned} f_{max}(\langle s \rangle) &= \begin{cases} 0 & \text{if } s \in S, \\ +\infty & \text{otherwise.} \end{cases} \\ f_{max}(\pi \cdot \langle s, t \rangle) &= \max\{f_{max}(\pi), d(s,t)\} \end{aligned} \tag{1}$$

in which $d(s,t)$ means the distance between ultrasonic signal features s and t, and a path π is defined as a sequence of adjacent features. As such, $f_{max}(\pi)$ computes the maximum distance between adjacent samples in π, when π is not a trivial path.

The OPF algorithm assigns one optimum path $P^*(s)$ from S to every ultrasonic signal feature $s \in Z_1$, originating an optimum path forest P (a function with no cycles, which assigns to each $s \in Z_1 \backslash S$ its predecessor $P(s)$ in $P^*(s)$ or a marker nil when $s \in S$. Let $R(s) \in S$ be the root of $P^*(s)$ that can be reached from $P(s)$. The OPF algorithm computes for each $s \in Z_1$ the cost $C(s)$ of $P^*(s)$, the label $L(s) = \lambda(R(s))$ and the predecessor $P(s)$.

The OPF classifier is composed of two distinct phases: (I) training; and (II) classification. The former step consists, essentially, of finding the prototypes and computing the optimum path forest, which is the union of all OPTs rooted at each prototype. After that, a sample is picked from the test sample, which connects it to all of the samples of the optimum path forest generated in the training phase. Notice that this test sample is not permanently added to the training set, *i.e.*, it is performed only once. The next sections describe this procedure in more detail.

2.3. Training

We say that S^* is an optimum set of prototypes when the OPF algorithm minimizes the classification errors for every $s \in Z_1$. S^* can be found by exploiting the theoretical relation between the minimum spanning tree (MST) and optimum path tree for f_{max} [29]. The training essentially consists of finding S^* and an OPF classifier rooted at S^*.

By computing an MST in the complete graph (Z_1, A), we obtain a connected acyclic graph whose nodes are all ultrasonic signal features of Z_1, and the arcs are undirected and weighted by the distances d between adjacent features. The optimum spanning tree is the tree that has the least sum of its arc compared to any other spanning tree in the complete graph. In the MST, every pair of ultrasonic signal features is connected by a single path that is optimum according to f_{max}. That is, the minimum spanning tree contains one optimum path tree for any selected root node.

The optimum prototypes are the closest elements of the MST with different labels in Z_1; *i.e.*, elements that fall in the frontier of the classes. By removing the arcs between different classes, their adjacent features become prototypes in S^*, and OPF can compute an optimum path forest with minimum classification errors in Z_1. It should be noted that a given class may be represented by multiple prototypes, *i.e.*, optimum path trees, and there must exist at least one prototype per class.

2.4. Classification

For any ultrasonic signal feature $t \in Z_2$, all arcs connecting t with samples $s \in Z_1$ are addressed, as though t were part of the training graph. Considering all possible paths from S^* to t, the optimum path $P^*(t)$ from S^* is found, and t is labeled with the class $\lambda(R(t))$ of its most strongly connected prototype $R(t) \in S^*$. This path can be identified incrementally by evaluating the optimum cost $C(t)$ as:

$$C(t) = \min\{\max\{C(s), d(s,t)\}\}, \ \forall s \in Z_1 \tag{2}$$

Let the node $s^* \in Z_1$ be the one that satisfies Equation (2), *i.e.*, the predecessor $P(t)$ in the optimum path $P^*(t)$. Given that $L(s^*) = \lambda(R(t))$, the classification simply assigns $L(s^*)$ as the class of t. An error occurs when $L(s^*) \neq \lambda(t)$.

2.5. Performance Evaluation Metrics

In order to analyze the performance of the machine learning technique used, three metrics were employed: accuracy, sensitivity and specificity.

Accuracy (Acc) is defined as the ratio of the total number of samples correctly classified and the number of total samples,

$$Accuracy = \frac{number\ of\ correctly\ classified\ samples}{number\ total\ of\ samples} \tag{3}$$

Sensitivity (Se) can be defined as the ratio of the total number of samples correctly classified of one class and the total number of samples classified as belong to that class, including the number of the missed classified samples,

$$Sensitivity = \frac{true\ positives}{true\ positives\ +\ false\ negatives} \tag{4}$$

in which true positives and false negatives stand for the number of samples of a given class correctly and incorrectly classified, respectively.

Specificity (Sp) stands for the ratio of the total number of samples correctly classified and the number of all samples classified as belonging to a specific class,

$$Specificity = \frac{true\ negatives}{true\ negatives\ +\ false\ positives} \tag{5}$$

in which true negatives stands for the number of samples not belonging to a given class classified as not belonging to the considered class, while false positives stands for the number of samples incorrectly classified as belonging to a given class. Observe that these last two measures are based on the data of each class.

Furthermore, we also propose the use of a harmonic average between sensitivity and specificity, that is the harmonic mean (HM):

$$HM = 2 \times \frac{Se \times Sp}{Se + Sp} \tag{6}$$

These evaluation metrics can be computed from a confusion matrix, which can be obtained by comparing the expected classification (reference data) with the ones predicted by the classifier. Besides these measures for evaluating and comparing the effectiveness performance of the classifier used, we also compute the training and testing times.

3. Results and Discussion

The original ultrasonic background echo and backscattered signals, acquired using 4- and 5-MHz transducers, were classified using the OPF classifier configured with the Euclidean, chi-square, Manhattan, Canberra, squared chi-squared and Bray–Curtis distances. The classification efficiency (processing time) and efficacy (accuracy rate, confusion matrices and harmonic mean) were analyzed. Thus, it is possible to evaluate the performance of the classifier to identify the microstructural classes. The original signals, for all distance metrics, were partitioned using the holdout method (50% for training and 50% for testing). The standard deviation for the mean accuracy, harmonic mean and processing time load over 10 iterations generated randomly were computed. The execution was performed on a personal computer with an Intel Core i3, at 3 GHz and with 3 GB of RAM using Linux Ubuntu as the operational system.

3.1. Efficiency and Effectiveness Analysis

The performance of each distance used by the OPF classifier was assessed through the accuracy rate, harmonic mean between specificity and sensitivity, processing time considering the training and testing phases and, finally, the confusion matrix.

Figure 1 shows the PCA (principal component analysis) corresponding to the spatial distribution of the samples of the 4-MHz backscattered signals at 650 °C (Figure 1a), 950 °C (Figure 1b) and both temperatures together (Figure 1c), as well as for the background pulse echo signals for 650 °C (Figure 1d), 950 °C (Figure 1e) and both temperatures together (Figure 1f). Accordingly, it is possible to analyze the complexity during the difficult assignment of classifying the samples.

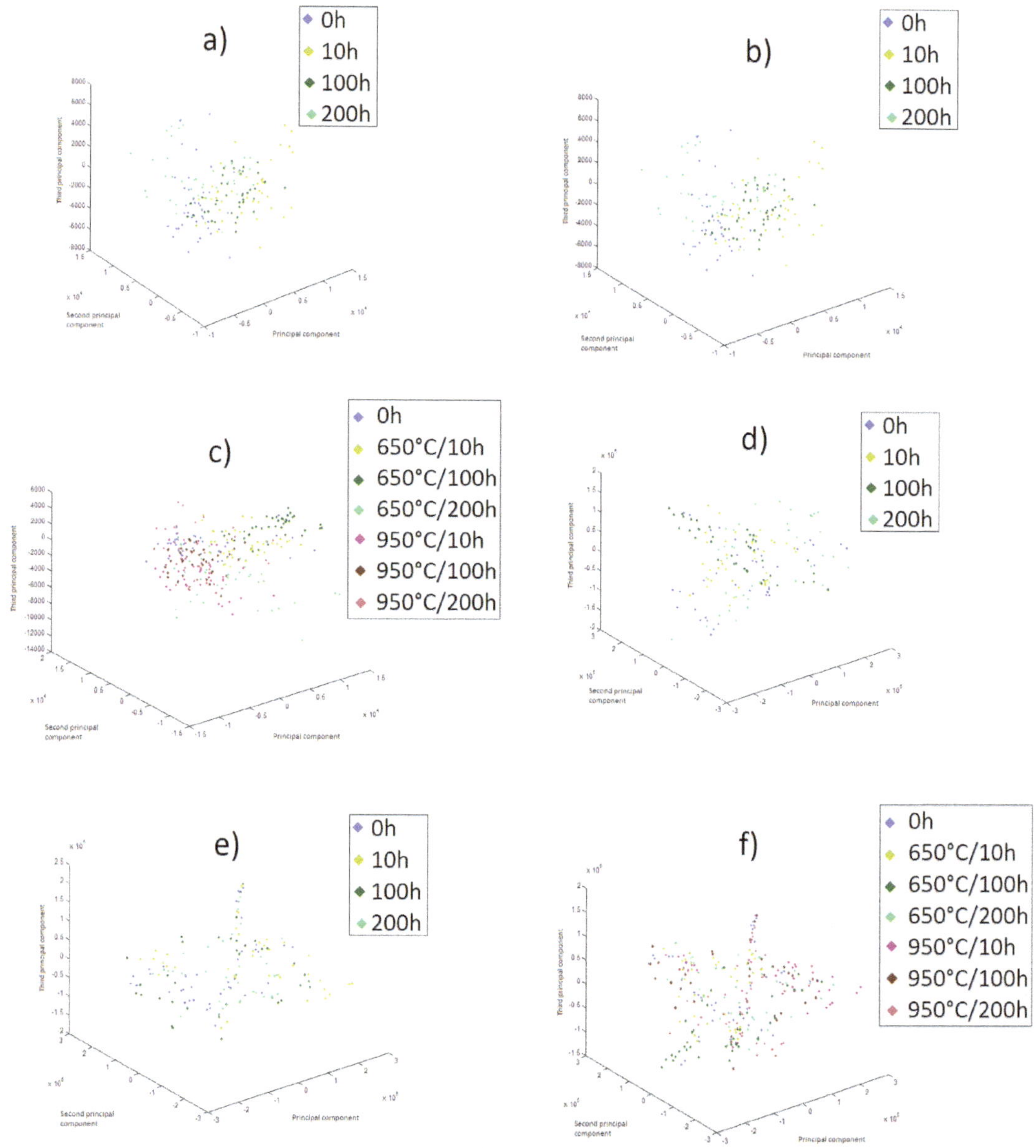

Figure 1. Spatial distribution of the samples of the (**a**) 4-MHz backscattered signals at 650 °C, (**b**) 950 °C and (**c**) of both temperatures together and of the (**d**) pulse echo signals at 650 °C, (**e**) 950 °C and (**f**) of both temperatures together.

3.1.1. Aged Samples at 650 °C

In Table 1, which shows the accuracy rate and harmonic mean for the 4- and 5-MHz frequency signals, the highest accuracy was achieved using the Manhattan distance (88.75%), highlighted in bold in the table. This value is around 3.75% higher than the second best result that was attained by the Euclidean distance metric (85%).

Table 1. Accuracy rate and harmonic mean for 4- and 5-MHz frequency signals at 650 °C.

Frequency(MHz)	**Signal**		**Manhattan**	**Euclidean**	**Squared Chi-Squared**
4	Backscattered	Acc	88.75 ± 2.11	85 ± 5.43	41.25 ± 4.22
		MH	61.9 ± 1.3	89.52 ± 4.23	44.97 ± 13.29
	Background echo	Acc	60.63 ± 3.54	58.13 ± 6.1	59.38 ± 4.92
		MH	67.89 ± 8.98	68.04 ± 10.66	69.34 ± 7.7
5	Backscattered	Acc	58.75 ± 1.29	60 ± 0	32.5 ± 3.65
		MH	66.31 ± 9.18	69.81 ± 10.27	46.47 ± 8.62
	Background echo	Acc	65.63 ± 4.1	64.38 ± 4.26	56.25 ± 4.72
		MH	74.86 ± 7.39	75.88 ± 9.08	67.19 ± 14.04
Frequency (MHz)	**Signal**		**Canberra**	**Bray–Curtis**	**Chi-Square**
4	Backscattered	Acc	30 ± 1.94	27.5 ± 2.84	18.75 ± 3.76
		MH	13.76 ± 4.94	15.08 ± 11.6	9.5 ± 6.16
	Background echo	Acc	53.75 ± 4.77	51.88 ± 3.87	45 ± 5.5
		MH	62.24 ± 9.81	61.87 ± 7.20	58.39 ± 8.03
5	Backscattered	Acc	29.38 ± 3.61	26.25 ± 3.23	31.88 ± 4.21
		MH	43.23 ± 20.08	37.8 ± 2.3	44.42 ± 10.36
	Background echo	Acc	48.13 ± 4.02	40.63 ± 4.49	26.25 ± 5.16
		MH	53.83 ± 16.29	51.01 ± 17.25	35.35 ± 10.92

In Table 2 is shown the confusion matrix of the classification means for the 4-MHz frequency backscattered signals; one can observe the difference between the Euclidean and Manhattan distances' performance compared to the other distances, since these two distances properly classified an average of 16–18 samples per class, while the other ones confused a large part of the samples, the majority being classified as belonging to the 200-h class.

In Table 3 is presented the confusion matrix of the classification means for the 4-MHz frequency pulse echo signals; one can conclude that the classification was more distributed across the classes, since 7–14 of the samples were classified correctly with the others distributed across the other classes.

Table 2. Confusion matrix of the classification means for the 4-MHz frequency backscattered signals at 650 °C.

True Class	Classified as	Manhattan	Euclidean	Squared Chi-Squared	Canberra	Bray Curtis	Chi Square
0 h	0 h	17	18	5	2	1	1
	10 h	1	1	2	0	1	1
	100 h	2	1	2	1	1	1
	200 h	0	0	11	17	17	17
10 h	0 h	1	0	1	1	0	0
	10 h	18	18	2	1	0	1
	100 h	1	2	1	0	0	0
	200 h	0	0	16	18	20	19
100 h	0 h	0	1	0.5	0	1	1
	10 h	2	3	1.5	0	0	0
	100 h	18	16	7	1	3	1
	200 h	0	0	11	19	16	18
200 h	0 h	1	0	0.5	0	1	1
	10 h	0	1	1.5	0	0	1
	100 h	1	1	1	1	1	4
	200 h	18	18	17	19	18	14

Table 3. Confusion matrix of the classification means for the 4-MHz frequency pulse echo signals at 650 °C.

True Class	Classified as	Manhattan	Euclidean	Squared Chi-Squared	Canberra	Bray Curtis	Chi Square
0 h	0 h	10	9	11.5	9.5	9	9.5
	10 h	4	6	3.5	4.5	6	4
	100 h	2	1	2	2	1	3
	200 h	2	4	3	4	4	3.5
10 h	0 h	3	3.5	3	3.5	3.5	4
	10 h	12	12	12	11.5	12	10
	100 h	2	2	2	2	2	1
	200 h	3	2.5	3	3	2.5	5
100 h	0 h	1.5	2.5	1.5	2	2.5	4
	10 h	2.5	4.5	1	1.5	4.5	2.5
	100 h	16	12	16	14.5	12	11.5
	200 h	0	1	1.5	2	1	2
200 h	0 h	3.5	5	4.5	5	5	5
	10 h	2.5	5	3	4	5	6.5
	100 h	3	2	3	1	2	1
	200 h	11	8	9.5	10	8	7.5

In Table 4 is presented the confusion matrix for the 5-MHz frequency backscattered signals. In this case, the performance of the Euclidean and Manhattan distances was around 25% below the one for the frequency backscattered signals, whereas the Euclidean distance has an accuracy between eight and 15 samples of each class, and the Manhattan distance had an accuracy from 10–14. The other distances classified correctly, on average, less than 10 samples per class.

Table 4. Confusion matrix of the classification means for the 5-MHz frequency backscattered signals at 650 °C.

True Class	Classified as	Manhattan	Euclidean	Squared Chi-Squared	Canberra	Bray Curtis	Chi Square
0 h	0 h	11	14	7	9	5	9
	10 h	9	6	6	6.5	6	2
	100 h	0	0	7	3	6	5
	200 h	0	0	0	1.5	3	4
10 h	0 h	2	0	8	7	5	5
	10 h	12	11	4	1	5	5
	100 h	6	9	6	5	5	3
	200 h	0	0	2	7	5	7
100 h	0 h	1	2	4	7	4	5
	10 h	9	3	5	6	5	3
	100 h	10	15	8	5	6	5
	200 h	0	0	3	2	5	7
200 h	0 h	1	2	3	1	5	4
	10 h	3	4	6	4	4	4
	100 h	2	6	5	6	6	4
	200 h	14	8	6	9	5	8

Table 5 shows the confusion matrix for the 5-MHz frequency pulse echo signals. Here, the Manhattan distance achieved a mean accuracy of 11–16 samples, reaching in the best case 70% accuracy and in the worst cases 58.75%, whereas the Euclidean distance achieved a mean accuracy between 10 and 17, reaching a maximum of 75% and a minimum of 61.25%. Notable was the squared chi-squared distance that in its best result achieved an accuracy rate of 63.75%; the other distances achieved a mean accuracy from 1–15.

In Table 6, the training and testing average times for the 4- and 5-MHz frequency signals are presented in milliseconds. All distances achieved analogous results regarding training and testing times for the backscattered signals, keeping a time of 0.2 ms for training and of 0.1 ms for testing. For the pulse echo signals, average times range from 0.5–1 milliseconds for training and 0.2–0.3 for testing.

Table 5. Confusion matrix of the classification means for the 5-MHz frequency pulse echo signals at 650 °C.

True Class	Classified as	Manhattan	Euclidean	Squared Chi-Squared	Canberra	Bray Curtis	Chi Square
0 h	0 h	14	14	15.5	9	9	4
	10 h	3	4	3.5	3	2	6
	100 h	1	1	0	3	4	4
	200 h	2	1	1	5	5.5	6
10 h	0 h	4	3	3	2.5	2	4
	10 h	11	13	8	6.5	4	4
	100 h	3	3	4	5	6	5
	200 h	2	1	5	6	8	7
100 h	0 h	0.5	1	3	1	2	3
	10 h	2	1	1	2	2	4
	100 h	16	17	16	16	13	7
	200 h	1.5	1	1	1	3	6
200 h	0 h	2	3	3.5	3.5	4.5	5
	10 h	2	3.5	4	5.5	5.5	4
	100 h	3	3	3	2	4	5
	200 h	13	10.5	9.5	9	6	6

Table 6. Average of training and testing times in milliseconds for the 4- and 5-MHz frequency signals at 650 °C.

Frequency (MHz)	Signal	Time	Manhattan	Euclidean	Squared Chi-Squared
4	Backscattered	Train	0.279 ± 0.159	0.355 ± 3.912	0.236 ± 0.092
		Test	0.105 ± 0.056	0.097 ± 0.031	0.090 ± 0.063
	Background echo	Train	0.512 ± 0.015	0.312 ± 0.063	1.050 ± 2.680
		Test	0.162 ± 0.035	0.126 ± 0.040	0.226 ± 0.293
5	Backscattered	Train	0.3 ± 0.179	0.288 ± 0.174	0.258 ± 0.131
		Test	0.095 ± 0.074	0.125 ± 0.049	0.143 ± 0.041
	Background echo	Acc	0.520 ± 0.097	0.446 ± 9.254	0.376 ± 0.283
		Test	0.285 ± 0.041	0.192 ± 0.054	0.191 ± 0.056
Frequency (MHz)	**Signal**	**Time**	**Canberra**	**Bray–Curtis**	**Chi-Square**
4	Backscattered	Train	0.205 ± 0.016	0.209 ± 0.030	0.216 ± 0.136
		Test	0.095 ± 0.018	0.113 ± 0.013	0.090 ± 0.021
	Background echo	Acc	0.593 ± 0.392	0.707 ± 0.359	0.398 ± 0.066
		Test	0.185 ± 0.047	0.247 ± 0.141	0.152 ± 0.035
5	Backscattered	Train	0.229 ± 0.010	0.329 ± 0.129	0.238 ± 0.116
		Test	0.091 ± 0.009	0.158 ± 0.080	0.113 ± 0.036
	Background echo	Acc	0.542 ± 0.449	0.388 ± 0.047	0.525 ± 0.166
		Test	0.324 ± 0.102	0.166 ± 0.036	0.237 ± 0.050

3.1.2. Samples Aged at 950 °C

In order to ascertain the best classification performance of the samples aged at 950 °C, we analyzed in detail the classification of each round. As shown in Table 7, which indicates the accuracy rates and harmonic mean for the 4- and 5-MHz frequency signals, the Manhattan distance achieved 60% accuracy for the 4-MHz pulse echo and backscattered signals and 60.63% for the 5-MHz pulse echo signals. However, the Euclidean distance used with the 5-MHz pulse echo signals may be considered more suitable for the classification, since in its best case, it achieved 70% accuracy and in its worst case 53.75%, whereas the results of the Manhattan distance for the 4-MHz backscattered signals ranged from 57.5%–60% and for the 4-MHz pulse echo signals from 48.75%–63.75%. All rounds using the 4-MHz backscattered signals with the Euclidean distance achieved an accuracy rate of 57.5%.

In general, the classifications involving the 5-MHz backscattered signals were very unsatisfactory, with accuracy rates below 40%.

Table 7. Accuracy rates and harmonic mean for the 4- and 5-MHz frequency at 950 °C.

Frequency (MHz)	Signal		Manhattan	Euclidean	Squared Chi-Squared
4	Backscattered	Acc	60 ± 1.05	57.5 ± 0	38.75 ± 1.44
		MH	71.36 ± 7.28	70.8 ± 16.65	50.86 ± 3.74
	Background echo	Acc	60 ± 4.75	56.88 ± 4.19	55 ± 4.02
		MH	69.92 ± 4.97	64.65 ± 6.2	62.03 ± 10.31
5	Backscattered	Acc	38.75 ± 4	36.88 ± 0.66	25 ± 2.77
		MH	53.95 ± 7.55	11.67 ± 22.99	34.64 ± 8.21
	Background echo	Acc	60.63 ± 5.24	60 ± 4.87	48.75 ± 4.42
		MH	69.94 ± 6.61	69.19 ± 8.68	59.75 ± 10.53
Frequency (MHz)	**Signal**		**Canberra**	**Bray–Curtis**	**Chi-Square**
4	Backscattered	Acc	30 ± 2.64	25 ± 0.53	34.38 ± 3.7
		MH	43.32 ± 4.38	33.78 ± 13.03	48.77 ± 6.74
	Background echo	Acc	50.63 ± 2.83	46.25 ± 3.59	43.13 ± 6.82
		MH	59.98 ± 8.54	56.9 ± 10.87	55.12 ± 7.58
5	Backscattered	Acc	23.75 ± 2.49	31.25 ± 2.13	32.5 ± 4.49
		MH	34.28 ± 11.93	47.36 ± 7.76	47.65 ± 8.96
	Background echo	Acc	39.38 ± 4.19	34.38 ± 3.76	28.13 ± 4.05
		MH	49.66 ± 13.32	44 ± 5.4	41.63 ± 2.66

Table 8 shows that for the 4-MHz frequency backscattered signals, all distances at some time could accurately classify more than 10 samples as 0 h and 10 h classes, whereas only the Euclidean and Manhattan distances could do the same with the 100 h and 200 h classes.

In Table 9 is presented the confusion matrix of the classification means of the 4-MHz frequency pulse echo signals. It can be observed that only the Euclidean and Manhattan distances achieved more than 50% of classification for each class, whereas the Euclidean distance achieved a maximum of 63.75% and the Manhattan distance a maximum of 60%.

The data in Table 10 show that for the 5-MHz frequency backscattered signals, a poor performance was achieved in general. The Manhattan distance achieved a maximum of 45% and a minimum of 31.25%

accuracy, whereas the Euclidean distance achieved a maximum and a minimum of 37.5% and 36.25%, respectively; the other ones kept below 35%.

Table 8. Confusion matrix of the classification means for the 4-MHz frequency backscattered signals at 950 °C.

True Class	Classified as	Manhattan	Euclidean	Squared Chi-Squared	Canberra	Bray Curtis	Chi Square
0 h	0 h	14	15	9	5	6	5
	10 h	3	1	3	2	6	3
	100 h	2	2	5	10	3	6
	200 h	1	2	3	3	5	6
10 h	0 h	0	1	2	3	7	2
	10 h	12	6	8	6	4	8
	100 h	7	10	6	8	5	5
	200 h	1	2	3	3	4	5
100 h	0 h	2	1	6	5.5	6	1
	10 h	7	5	5	4.5	3	4
	100 h	10	13	7	6	3	9
	200 h	1	1	2	4	8	6
200 h	0 h	4	1.5	4	2.5	3	2
	10 h	1	2.5	5	4	5	4.5
	100 h	3	4	3	7	4	6
	200 h	12	12	8	6.5	8	7.5

Table 9. Confusion matrix of the classification means for the 4-MHz frequency pulse echo signals at 950 °C.

True Class	Classified as	Manhattan	Euclidean	Squared Chi-Squared	Canberra	Bray Curtis	Chi Square
0 h	0 h	12	10.5	10	9.5	9.5	10
	10 h	2	2	2.5	5	4	4.5
	100 h	4	4	3	2	3	4.5
	200 h	2	3.5	4.5	3.5	3.5	1
10 h	0 h	2	1	1	2	2	3.5
	10 h	14.5	13.5	15	15	14	11.5
	100 h	2	3	2	2	2	4
	200 h	1.5	2.5	2	1	2	1
100 h	0 h	4.5	3.5	5	5.5	6	6
	10 h	3	3.5	5	5.5	6	5
	100 h	12	11	8	8	7	7
	200 h	0.5	2	2	1	1	2
200 h	0 h	3.5	4.5	4.5	4.5	5	4
	10 h	3	2	3	3	3	4
	100 h	2	3	2	2	3	4
	200 h	11.5	10.5	10.5	11.5	8	8

Table 10. Confusion matrix of the classification means for the 5-MHz frequency backscattered signals at 950 °C.

True Class	Classified as	Manhattan	Euclidean	Squared Chi-Squared	Canberra	Bray Curtis	Chi Square
0 h	0 h	6	13	4	7	7	7
	10 h	1	2	7.5	3.5	5	6
	100 h	2	2	5	4.5	4	4
	200 h	11	3	3.5	5	4	3
10 h	0 h	1	2.5	3	6	5	3
	10 h	8	6.5	7	6	7	8
	100 h	4	6	6	5	4	5
	200 h	7	5	4	3	4	4
100 h	0 h	3	3.5	5	5	5	4
	10 h	3	5.5	6	7	6	5
	100 h	8	5	5	3	7	9
	200 h	6	6	4	5	2	2
200 h	0 h	2	3	4	5	4	3
	10 h	6	9.5	7	3	7	7
	100 h	2	2	6	9	5	6
	200 h	10	5.5	3	3	4	4

Table 11 shows the confusion matrix of the classification means of the 5-MHz frequency pulse echo signals. The data presented show that the best performance for the temperature of 950 °C was achieved by the Euclidean distance with a maximum of 70%, followed by the Manhattan distance with 68.75%; the others remained between 57.5% and 22.5%.

Table 11. Confusion matrix of the classification means for the 5-MHz frequency pulse echo signals at 950 °C.

True Class	Classified as	Manhattan	Euclidean	Squared Chi-Squared	Canberra	Bray Curtis	Chi Square
0 h	0 h	15	17	14	12	9	5.5
	10 h	2	1	1	1	1	3.5
	100 h	2	1	3	4	5	5
	200 h	1	1	2	3	5	6
10 h	0 h	1.5	1	1	3.5	3	5
	10 h	13.5	12.5	8	5.5	6	6
	100 h	4	4	8	7	7	6
	200 h	1	2.5	3	4	4	3
100 h	0 h	3	4	3	5.5	7	5
	10 h	5	5	5	4.5	4	7
	100 h	10	10	10	7	5	5
	200 h	2	1	2	3	4	3
200 h	0 h	3	4	2.5	5	5.5	6
	10 h	4	3	6	4	4.5	2.5
	100 h	2	1	2	3	3	5
	200 h	11	12	9.5	8	7	6.5

In Table 12 is shown the training and testing average times in milliseconds for the 4- and 5-MHz frequency signals. All distances achieved analogous results for the training and testing times for the backscattered signals, keeping a time of 0.2 ms for training and of 0.1 for testing. For the pulse echo signals, the average times range from 0.2–0.4 ms for training and 0.1 for testing.

Table 12. Average of training and testing times in milliseconds for the 4- and 5-MHz frequency signals at 950 °C.

Frequency (MHz)	Signal	Time	Manhattan	Euclidean	Squared Chi-Squared
4	Backscattered	Train	0.254 ± 0.123	0.232 ± 0.007	0.233 ± 0.152
		Test	0.098 ± 0.040	0.084 ± 0.016	0.091 ± 0.035
	Background echo	Train	0.282 ± 0.055	0.252 ± 0.029	0.629 ± 0.394
		Test	0.092 ± 0.054	0.092 ± 0.003	0.221 ± 0.075
5	Backscattered	Train	0.271 ± 0.097	0.240 ± 0.107	0.284 ± 0.122
		Test	0.090 ± 0.032	0.106 ± 0.033	0.118 ± 0.060
	Background echo	Acc	0.397 ± 0.086	0.415 ± 0.132	0.393 ± 0.240
		Test	0.195 ± 0.090	0.154 ± 0.040	0.189 ± 0.030
Frequency (MHz)	**Signal**	**Time**	**Canberra**	**Bray–Curtis**	**Chi-Square**
4	Backscattered	Train	0.234 ± 0.188	0.322 ± 0.151	0.219 ± 0.087
		Test	0.095 ± 0.037	0.100 ± 0.032	0.092 ± 0.024
	Background echo	Acc	0.262 ± 0.047	0.478 ± 0.146	0.255 ± 0.024
		Test	0.100 ± 0.114	0.242 ± 0.100	0.095 ± 0.013
5	Backscattered	Train	0.231 ± 0.105	0.339 ± 0.120	0.224 ± 0.133
		Test	0.096 ± 0.055	0.113 ± 0.044	0.082 ± 0.078
	Background echo	Acc	0.379 ± 0.061	0.399 ± 0.070	0.369 ± 0.070
		Test	0.201 ± 0.058	0.186 ± 0.038	0.182 ± 0.072

3.1.3. Samples Aged at 650 and 950 °C

For this dataset, as can be seen in Table 13, the best accuracy was achieved by the Euclidean distance, with a value of 65.86%, but the highest harmonic mean belongs to the Manhattan metric, with 83.5%. This is due to the fact that in some rounds, the classification performance achieved by the Manhattan distance was higher than the one achieved by the Euclidean distance. The best classification of the Manhattan distance was 71.4%, whereas the best accuracy achieved by the Euclidean distance was equal to 67.86%. The other distances achieved accuracy rates lower than 30%, with the chi-squared distance achieving in its worst classification a value of 10.71%.

Regarding the processing times, indicated in Table 14, the distances that classified the samples correctly more often were those that took longer to do the training, with the Manhattan distance taking between 0.5 and 1 ms and the Euclidean distance from 0.4–0.6 ms. For the test, the Manhattan distance took from 0.16–0.3 ms, whereas the Euclidean distance took from 0.18–0.24 ms. The pulse echo signals were those that took longer to train the classifier, taking around 0.8 ms for all distances.

For the 4-MHz backscattered signals, as shown in Table 15, the Manhattan distance could classify correctly many of the samples related to the signals associated with the temperature of 650 °C, with an average from 15–18.5 samples classified correctly, whereas for the signals associated with the temperature of 950 °C, it achieved an average of correct classifications from 7.5–11 samples. The same applies to the Euclidean distance, that classified correctly from 10–15 samples of the classes associated with the temperature of 650 °C and 7–13 samples for the classes related to the temperature of 950 °C. The remaining distances confused considerably the data associated with the 650 °C/200 h class, with an average of 11–18 samples in all classes classified as belonging to this class.

Table 13. Accuracy rates and harmonic mean for the 4- and 5-MHz frequency signals at 650 and 950 °C.

Frequency (MHz)	Signal		Manhattan	Euclidean	Squared Chi-Squared
4	Backscattered	Acc	65.71 ± 2.94	67.86 ± 3.32	27.14 ± 1.74
		MH	83.50 ± 16.26	80.61 ± 14.81	25.91 ± 14.31
	Background echo	Acc	46.43 ± 3.33	46.43 ± 3.28	46.43 ± 3.9
		MH	56.47 ± 7.59	60.09 ± 8.65	55.77 ± 8.92
5	Backscattered	Acc	30.71 ± 5.9	39.29 ± 1.51	19.29 ± 2.82
		MH	46.15 ± 10.03	50.33 ± 7.25	25.34 ± 13.79
	Background echo	Acc	55 ± 4.31	53.21 ± 3.54	41.07 ± 3.52
		MH	66.76 ± 8.48	67.2 ± 7.97	50.33 ± 13.58
Frequency (MHz)	**Signal**		**Canberra**	**Bray–Curtis**	**Chi-Square**
4	Backscattered	Acc	18.57 ± 1.32	16.43 ± 1.59	13.21 ± 1.73
		MH	13.93 ± 9.69	9.5 ± 13.16	9.51 ± 11.67
	Background echo	Acc	39.29 ± 3.35	36.07 ± 2.79	28.93 ± 3.46
		MH	50.78 ± 12.36	46.65 ± 9.67	45.26 ± 10.59
5	Backscattered	Acc	17.14 ± 1.84	26.43 ± 1.38	21.07 ± 2.18
		MH	25.18 ± 9.42	39.44 ± 10.73	35.36 ± 13.35
	Background echo	Acc	35 ± 2.46	27.5 ± 2.69	16.43 ± 3.17
		MH	39.29 ± 16.7	36.16 ± 14.77	25.76 ± 6.68

Table 14. Average of training and testing times in milliseconds for the 4- and 5-MHz frequency signals at 650 and 950 °C.

Frequency (MHz)	Signal	Time	Manhattan	Euclidean	Squared Chi-Squared
4	Backscattered	Train	0.512 ± 0.015	0.528 ± 0.204	0.411 ± 0.014
		Test	0.162 ± 0.035	0.185 ± 0.048	0.192 ± 0.075
	Background echo	Train	0.945 ± 0.211	0.599 ± 0.099	0.888 ± 6.323
		Test	0.333 ± 0.108	0.243 ± 0.056	0.290 ± 0.080
5	Backscattered	Train	0.555 ± 0.182	0.454 ± 0.218	0.456 ± 0.197
		Test	0.192 ± 0.050	0.188 ± 0.027	0.219 ± 0.063
	Background echo	Acc	0.638 ± 0.075	0.492 ± 0.017	0.751 ± 0.136
		Test	0.265 ± 0.076	0.182 ± 0.070	0.360 ± 0.205
Frequency (MHz)	**Signal**	**Time**	**Canberra**	**Bray–Curtis**	**Chi-Square**
4	Backscattered	Train	0.386 ± 0.019	0.387 ± 0.047	0.387 ± 0.022
		Test	0.192 ± 0.015	0.189 ± 0.016	0.197 ± 0.035
	Background echo	Acc	0.957 ± 0.285	0.804 ± 0.222	0.899 ± 7.768
		Test	0.371 ± 0.105	0.411 ± 0.100	0.303 ± 0.160
5	Backscattered	Train	0.453 ± 0.126	0.410 ± 0.032	0.410 ± 0.080
		Test	0.205 ± 0.022	0.198 ± 0.040	0.205 ± 0.026
	Background echo	Acc	0.581 ± 0.054	0.800 ± 0.445	0.437 ± 0.040
		Test	0.264 ± 0.044	0.380 ± 0.189	0.195 ± 0.036

Table 15. Confusion matrix of the classification means for the 4-MHz frequency backscattered signals at 650 and 950 °C.

True Class	Classified as	Manhattan	Euclidean	Squared Chi-Squared	Canberra	Bray Curtis	Chi Square
0 h	0 h	15	18	3.5	2.5	1	1
	650 °C/10 h	1	1	2	1	1	0
	650 °C/100 h	0.5	1	1	1	1	1
	650 °C/200 h	0	0	11	13	16	13
	950 °C/10 h	2	0	1	0	0	1
	950 °C/100 h	0	0	1	1	0	2.5
	950 °C/200 h	2	0	1	0.5	0	1
650 °C/10 h	0 h	0	0	0	0	0	0.5
	650 °C/10 h	15	14	3	1	0	1
	650 °C/100 h	1	3	2	1	0	0
	650 °C/200 h	0	0	13	16	17	15
	950 °C/10 h	1.5	1	0	1	0	1
	950 °C/100 h	1	2	1.5	0	2	1
	950 °C/200 h	0.5	0	0	1	0	2
650 °C/100 h	0 h	0	0	0	0	0	0.5
	650 °C/10 h	3	2	1	0	0	1
	650 °C/100 h	16.5	18	7	2	4	1
	650 °C/200 h	0	0	11.5	16	15	14
	950 °C/10 h	0	0	0	0	0	1
	950 °C/100 h	0	0	0	0	0	2.5
	950 °C/200 h	0.5	0	0	0	0	2
650 °C/200 h	0 h	0	0	0	1	0	1
	650 °C/10 h	1	1	1	0	1	1.5
	650 °C/100 h	0.5	1	1	0	2	2
	650 °C/200 h	18.5	16	16	17	16	8.5
	950 °C/10 h	0	0	0	0	0	2
	950 °C/100 h	0	0	1	1	0	2.5
	950 °C/200 h	0	2	0	1	0	3.5
950 °C/10 h	0 h	2	2	0	0	1	1
	650 °C/10 h	1	0	0	0	0	1
	650 °C/100 h	0	0	0	0	1	0
	650 °C/200 h	0	0	12	15	18	11.5
	950 °C/10 h	7.5	8	2.5	2	0	3.5
	950 °C/100 h	8	9	2	1	1	2
	950 °C/200 h	2	1	1	0.5	0	1.5
950 °C/100 h	0 h	0	2	2	1	0	0
	650 °C/10 h	0.5	1	1	0	0	0.5
	650 °C/100 h	0	0	0	0	0	1
	650 °C/200 h	0	0	12	17	18	13
	950 °C/10 h	7	6	2	0	0.5	1
	950 °C/100 h	11	10	2	1.5	1	3
	950 °C/200 h	2	1	1	0	0	1
950 °C/200 h	0 h	3	2	2	2	0	1
	650 °C/10 h	1	1	0	0	0	2.5
	650 °C/100 h	0.5	0	0	0	1	1
	650 °C/200 h	0	0	13	16	17	15
	950 °C/10 h	4	3	1	1	1	0.5
	950 °C/100 h	2.5	3	1	1	0.5	1
	950 °C/200 h	9.5	11	3	0	0.5	1

With the 4-MHz frequency pulse echo signals, Table 16, the classification was more confusing, with wrong classifications for all classes and distances. The Manhattan distance achieved an average accuracy in most of its classes from 8–13.5 samples, whereas the Euclidean distance achieved an average accuracy from 6.5–13 samples. The squared chi-squared and Canberra distances achieved similar results, with an average of samples classified correctly between 5 and 12.5, whereas the Bray–Curtis distance from 5–10.5 and the chi-squared distance below 10 samples.

Table 16. Confusion matrix of the classification means for the 4-MHz frequency pulse echo signals at 650 and 950 °C.

True Class	Classified as	Manhattan	Euclidean	Squared Chi-Squared	Canberra	Bray Curtis	Chi Square
0 h	0 h	9.5	10.5	7.5	6	5	7
	650 °C/10 h	2	2.5	3.5	2.5	3	2.5
	650 °C/100 h	1.5	2	1	2.5	2	2.5
	650 °C/200 h	1.5	2	2	3	3	3
	950 °C/10 h	2	1	1.5	3.5	4	1.5
	950 °C/100 h	1	1	3.5	1.5	2	2.5
	950 °C/200 h	2.5	1	1	2	2	1
650 °C/10 h	0 h	2	2	2.5	3	2	2
	650 °C/10 h	9	7.5	10	9	8	4
	650 °C/100 h	2	1.5	2.5	2.5	1.5	1
	650 °C/200 h	1	2	1	1.5	2	5
	950 °C/10 h	3	4	3.5	2	3.5	3
	950 °C/100 h	2	2	0	1	1.5	2
	950 °C/200 h	1	1	1.5	1	1.5	2
650 °C/100 h	0 h	1	1.5	2	1	1	1
	650 °C/10 h	1	1	1	2	2	2.5
	650 °C/100 h	13.5	13	11	12.5	9	9
	650 °C/200 h	1	0	1	1	2	2
	950 °C/10 h	1.5	2	2	1	3.5	2
	950 °C/100 h	1	1	1.5	1.5	1	2
	950 °C/200 h	1	1	1.5	1	1	1.5
650 °C/200 h	0 h	2	2	1.5	3	4	3
	650 °C/10 h	3	2	2	2.5	2	3
	650 °C/100 h	1	1	1	2	1	2
	650 °C/200 h	8.5	9	8.5	5.5	5	5
	950 °C/10 h	2	2	3	3	3	3.5
	950 °C/100 h	2	2	2	2	3	1
	950 °C/200 h	1.5	2	2.5	2	1	2.5
950 °C/10 h	0 h	1	1	1	1	1.5	1.5
	650 °C/10 h	2	1.5	1	1.5	2	3
	650 °C/100 h	2.5	1.5	1.5	1.5	1	3
	650 °C/200 h	1.5	2	1	1.5	2.5	2.5
	950 °C/10 h	10.5	10.5	12.5	11	11.5	8
	950 °C/100 h	1.5	2.5	2	1.5	1	2
	950 °C/200 h	1	1	1	1	0.5	0
950 °C/100 h	0 h	3.5	2	4.5	3	3.5	4
	650 °C/10 h	2	2	1.5	2.5	2.5	3
	650 °C/100 h	1	2	1.5	1.5	2	1.5
	650 °C/200 h	1.5	1.5	1.5	1.5	2	1.5
	950 °C/10 h	3	3.5	4	3	4	4
	950 °C/100 h	8.5	8.5	7	7.5	5.5	4
	950 °C/200 h	0.5	0.5	0	1	1	2
950 °C/200 h	0 h	2.5	3	2.5	3	3.5	3.5
	650 °C/10 h	2	1.5	1.5	2.5	3.5	2.5
	650 °C/100 h	2	2	2	1	1.5	1
	650 °C/200 h	2	2	2	2	2	3
	950 °C/10 h	1.5	1.5	2	2.5	3	2
	950 °C/100 h	1.5	1.5	2	2	0.5	1
	950 °C/200 h	8.5	8.5	8	7	6	7

Table 17 presents the result of the confusion matrix for the 5-MHz frequency backscattered signals, and it can be observed that all distances had bad performances; with the exception of the 650 °C/10 h class, for which the Manhattan and Euclidean distances have the best accuracy rate of 42.14% and 40.71%, respectively. The distance for others had the best results below 30%.

From Table 18, which shows the confusion matrix for the 5-MHz frequency pulse echo signals, it can be confirmed that the performance of all distances was better than the ones presented in Table 17, but still lower than the ones shown in Table 15. The Manhattan distance achieved an average of classification varying from 9–15 correct samples per class, with a maximum accuracy of 64.29% and a minimum accuracy of 48.57%. The Euclidean distance achieved an average of accurate correct classification varying from 9–14.5 samples per class, with a maximum accuracy of 59.29% and a minimum accuracy of 47.14%. The squared chi-squared metric achieved a maximum accuracy of 47.86% and a minimum accuracy of 34.29%. The Canberra distance achieved a maximum accuracy of 36.42% and a minimum accuracy of 29.29%; the Bra–Curtis distance achieved a maximum accuracy of 30% and a minimum accuracy of 22.14%; and finally, the chi-squared distance achieved a maximum accuracy of 22.14% and a minimum accuracy of 11.43%.

4. Conclusions

This work evaluated the efficiency and efficacy of the OPF classifier configured with six distance functions to classify ultrasonic signals, raw data background echo and backscattered signals acquired at frequencies of 4 and 5 MHz, to characterize the phase transformations on a Nb-base alloy, thermally aged at 650 and 950 °C for 10, 100 and 200 h, as well as in the as-welded condition.

In regard to this work, the following conclusions can be pointed out:

(1) The results revealed that the classification of the ultrasonic signals using the OPF classifier was sensitive to the microstructural changes occurring in the inconel 625 alloy and that the formation of the secondary phases during the welding process, as well as the phase transformation kinetics due to the different thermal aging times can be efficiently identified;

(2) The best accuracy rates for the thermal aging at 650 °C were obtained using the OPF configured with the Manhattan distance on the backscattered signals acquired with a 4-MHz transducer (accuracy of 88.75%, in 0.384 ms, with a harmonic mean of 61.9);

(3) For the thermal aging at 950 °C, the best results were obtained using the OPF with Euclidean distance on the background echo signals acquired with the 5-MHz transducer (accuracy of 60%, in 0.569 ms, with a harmonic mean of 69.19);

(4) For the thermal aging at 650 and 950 °C, the best results were obtained using the OPF with the Euclidean distance on the backscattered signals acquired with the 4-MHz transducer (accuracy of 67.86%, in 0.713 ms, with a harmonic mean of 80.61);

Based on the results obtained for accurately classifying the ultrasonic signals, it is possible to confirm that the OPF classifier is able to assess the aging conditions to which the inconel 625 alloy is submitted, making it possible to detect the best moment to carry out maintenance services, reducing the costs and maintenance time.

Table 17. Confusion matrix of the classification means for the 5-MHz frequency backscattered signals at 650 and 950 °C.

True Class	Classified as	Manhattan	Euclidean	Squared Chi-Squared	Canberra	Bray Curtis	Chi Square
0 h	0 h	6	7	3.5	3	5	6
	650 °C/10 h	4	3.5	2.5	2	0	2
	650 °C/100 h	2	1	3	1	2	2
	650 °C/200 h	0	0	1	0	0	3
	950 °C/10 h	2	1.5	4	5	4	3
	950 °C/100 h	1	2	3	5	5	3
	950 °C/200 h	5	5	3	4	2	1
650 °C/10 h	0 h	3	5	3	4	1	2
	650 °C/10 h	11.5	10	2	3	5	2
	650 °C/100 h	4.5	4	4.5	0.5	2	2.5
	650 °C/200 h	0	0	1	4	4	4
	950 °C/10 h	0	0	4	3	2	4
	950 °C/100 h	0	0.5	3	2.5	3	3.5
	950 °C/200 h	1	0.5	2.5	3	3	3
650 °C/100 h	0 h	3	2	3	2.5	1.5	2
	650 °C/10 h	9	9	2	3	1.5	1
	650 °C/100 h	8	8.5	3.5	2.5	3	3
	650 °C/200 h	0	0	0	1	3	4
	950 °C/10 h	0	0	5	6	4	4
	950 °C/100 h	0	0	3	2	4	3
	950 °C/200 h	0	0.5	3.5	3	3	3
650 °C/200 h	0 h	1.5	0.5	1	1.5	2.5	3
	650 °C/10 h	5	5	5	5	2	3
	650 °C/100 h	5	3.5	3	3	1	3
	650 °C/200 h	8	9.5	5	4	8.5	6.5
	950 °C/10 h	0.5	0	2	3	1	1
	950 °C/100 h	0	0.5	2	2.5	2	2
	950 °C/200 h	0	1	2	1	3	1.5
950 °C/10 h	0 h	1	1	4	2.5	3	2
	650 °C/10 h	0.5	0	1.5	1.5	0	1
	650 °C/100 h	0.5	0	2.5	2	3	2
	650 °C/200 h	0	0	0	1	1.5	2.5
	950 °C/10 h	6	6.5	6	6	5	5.5
	950 °C/100 h	6	5.5	3	4	3.5	5
	950 °C/200 h	6	7	3	3	4	2
950 °C/100 h	0 h	2	3.5	5	3	3.5	4
	650 °C/10 h	0	1	1.5	1	5	1
	650 °C/100 h	0	0	2	2	2	1
	650 °C/200 h	0	0	0	0	0.5	1
	950 °C/10 h	6	4.5	3	4.5	2	3.5
	950 °C/100 h	8	7	3.5	3.5	6	6
	950 °C/200 h	4	4	5	5	1	2.5
950 °C/200 h	0 h	3.5	4	4.5	3	5	3
	650 °C/10 h	0.5	1	2.5	1	1	1.5
	650 °C/100 h	1	0.5	2	1	0.5	1.5
	650 °C/200 h	0	0	0	0	2	4
	950 °C/10 h	5	4	6	6	5	4.5
	950 °C/100 h	3	4	4	7	3.5	3
	950 °C/200 h	7	6.5	1	2	3	2.5

Table 18. Confusion matrix of the classification means for the 5-MHz frequency pulse echo signals at 650 and 950 °C.

True Class	Classified as	Manhattan	Euclidean	Squared Chi-Squared	Canberra	Bray Curtis	Chi Square
0 h	0 h	14	13	12.5	8	5	3.5
	650 °C/10 h	2	3	1.5	3	2	3.5
	650 °C/100 h	1	1	1.5	2.5	2	3
	650 °C/200 h	1	1.5	0	3.5	4	4
	950 °C/10 h	1	1	1	0	2.5	1
	950 °C/100 h	0.5	0.5	2	3	2	2
	950 °C/200 h	0.5	0	1.5	1.5	2.5	3
650 °C/10 h	0 h	3	3	3	1.5	1.5	3
	650 °C/10 h	10.5	10	6	5	4	2.5
	650 °C/100 h	2.5	2	3.5	5	5.5	4
	650 °C/200 h	1	0	2.5	5.5	5.5	5
	950 °C/10 h	1	1	1.5	0	0	0.5
	950 °C/100 h	1	2.5	2	2	2	2
	950 °C/200 h	1	1.5	1.5	1	1.5	3
650 °C/100 h	0 h	0.5	1	1.5	1	1.5	1.5
	650 °C/10 h	1.5	1	1	1	1.5	3
	650 °C/100 h	15	14.5	15	14.5	12	6
	650 °C/200 h	1.5	1	1	2	2	5
	950 °C/10 h	0.5	0.5	0	0	1	1
	950 °C/100 h	1	2	1.5	0	0	2
	950 °C/200 h	0	0	0.5	1.5	2	1.5
650 °C/200 h	0 h	1.5	2	3	2	3	4
	650 °C/10 h	2	2.5	2.5	4	3	3
	650 °C/100 h	2.5	2.5	1.5	2	2.5	2.5
	650 °C/200 h	9.5	9	6.5	8	5	4.5
	950 °C/10 h	2	1	2	1	2	1
	950 °C/100 h	2	3	3	1.5	3	2
	950 °C/200 h	0.5	0	1.5	1.5	1.5	3
950 °C/10 h	0 h	0	1	0.5	1	1	2
	650 °C/10 h	2	2	0.5	1	1	1.5
	650 °C/100 h	1	1	1.5	3	4	3
	650 °C/200 h	1	0.5	2	4.5	3.5	4
	950 °C/10 h	11	10.5	7	4.5	5.5	4
	950 °C/100 h	3	3.5	5.5	4	4	2.5
	950°C/200 h	2	1.5	3	2	1	3
950 °C/100 h	0 h	2	2.5	2	2	2	3
	650 °C/10 h	2	2	4	2	3	2
	650 °C/100 h	2	2	2.5	5	4	3
	650 °C/200 h	1	1	1.5	3.5	5	5
	950 °C/10 h	3	2.5	3.5	3	1.5	3
	950 °C/100 h	9	9	5.5	3	3.5	3
	950 °C/200 h	1	1	1	1.5	1	1
950 °C/200 h	0 h	2	2	2.5	1.5	2.5	2
	650 °C/10 h	2	1	1.5	2.5	3	3
	650 °C/100 h	0.5	2	2	4.5	4.5	2.5
	650 °C/200 h	1	1	1	5	4.5	7
	950 °C/10 h	3	2.5	2.5	0.5	2.5	2.5
	950 °C/100 h	1.5	1.5	1	1	2	2
	950 °C/200 h	10	11	7.5	5	2.5	3

Acknowledgments

All authors are grateful for the support given by the following laboratories of the Federal University of Ceará, in Brazil: Welding Engineering Laboratory (ENGESOLDA), Materials Characterization Laboratory (LACAM), Center of Non-Destructive Testing (CENDE), in particular Paulo G. Normando. All authors are grateful, as well as, for the financial support given by the Research and Projects Financing (FINEP), the Coordination for the Improvement of People with Higher Education (CAPES) and, finally, Petróleo Brasileiro S.A.

The last author acknowledges sponsorship from the Brazilian National Council for Research and Development (CNPq) via Grants 470501/2013-8 and 301928/2014-2.

Author Contributions

Victor Hugo C. de Albuquerque, Cleisson V. Barbosa, João P. Papa, Pedro P. Rebouças Filho and João Manuel R.S. Tavares contributed to the design of the work, assessment of the OPF classifier performance, configured with six distance functions, for the microstructural characterization, as well as thermal aging treatment. Cleiton C. Silva and Elineudo P. Moura carried out the welding of the samples and acquisition of the ultrasound signals, respectively. All authors contributed to the writing of the manuscript.

Conflicts of Interest

The authors declare no conflict of interest.

References

1. Cieslak, M.J.; Headley, T.J.; Romig, A.D. The welding metallurgy of HASTELLOY alloys C-4, C-22 and C-276. *Metall. Mater. Trans. A* **1986**, *17A*, 2035–2047.
2. Cieslak, M.J. The welding and solidification metallurgy of alloy 625. *Weld. J.* **1981**, *70*, 49–56.
3. Yang, J.X.; Zheng, Q.; Sun, X.F.; Guan, H.R.; Hu, Z.Q. Formation of μ phase during thermal exposure and its effect on the properties of K465 superalloy. *Scr. Mater.* **2006**, *55*, 331–334.
4. Dupont, J.N.; Banovic, S.W.; Marder, A.R. Microstructural evolution and weldability of dissimilar welds between a super austenitic stainless steel and nickel-based alloys. *Weld. Res.* **2003**, *82*, 125–156.
5. de Araújo Freitas, V.; Normando, P.G.; Albuquerque, V.H.C.; de Macedo Silva, E.; Almeida, A.; ao Manuel, R.S.; Tavares, J. Nondestructive Characterization and Evaluation of Embrittlement Kinetics and Elastic Constants of Duplex Stainless Steel SAF 2205 for Different Aging Times at 425 °C and 475 °C. *J. Nondestruct. Eval.* **2011**, *30*, 130–136.
6. De Albuquerque, V.H.C.; de Macedo Silva, E.; Leite, J.P.; de Moura, E.P.; de Araújo Freitas, V.L.; ao Manuel, R.S.; Tavares, J. Spinodal decomposition mechanism study on the duplex stainless steel UNS S31803 using ultrasonic speed measurements. *Mater. Des.* **2010**, *31*, 2147–2150.

7. De Albuquerque, V.H.C.; de A. Melo, T.A.; de Oliveira, D.F.; Gomes, R.M.; Tavares, J.M.R. Evaluation of grain refiners influence on the mechanical properties in a CuAlBe shape memory alloy by ultrasonic and mechanical tensile testing. *Mater. Des.* **2010**, *31*, 3275–3281.
8. Normando, P.G.; Moura, E.P.; Souza, J.A.; Tavares, S.S.; Padovese, L.R. Ultrasound, eddy current and magnetic Barkhausen noise as tools for sigma phase detection on a UNS S31803 duplex stainless steel. *Mater. Sci. Eng. A* **2010**, *527*, 2886–2891.
9. Vieira, A.; de Moura, E.; Gonçalves, L.; Rebello, J. Characterization of welding defects by fractal analysis of ultrasonic signals. *Chaos Solitons Fractals* **2008**, *38*, 748–754.
10. De Moura, E.P.; Normando, P.G.; Gonçalves, L.L.; Kruger, S.E. Characterization of cast iron microstructure through the statistical fluctuation and fractal analyses of ultrasonic backscattered signals. *AIP Conf. Proc.* **2011**, *1335*, 1104–1111.
11. Vieira, A.P.; de Moura, E.P.; Gonçalves, L.L. Fluctuation Analyses for Pattern Classification in Nondestructive Materials Inspection. *EURASIP J. Adv. Signal Process* **2010**, *2010*, doi:10.1155/2010/262869.
12. De Araújo Freitas, V.L.; de Albuquerque, V.H.C.; de Macedo Silva, E.; Silva, A.A.; ao Manuel, R.S.; Tavares, J. Nondestructive characterization of microstructures and determination of elastic properties in plain carbon steel using ultrasonic measurements. *Mater. Sci. Eng. A* **2010**, *527*, 4431–4437.
13. De Macedo Silva, E.; de Albuquerque, V.H.C.; Leite, J.P.; Varela, A.C.G.; de Moura, E.P.; Tavares, J.M.R. Phase transformations evaluation on a UNS S31803 duplex stainless steel based on nondestructive testing. *Mater. Sci. Eng. A* **2009**, *516*, 126–130.
14. Nunes, T.M.; Coelho, A.L.; Lima, C.A.; Papa, J.P.; de Albuquerque, V.H.C. {EEG} signal classification for epilepsy diagnosis via optimum path forest—A systematic assessment. *Neurocomputing* **2014**, *136*, 103–123.
15. Da S. Luz, E.J.; Nunes, T.M.; de Albuquerque, V.H.C.; Papa, J.P.; Menotti, D. {ECG} arrhythmia classification based on optimum-path forest. *Expert Syst. Appl.* **2013**, *40*, 3561–3573.
16. Papa, J.P.; Nakamura, R.Y.; de Albuquerque, V.H.C.; Falcão, A.X.; Tavares, J.M.R. Computer techniques towards the automatic characterization of graphite particles in metallographic images of industrial materials. *Expert Syst. Appl.* **2013**, *40*, 590–597.
17. Pereira, C.R.; Nakamura, R.Y.; Costa, K.A.; Papa, J.P. An Optimum-Path Forest framework for intrusion detection in computer networks. *Eng. Appl. Artif. Intell.* **2012**, *25*, 1226–1234.
18. Pereira, L.A.; Nakamura, R.Y.; de Souza, G.F.; Martins, D.; Papa, J.P. Aquatic weed automatic classification using machine learning techniques. *Comput. Electron. Agric.* **2012**, *87*, 56–63.
19. Iliev, A.I.; Scordilis, M.S.; Papa, J.P.; Falcão, A.X. Spoken emotion recognition through optimum-path forest classification using glottal features. *Comput. Speech Lang.* **2010**, *24*, 445–460.
20. Nunes, T.M.; de Albuquerque, V.H.C.; Papa, J.P.; Silva, C.C.; Normando, P.G.; Moura, E.P.; Tavares, J.M.R. Automatic microstructural characterization and classification using artificial intelligence techniques on ultrasound signals. *Expert Syst. Appl.* **2013**, *40*, 3096–3105.

21. Albuquerque, V.H.C.; Silva, C.C.; Normando, P.G.; Moura, E.P.; Tavares, J.M.R.S. Thermal aging effects on the microstructure of Nb-bearing nickel based superalloy weld overlays using ultrasound techniques. *Mater. Des.* **2012**, *36*, 337–347.
22. Albuquerque, V.H.C.; Silva, E.M.; Leite, J.P.; Moura, E.P.; Freitas, V.L.A.; Tavares, J.M.R.S. Spinodal decomposition mechanism study on the duplex stainless steel UNS S31803 using ultrasonic speed measurements. *Mater. Des.* **2010**, *31*, 2147–2150.
23. Silva, E.M.; Albuquerque, V.H.C.; Leite, J.P.; Varela, A.C.G.; Moura, E.P.; Tavares, J.M.R.S. Phase transformations evaluation on a UNS S31803 duplex stainless steel based on nondestructive testing. *Mater. Sci. Eng. A* **2009**, *516*, 126–130.
24. Normando, P.G.; Moura, E.P.; Souza, J.A.; Tavares, S.S.M.; Padovese, L.R. Ultrasound, eddy current and magnetic Barkhausen noise as tools for sigma phase detection on a UNS S31803 duplex stainless steel. *Mater. Sci. Eng. A* **2010**, *527*, 2886–2891.
25. Normando, P.G.; Nascimento, R.A.S.; Moura, E.P.; Vieira, A.P. Microstructure identification via detrended fluctuation analysis of ultrasound signals. *Phys. Rev. E* **2013**, *87*, 043304.
26. Papa, J.P.; Falcão, A.X.; Suzuki, C.T.N. Supervised Pattern Classification based on Optimum-Path Forest. *Int. J. Imaging Syst. Technol.* **2009**, *19*, 120–131.
27. Dijkstra, E.W. A note on two problems in connexion with graphs. *Numer. Math.* **1959**, *1*, 269–271.
28. Falcão, A.X.; Stolfi, J.; Lotufo, R.A. The Image Foresting Transform Theory, Algorithms, and Applications. *IEEE Trans. Pattern Anal. Mach. Intell.* **2004**, *26*, 19–29.
29. Allène, C.; Audibert, J.Y.; Couprie, M.; Cousty, J.; Keriven, R. Some Links between Min-cuts, Optimal Spanning Forests and Watersheds. In Proceedings of the Mathematical Morphology and its Applications to Image and Signal Processing (ISMM'07), MCT/INPE, Rio de Janeiro, Brazil, 10–13 October 2007; pp. 253–264.

6

Development of a Portable Non-Invasive Swallowing and Respiration Assessment Device †

Wann-Yun Shieh [1,‡,*], Chin-Man Wang [2,3,‡] and Chia-Shuo Chang [1]

1. Department of Computer Science and Information Engineering, Chang Gung University, No. 259, Wen-Hwa 1st Road, Kwei-Shan, Tao-Yuan 333, Taiwan; E-Mail: johnny7802@hotmail.com
2. Department of Physical Medicine and Rehabilitation, Chang Gung Memorial Hospital, Taoyuan 333, Taiwan; E-Mail: cmw1314@cgmh.org.tw
3. Department of Medicine, Chang Gung University, No. 5, Fu-Hsing Street, Kwei Shan, Tao-Yuan 333, Taiwan

† A preliminary version concept of this work was first published in "2014 International Symposium on InfoComm and Media Technology in Bio-Medical and Healthcare Application", Shanghai, China, 21 December 2014.

‡ These authors contributed equally to this work.

* Author to whom correspondence should be addressed; E-Mail: wyshieh@mail.cgu.edu.tw

Academic Editor: Stefano Mariani

Abstract: Dysphagia is a condition that happens when a person cannot smoothly swallow food from the mouth to the stomach. It causes malnourishment in patients, or can even cause death due to aspiration pneumonia. Recently, more and more researchers have focused their attention on the importance of swallowing and respiration coordination, and the use of non-invasive assessment systems has become a hot research trend. In this study, we aimed to integrate the timing and pattern monitoring of respiration and swallowing by using a portable and non-invasive approach which can be applied at the bedside in hospitals or institutions, or in a home environment. In this approach, we use a force sensing resistor (FSR) to detect the motions of the thyroid cartilage in the pharyngeal phase. We also use the surface electromyography (sEMG) to detect the contraction of the submental muscle in the oral phase, and a nasal cannula to detect nasal airflow for respiration monitoring during the swallowing process. All signals are received and processed for swallowing event recognition. A total of

19 volunteers participated in the testing and over 57 measurements were made. The results show that the proposed approach can effectively distinguish the swallowing function in people of different ages and genders.

Keywords: swallowing; respiration; dysphagia; thyroid cartilage motion; non-invasive detection; FSR

---

1. Introduction

The difficulty in swallowing food through the esophagus to the stomach, is a condition known as dysphagia. Many acute or chronic illnesses may cause dysphagia, majorly of two types: (I) structural dysphagia, when a structural tissue abnormality like in a cleft lip and palate born child, or oral cancer affects the swallowing function; (II) nervous system disorder, when some nervous system abnormality such as cerebral palsy, myasthenia gravis, stroke or another disease causes a swallowing dysfunction (for stroke abnormalities, the proportion of the patients with the oropharyngeal dysphagia may reach 32% to 45% [1])

The study of Greco *et al.* [1] showed that over 12% of patients in an acute medical institute had dysphagia, while in a long-term healthcare organization, the proportion may be more than 50%. If the symptoms of oropharyngeal dysphagia are not properly treated, it can lead to many complications. The most important one is that the eating behavior of patients will be affected and changed, which leads to dehydration, malnutrition, choking injuries, aspiration pneumonia and even death [2–4]. Typically, the swallowing process can be divided into four stages:

(1) Oral preparation stage: the mouth starts to chew the food and mix it with saliva.
(2) Oral stage: the bolus is pushed backwards by the tongue into the pharynx.
(3) Pharyngeal stage: many swallowing reflex motions occur, including a series of complex throat neuromuscular reactions, which can push the bolus into the esophagus.
(4) Esophagus stage: the bolus is pushed into the stomach through the esophagus.

Dysphagia can happen at any one of above stages, but the general assessment and therapy for neurogenic dysphagia particularly emphasizes the oral and the pharyngeal stage. This is because the entrance of the esophagus is in the close proximity to the larynx and both air and the swallowed bolus will share a common pathway through the pharynx. Many researchers have mentioned that breathing and swallowing are physiologically linked to ensure smooth gas exchange during oronasal breathing, and to prevent aspiration during swallowing [2,5–8]. When the food goes through the pharynx to the esophagus, swallowing and respiration cannot happen at the same time. This physiological mechanism is to ensure that the food can be swallowed through the esophagus smoothly and safely without getting into the trachea and lungs. It can also avoid suffocation, aspiration pneumonia and severe respiratory failure. In fact, most of the oropharyngeal dysphagia cases actually result from uncoordinated respiration and swallowing in the oropharyngeal stage. Therefore, bedside clinical evaluation is very important. Currently most oropharyngeal dysphagia evaluations are performed by well-trained physicians or speech therapists. A physician will ask the subjects to swallow food or liquid, and the physician record the strength and the speed of vomiting, coughing, swallowing reflex motions, eating posture, the difficulty of eating different kinds of food and nutritional status, *etc.*

Most instrumental methods for measuring oropharyngeal dysphagia symptoms are invasive approaches. The most widely used method is the video fluoroscopic swallowing study (VFSS), which is the gold standard of the oropharyngeal dysphagia examination in many medical organizations [9,10]. It uses an X-ray photography instrument. Before the test, the physician will ask the subject to swallow food containing a contrast material. By observing the distribution of the remains of the contrast material in the mouth and throat, the physician can adjust the degree of the oropharyngeal dysphagia. It is an accurate method. However, it will bring a risk of radiation exposure.

The second instrumental method uses a fiber optic endoscope to evaluate oropharyngeal dysphagia [11]. A subject will be asked to eat a colorful food or liquid, then the physician will use an endoscope with optical fiber to observe the motion of the nasopharynx in the mouth and throat. This approach can help the physician observe the reflected swallowing flow in the rest state of the oropharynx and hypopharynx before and after the subject eats the food. However, this method needs an endoscope placed in the mouth and throat during the test, making the subject feel uncomfortable and swallowing the food difficult.

Both of the above approaches require the patients to undergo the testing in a hospital or lab. Thus, the measurements will be limited in a specific time and place. For patients with poor mobility, it is very inconvenient to do the testing with the risk of radioactive exposure or nosocomial infection. Therefore there is a strong need to develop a wearable and portable device, as well as the corresponding assessment software, so physicians can perform the measurements directly near the bedside.

Some researchers have applied the Micro-Electro-Mechanical Systems (MEMs) sensors to measure oropharyngeal dysphagia [5,12]. The most widely used sensors for this purpose are accelerometers and pressure sensors, especially those based on the piezoelectric effect. We take the sensor in Figure 1a for example. Figure 1a shows a three-axis capacitive accelerometer which can perform static or dynamic acceleration measurements. The study of Lee *et al.* [12] pasted an accelerometer on a subject's throat to measure the upward and downward motions of the thyroid cartilage when the subject swallowed water (or food). The studies reported in [1,5,12] proposed similar approaches. Using an accelerometer, however, has an obvious drawback. That is, the measurement results will suffer signal interference from the subject's head or trunk shaking. Although there are other accelerometers which only measure vibrations, they cannot measure the force exerted by the thyroid cartilage.

(a) (b) (c)

Figure 1. Micro-Electro-Mechanical sensors: (**a**) accelerometer; (**b**) piezoelectric sensors; (**c**) force sensing resistor.

The piezoelectric sensor (shown in Figure 1b) is the kind of pressure sensors used in [13]. It is made by solid materials, which can accumulate electric charges in response to applied mechanical stress. The authors pasted a piezoelectric sensor tightly on a subject's throat. When a subject swallows water or food, the thyroid cartilage will press the sensor, making the sensor respond with a voltage. The voltage can be translated into a pressure, and a physician can determine the force exerted by the thyroid cartilage on the sensor from that pressure. Most piezoelectric sensors, however, are inflexible materials, which means that a subject may feel uncomfortable wearing them on the skin surface of the thyroid cartilage. In addition, pasting a solid sensor on the throat will affect a subject's swallowing function, which causes measurement bias.

There is a type of sensor which was not used in previous research, the force-sensing resistor (FSR), shown in Figure 1c. It is a kind of piezoresistive sensor. Its use principle requires contact with both sides of the sensor such that its resistance will be changed. Applying a larger pressure on both sides of the FSR will cause better conductivity, therefore we can connect a small power source to the sensor and detect a force applied on the sensor by measuring changes in the output voltage. Because the FSR can be manufactured into a very small thin-film structure, it is often used to measure the foot pressure [14,15]. There are advantages if we use a FSR sensor to detect the thyroid cartilage motions in the pharyngeal stage. First, it is a soft and flexible material. Thus we can paste the FSR sensor on the throat without affecting the swallowing function. Second, we can measure the output voltages of the sensor to monitor the force the throat exerts on the surface of the sensor during the swallowing. This can help us to develop an easy-to-use, lightweight, and wearable belt device for healthcare applications or bedside assessment.

This paper aims to use the FSR to design a wearable and portable monitoring system for oropharyngeal swallowing. Such a system can be applied to bedside assessments or homecare applications. The diagram of this system is illustrated in Figure 2. The system consists of a FSR throat belt, a holter and a remote monitoring system. The FSR throat belt is made by soft and flexible materials with a FSR embedded in the middle of the belt. The signals measured by the FSR will be collected in the holter. The holter performs basic signal processing with a 1 kHz sample rate, 20 Hz low-pass filtering, and 8-bit analog-to-digital (A/D), *etc.* The data afterwards can be displayed immediately on a smart phone or be sent to a remote monitoring system for further physiological analysis. Later we will show the detailed design of the FSR belt and its test results. An implementation example of the holter system will be shown before the end of this paper.

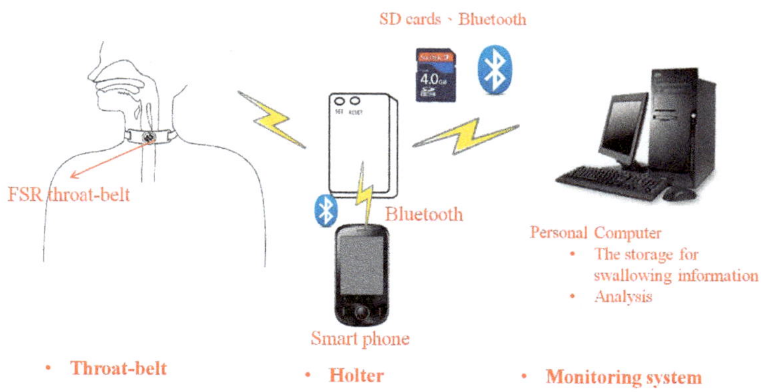

Figure 2. The diagram of the wearable swallowing monitoring system.

The concept of this work was first reported in [16] with less than 10% overlap with the current pape, where we clarify the design in more details and include more test results. The remainder of this paper is organized as follows: in Section 2 we survey the previous research on approaches for monitoring dysphagia. In Section 3 we present the design of the proposed FSR belt and the signal analysis method. In Section 4 we first present the testing method, and then the test results with discussion. Finally, Section 5 gives the conclusions.

2. Related Works

Many researchers have mentioned that the evaluation of the swallowing function before treatment is very important. According to the studies in [2–4], a patient's quality of life can be greatly affected by dysphagia. To observe swallowing behavior, Logemann *et al.* [10] used VFSS equipment to monitor the swallowing motion during the oral and pharyngeal stage. Ertekin and Pehlivan *et al.* combined the electrophysiological and mechanical approaches to measure the laryngeal motions with submental sEMG during swallowing [3]. They proved that the study of the physiology of deglutition is very useful in the clinical evaluation of dysphagia patients.

Some researchers [17–19] have mentioned that the tongue pressure will also affect the swallowing function. They designed an apparatus based on a three-bulb silicon array to measure the tongue pressure. The apparatus is a handheld device, and can be put into fixed positions above the tongue for measurement. When the subject's tongue touches the silicon array, each bulb will reflect the pressure during the swallowing process. This approach, however, cannot reflect the swallowing status in the pharyngeal stage.

Martin-Harris and Brodsky [6,7] developed a normative model to integrate breathing and swallowing patterns, using concomitant videofluoroscopic images and nasal respiratory airflow recordings. Catiuscia and Luiz have mentioned that the methods currently used for diagnosis, however, are only qualitative, and need expensive equipment [1]. Also, most current clinical bedside evaluations are based on physician experience. Therefore, in Catiuscia and Luiz's study [1], they used the three-axis accelerometer to measure the neck vibrations associated with deglutition, and used a PDA as a portable device for bedside measurements. Guilherme and Evert *et al.* [5] used the nasal airflow measurement subsystem with the accelerometers to measure the correlations between breathing and swallowing.

There are other researches which evaluated how different factors of bolus affect the swallowing physiology, including the temperature or the volume [13]. From the research in [13], we find that if the volume of water was larger than 20 mL, some subjects failed to swallow the bolus after the first try and they had to divide the bolus into two or more pieces as piecemeal deglutition in a hotter temperature range.

In our approach, we use not only the FSR throat-belt to measure swallowing motions, but also a nasal airflow sensor to monitor the association with respiration. By integrating the signals from the FSR throat-belt and the nasal airflow sensor, we can compare the correlations between swallowing and breathing. Moreover, we have developed a portable holter to continuously collect the data. The data can be transmitted to computers via wireless communication, making our approach is very suitable for bedside and homecare measurement applications.

3. Methods

In this section we show the design of the proposed FSR throat belt. Then we present the algorithm by which the swallowing and breathing events can be identified automatically from different signals (*i.e.*, FSR, sEMG, and nasal airflow).

3.1. FSR Throat-Belt Design

The major component of this work is the FSR throat-belt design. Recall that the FSR should be used by contacting both sides of the sensor to change the internal conductivity. We found that if we only use medical tape to paste the FSR on a subject's throat, the signals could not be obviously detected because medical tape cannot provide a sufficient reaction force on the FSR as the larynx or the thyroid cartilage does during swallowing. To resolve this problem, we design a throat-belt where the FSR sensor is fixed on the center of the belt and the subject can wear it around the neck, as shown in Figure 3. The belt has good elasticity and we use Velcro straps to close it. The maximal width of the belt is 5 cm, therefore it does not obstruct the natural swallowing motions. Particularly, we insert a small airbag between the belt and FSR such that FSR can be fixed on the center of the thyroid cartilage without any movement during the testing. If FSR is not fixed on that point, the measurement accuracy will be affected. When the belt is tied around the neck, the airbag will provide a stable initial pressure on the FSR which will be considered as the baseline in each measurement. During swallowing, the larynx and the thyroid cartilage will retract, resulting in released and changed pressure on the FSR, as shown in Figure 4.

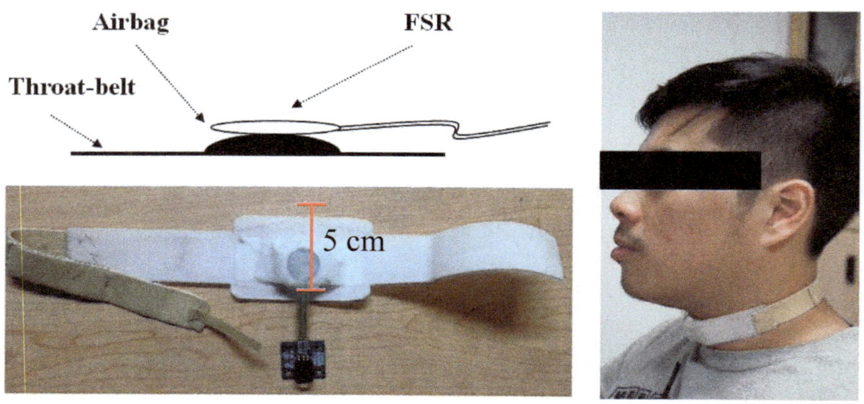

Figure 3. The FSR throat-belt.

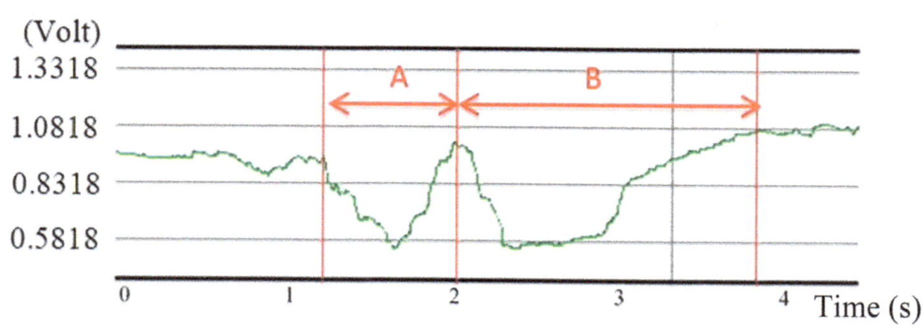

Figure 4. A typical swallowing signal measured by FSR on a healthy subject.

We let a healthy subject (21 years old) swallow 10 mL of room temperature water as an example. Figure 4 shows the swallowing signal measured by FSR. At first, the FSR detected the initial pressure from the airbag. Then two significant responses followed during time periods "A" and "B". The first valley in time period "A" represents that the larynx starting to move upward and forward such that the bolus can be pushed on the top of the esophagus. After that, the second valley in time period "B" appears, representing when the thyroid cartilage retracts and the bolus is pushed into the esophagus. Finally the thyroid cartilage returns back to the original position. Because the belt is made of all soft materials, such results can reflect real motions of the thyroid cartilage without affecting the subject's swallowing function.

3.2. Signal Analysis for the Physiology of Swallowing

We are interested in finding the correlations between swallowing and respiration. Therefore we attach the FSR belt around the neck, paste a pair of sEMG electrode pads on the submental muscle, and put the nasal airflow cannula in front of the nose. To synchronize those three different signal sources, in the following measurements we connect all of them into a BIOPAC MP100 system for data acquisition and analysis. Also, here we mostly focus on detecting the timing of swallowing events. We do not calibrate the voltage domain (Volts) of signals to the force domain (N); *i.e.*, we only display the voltages of signals. Later we will discuss the implementation of the holter (*cf.* Figure 2) such that the FSR belt and other signals can be collected for practical applications.

We use the sEMG sensor to detect the signals from the surface of the submental muscle. Figure 5a,c show the submental sEMG and the FSR signals from a healthy subject swallowing 10 mL of room temperature water. The sEMG signals have been amplified by 1000 times on a scale between −1 Volts and 1 Volts. In the figure we find that the onset time of sEMG covered most of the swallowing process. This is reasonable because when we swallow the bolus, the submental muscle will continue to squeeze the bolus into the top of esophagus. Thus we can use the responses of the submental sEMG to identify the beginning and the ending of the pharyngeal stage in each swallowing test.

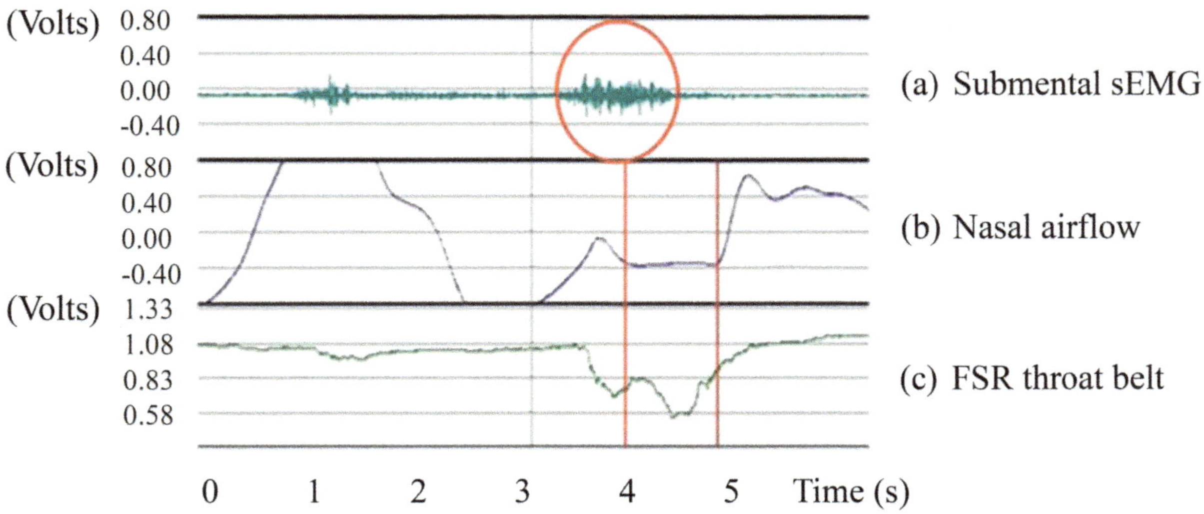

Figure 5. Signals from three sensors: (**a**) Submental sEMG; (**b**) Nasal airflow; and (**c**) FSR throat belt (the Y-axis preserves the original scaling in the BIOPAC MP100 for each signal).

We also used the nasal cannula to measure the airflow changes caused by respiration. The amount of nasal airflow through the nasal cannula will be translated into digital signals by a pressure transducer. Figure 5b,c show the nasal cannula signals and the FSR signals in the same 10 mL room temperature water swallowing test. The respiration of healthy participants, as shown in the figure, will display an apnea episode during the swallowing. It is because our the human physiological mechanism will protect us reflectively from choking injuries. According to the signals from the nasal airflow and the FSR throat-belt, a physician can immediately evaluate the coordination between the swallowing and the respiration.

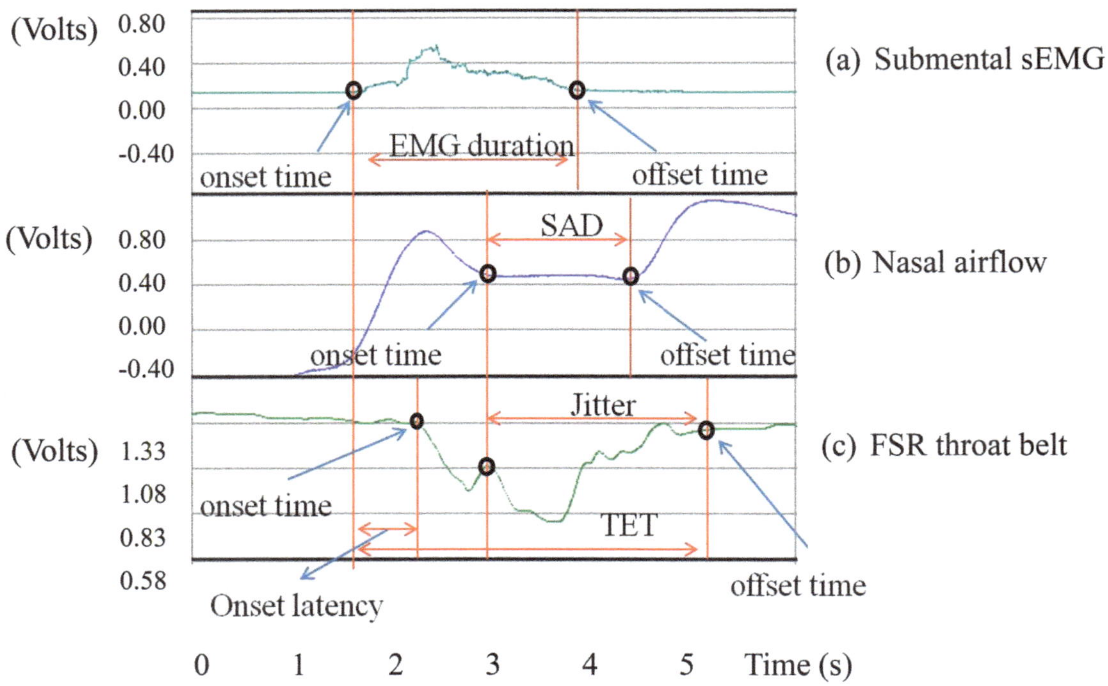

Figure 6. Representative non-invasive swallowing measurements: (**a**) Submental sEMG; (**b**) Nasal airflow; and (**c**) FSR throat belt.

By analyzing the signals from these three sensors (*i.e.*, Figure 5), we can find that the onset time and the offset time of each signal have different physiological meanings, which can be used to partition a normal swallowing process into several periods:

(1) EMG duration: This is the duration of the submental muscle motion, which can be obtained by measuring the time interval between the onset time and the offset time of the submental sEMG waveform, as shown in Figure 6a.

(2) Onset latency: This is the duration from the onset time of the submental muscle "starting to contract", to the onset time of the thyroid cartilage "starting to move upwards". It can be obtained by measuring the time interval between those two onset time points of the submental muscle sEMG and the throat-belt waveforms, respectively, as shown in Figure 6a,c.

(3) Total excursion time (TET): This is the time from when the thyroid cartilage starts to move upwards to the thyroid cartilage returning to the original status. It can be obtained by measuring the time interval between the onset time and the offset time of the FSR throat belt waveform, as shown in Figure 6c.

(4) Swallowing apnea duration (SAD): This is the time of apnea, which can be obtained by measuring the time interval between the onset time and the offset time of the nasal airflow waveform, as shown in Figure 6b.

(5) Jitter: The time while the throat is pushing the bolus into the esophagus. It can be obtained by measuring the time interval between the turn-around point and the offset time from the throat belt waveform (*i.e.*, the period "B" in Figure 4), as shown in Figure 6c.

The time intervals mentioned above all involve event identification among different signals. In order to identify the timing of those events automatically for analysis, we apply the following pre-processing steps to each signal:

(1) For submental muscle sEMG: We use the BIOPAC MP100 system [20] to record the sEMG signals (e.g., Figure 7a). The sEMG signals were filtered through a Butterworth digital filter (second-order band-pass filter) with a passband of 20–400 Hz, and smoothed using a 15-point moving average [21,22], as shown in Figure 7b. To identify the sEMG onset time, we fetch the data from the first three seconds and calculate the mean (M) with the standard deviation (SD). For the data (say: sEMG[i]) after the third second, we set a threshold ($M + \alpha \cdot SD$), where α is a factor ranged from 1 to 3 (the value of α should be determind by real signal patterns; in this work, we set the value of α as one.) If sEMG[i] is the first point of which the value is over the threshold, we mark it as the onset time point. Typically a swallow will finish in a few seconds. Therefore we fetch totally four-second data after the oneset time point for analysis. If sEMG[j] is the last point of which the value is over the threshold in this period, we mark it as the offset time point. The time between sEMG[i] and sEMG[j] is the "EMG duration".

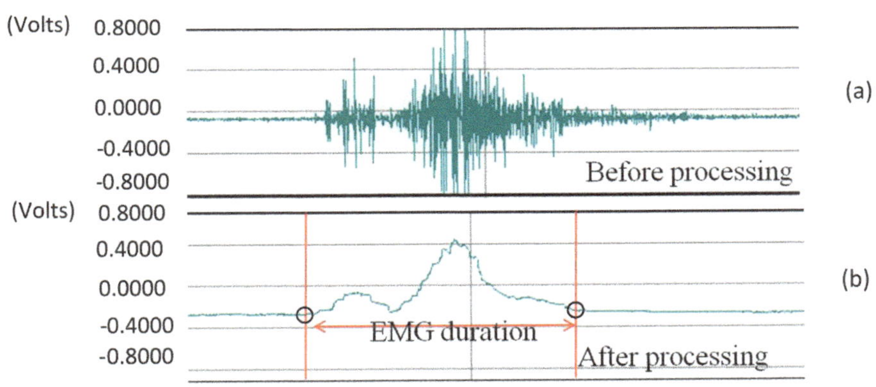

Figure 7. The processing of the submental muscle sEMG: (**a**) Before; (**b**) After.

(2) Nasal airflow sensor: We also use the BIOPAC MP100 system [20] to record nasal airflow signals (e.g., Figure 8a). In order to enhance the detection accuracy, we differentiate between the inspiration and the expiration phases first. That is, if the data (say nasal[k]) is larger than a threshold ($M + \beta \cdot SD$), we amplify the value of nasal[k] by N times. Here, M is the mean of the first three-second data (baseline), SD is the standard deviation, and in the following testing we let $\beta = 1$ and $N = 3$. Figure 8b shows the results after the amplification. Then we can apply the similar method to detect the onset time and the offset time. The interval between both time points is SAD.

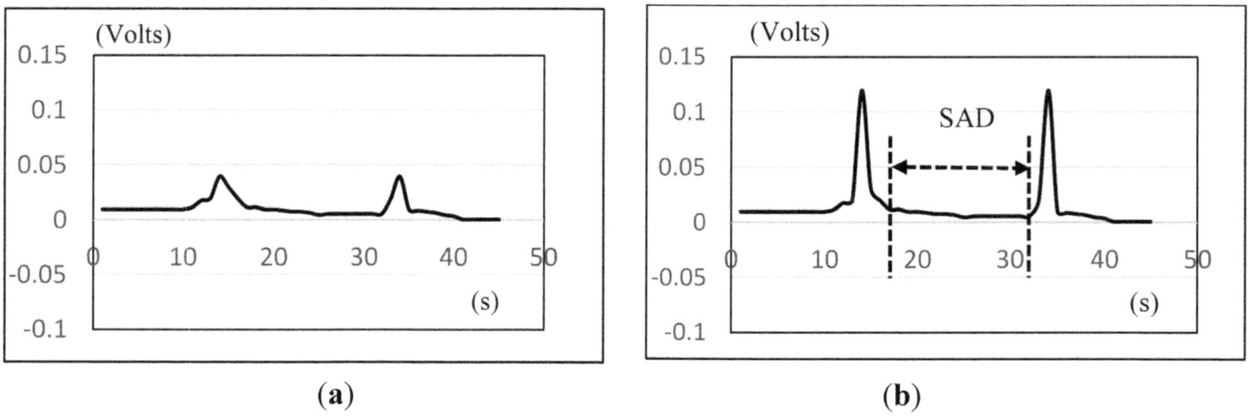

Figure 8. The amplification and the detection of the nasal airflow signals: (**a**) Before; (**b**) After.

(3) Throat-belt: To detect the onset time and the offset time of the FSR signals, we apply the same approach used in the sEMG detection. We calculate the mean of data in the first three seconds as the baseline, and set a threshold by subtracting a ratio of the standard deviation from the mean (*i.e.*, ($M - r \cdot SD$, $0.5 \leq r \leq 1$), The value of r should be determined by real signal patterns (in this work, we set the value of r as one). The onset time and the offset time can then be identified as the first and the last points less than the threshold, respectively. Particularly, we need to detect the middle turn-around point between the onset and the offset points for calculating the jitter (*cf.* Figure 6). The turn-around time detection can be done by scanning the slope variation along the curve between the onset and the offset time points.

4. Results and Discussion

We used the same equipment described in [21,23] to verify the proposed approach, including the nasal cannula to measure the nasal airflow, the sEMG electrode pad to measure the submental muscle motions, and finally using the BIOPAC MP100 system to record all signals. Instead of using the piezoelectric sensor as in [21,23], in this study we used the proposed FSR throat belt to detect the motions of the thyroid cartilage.

We recruited nineteen healthy subjects to participate in the testing, and divided them into three age groups: 21–30 years old, 31–50 years old and 51–60 years old, as shown in Table 1. The exclusion criteria were any known history of dysphagia, cardiopulmonary disease, neurological disease, hiatal hernia, chronic indigestion disorder, gastroesophageal reflux disease, cancer or disease of the head and neck, current use of medications with known effects on swallowing or breathing, tobacco use in the past 10 years, or age of more than 70 years. We excluded participants aged more than 70 years, because this age group is more likely to have comorbidities that influence swallowing and respiration. The Ethics Committee of Chang Gung Medical Foundation approved this study. All participants signed informed consent forms before participation.

Table 1. Characteristics of participants in the three age groups.

Group	Age Range	Mean ± SD	Male	Female
Young	21–30	24 ± 3	4	3
Middle	31–50	37 ± 5	4	4
Old	51–60	55 ± 4	2	2

During the experiment, the participant sat on a chair with a natural posture. We cleaned the surface skin of the jaw and throat using medicinal alcohol. After that we pasted the sEMG electrode on the submental muscle. Also, we helped the participants fit the throat-belt such that the FSR can be placed on the surface of the thyroid cartilage, and the nasal airflow cannula is placed in front of the nose.

The experiment requires consumption of five different volumes (1, 3, 5, 10 and 20 mL) of room temperature water. Each subject performed three trials for each volume. After each trial, we let the subjects rest for three minutes. Most importantly, we remind the subjects to swallow normally and within their ability. The standard operating procedure (SOP) is shown in Figure 9. For safety considerations, we did not ask the participants to do the counterbalancing measurement. We asked the participants to start the test from a small volume of water, and increase the water volume step by step. If we do not know the participant's swallowing limit in advance and start the test with a large volume of water, he or she may have choking risk which affects the subsequent testing.

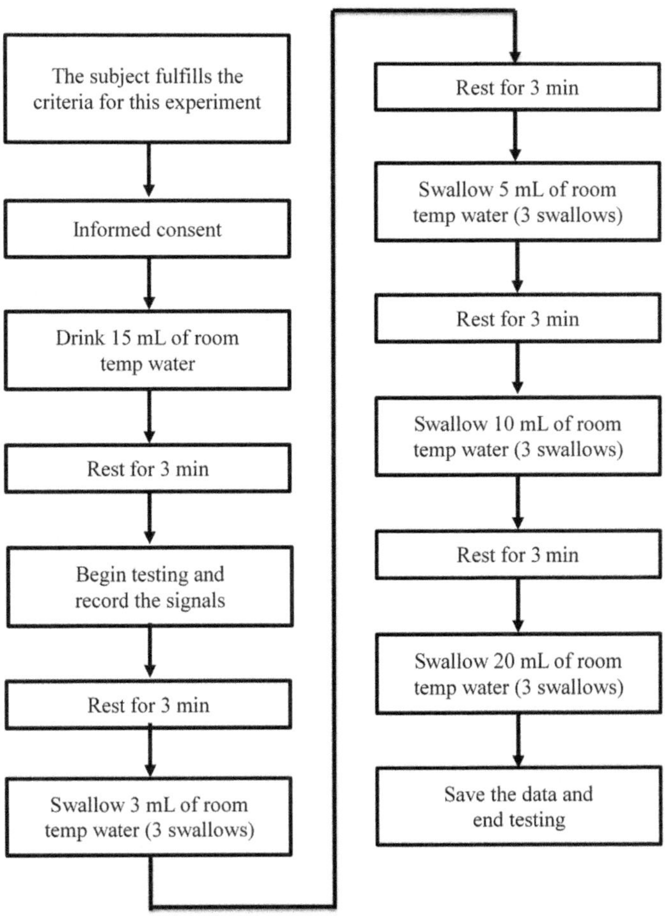

Figure 9. The standard operating procedure in our testing.

4.1. Experimental Results

In this section, we will show the results measured by the submental muscle sEMG, the nasal airflow, and the FSR throat belt, as well as the correlation analysis between them. Our goal is to show that the proposed belt can have good sensitivity to the effects of age, the volume of water swallowed, or the gender of the subjects performing the swallowing functions.

4.1.1. The Results of sEMG and Nasal Air Flow

Figure 10 shows the comparison results of sEMG in different age groups, where (a) is for male participants and (b) is for female participants. The first point we found is, although 1 mL water is the smallest water volume, it does not mean that the participants (no matter whether male or female) needed the least time to swallow it. Also, as the age of the participants increases, the sEMG duration will have a slight increase. This fact was also found in other previous studies [21,23,24]. A more acceptable reason is that 1 mL water is too small a sample so most participants need to use more strength and time to swallow it. The other point is, for 3, 5 or even 10 mL water, there are no obvious differences among different age groups. This result shows that using sEMG alone would not be a good approach to evaluate the swallowing function, because the duration time of sEMG actually cannot represent the whole swallowing process (however, in some previous studies, like [18,24], they evaluated the swallowing function by only observing the strength or the duration of sEMG).

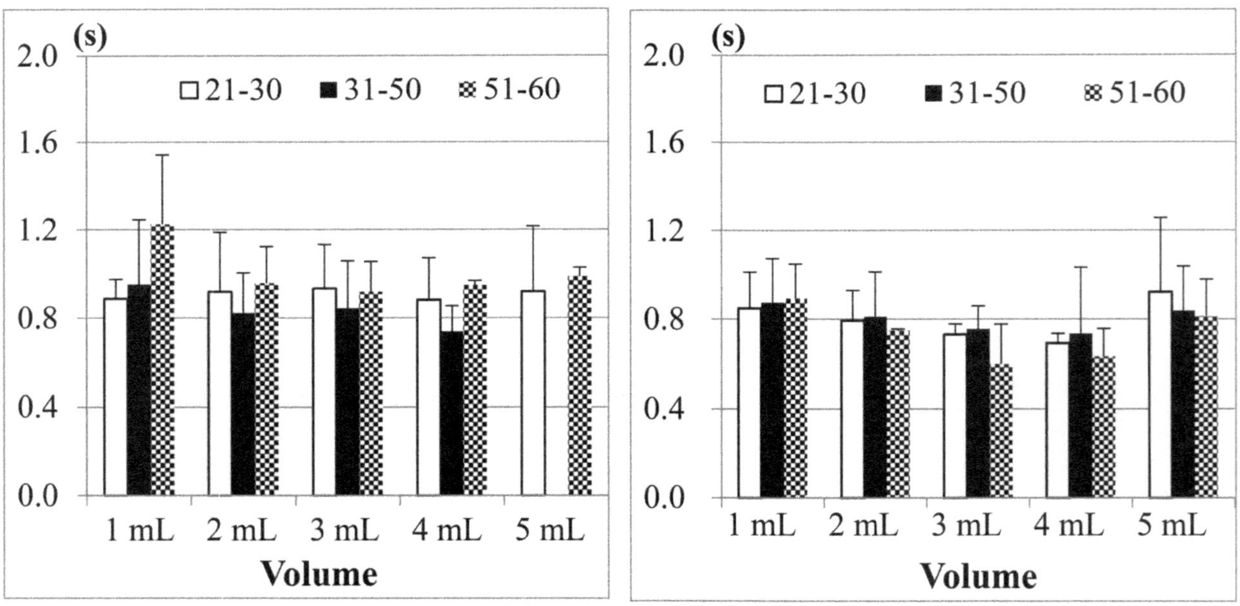

(a) (b)

Figure 10. Comparison of the average duration time of the submental sEMG in different age groups for (**a**) male; and (**b**) female participants. The X-axis lists the water volumes and the Y-axis shows the mean duration time.

Note that the results of the 31 to 50 years old male group in the 20 mL trials are not included in Figure 10a. This is because the participants in this group peformed piecemeal swallowing (also known as "piecemeal deglutition" [13]), where they divided the water into smaller volumes, and finished the swallowing in several times. In this case, the volume represents a dysphagia limit and the duration time of swallowing fluctuates, which makes measurement and analysis difficult. Thus we treat piecemeal swallowing as a special case and will discuss it later.

Figure 11 shows the SAD results, calculated from the nasal airflow, where (a) is for male participants; and (b) is for female participants. Here also the 20 mL water results of the male group aged from 31 to 50 years old are not included due to the piecemeal swallowing mentioned above.

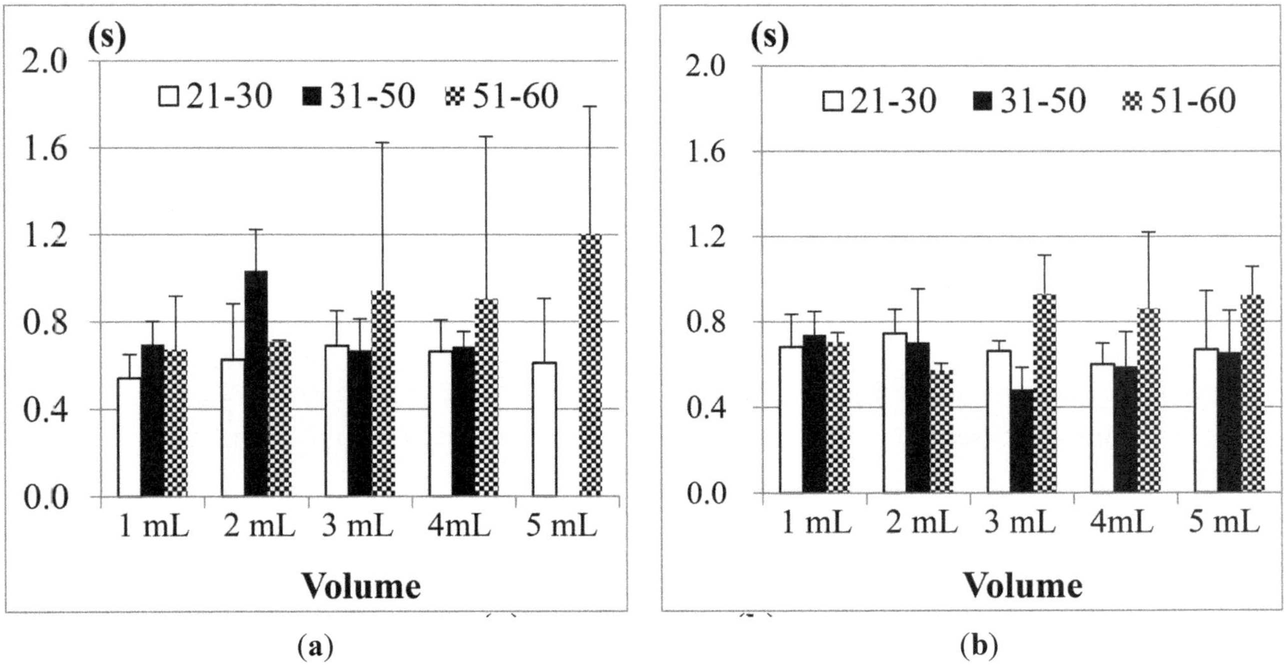

Figure 11. Comparison of the average duration of SAD in different age groups for: (**a**) male; and (**b**) female participants. The X-axis lists the water volumes and the Y-axis shows the mean duration time.

As we can see in Figure 11, the male and the female participants aged from 51 to 60 years old have obviously longer SAD duration in the 5, 10, and 20 mL tests than the other age participants, but for the participants aged from 21 to 30 and from 31 to 50, there are no significant differences for different volumes of water. We know that when the throat pushes the water or bolus into the stomach via the esophagus, the breathing will undergo a brief apnea period. The duration of this period (*i.e.*, SAD in Figure 6) will differ, mostly depending on each one's swallowing behavior. In our tests, some participants stopped breathing before swallowing for a very short time. Even after swallowing the water, some participants will continue to have this apnea period for a while. Figure 12 illustrates both of these cases if we compare the nasal airflow signals with the FSR signals from our throat belt.

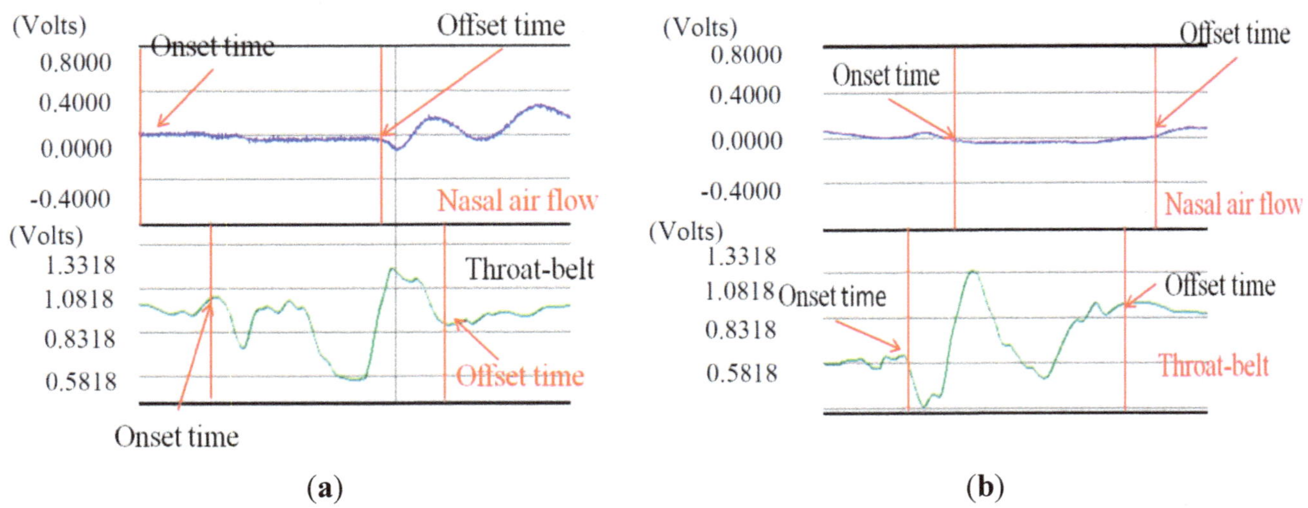

Figure 12. Two cases of SAD: (**a**) the onset time of SAD is earlier than the onset time of FSR; (**b**) the offset time of SAD is later than the time of FSR.

Figure 12a is the case of which the participant is a 21 year-old male in the 5 mL test. The nasal air flow signals show that the onset time of SAD (*i.e.*, apnea period) is earlier than the onset time of FSR (*i.e.*, the time of the thyroid cartilage starts to move). Different from Figure 12a, Figure 12b is another case of which the subject is a 40 year-old male in the 10 mL test. In this case, the offset time of SAD happens later than the time the thyroid cartilage stops the motions. This is why in in Figure 11a there are some cases where the standard deviations are particularly large. In our testing, we find that the swallowing function cannot be distinguished clearly only by sEMG or only by SAD. Thus in the next section, we will add the FSR throat belt to the testing for swallowing function evaluation.

4.1.2. The Results of Jitter and TET by the Throat Belt

Recall that the duration of Jitter and TET (in Figure 6) can be used to measure the motions of the thyroid cartilage. According to the research of Wang and Chuang *et al.* [21,23], the duration of Jitter and TET will be affected by the bolus volume. Here we use the FSR throat belt to measure Jitter and TET, and compare the effects of the bolus volumes on different genders and age groups.

(1) Comparison between men and women

Figure 13a shows the results of Jitter, and Figure 13b shows the results of TET for the participants aged from 21 to 30 years old in tests with different water volumes. The results show that both male and female participants have longer Jitter and TET durations as the volumes of water increase, except for the case of 1 mL of water. The reason is the same as what we mentioned in Figure 10.

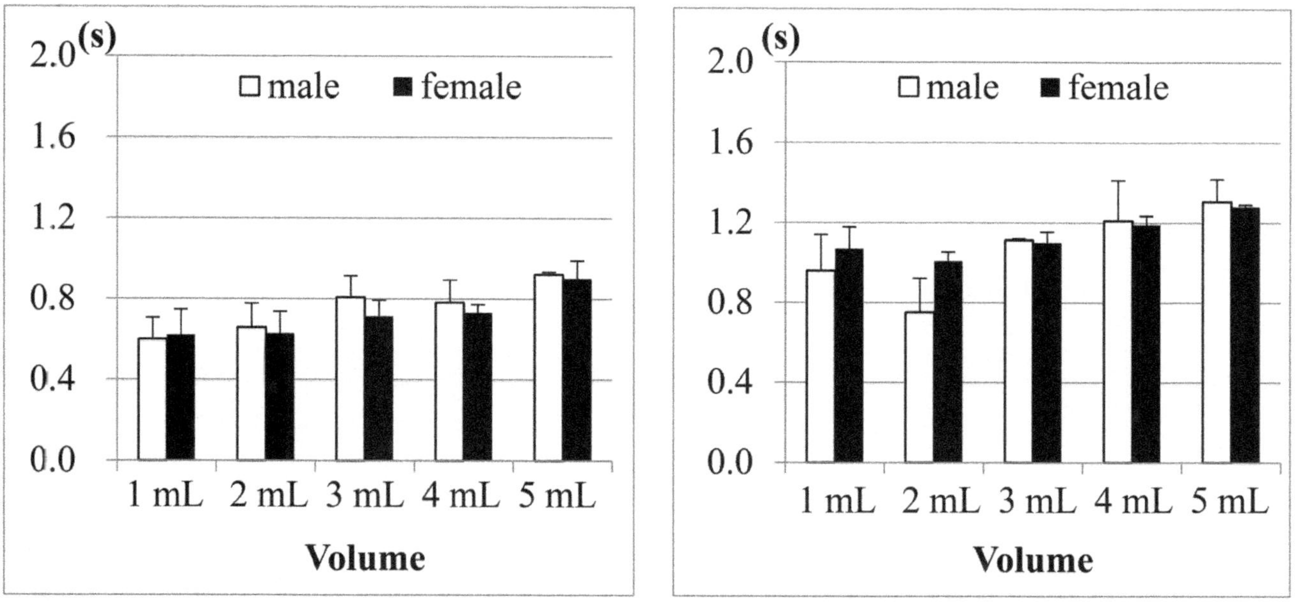

Figure 13. Average duration of (**a**) Jitter and (**b**) TET for the age group 21–30 years old.

The results in Figure 13 also show that both the Jitter and TET of young male participants (aged from 21 to 30), on average, are slightly longer than those of the female participants (aged from 21 to 30) when the volume of water is larger than 5 mL, but for the volumes smaller than 5 mL, the male participants seem to have shorter Jitter and TET. We use pairwise *T*-test to evaluate if the male and female participants in this age group have obvious differences (*i.e.*, research hypothesis). Table 2 shows the statistical analysis of average Jitter and average TET between the male and female groups aged from 21 to 30 years old for different water volumes. The research hypothesis will be accepted if (*T*-value > *F*-value) or (*T*-value < −*F*-value) at the $\alpha = 0.05$ level. From Table 2 we can find that in this age group of the participants, the average Jitter or TET between different genders do not have statistically significant differences.

Table 2. Statistical analysis of average Jitter and average TET between the male and female groups aged from 21 to 30 years old for different water volumes.

(α = 0.05)		1 mL		3 mL		5 mL		10 mL		20 mL	
		Jitter	**TET**	**Jitter**	**TET**	**Jitter**	**TET**	**Jitter**	**TET**	**Jitter**	**TET**
Male's mean		0.599	0.959	0.656	0.751	0.807	1.112	0.781	1.210	0.921	1.305
Female's mean		0.620	1.068	0.627	1.006	0.713	1.100	0.731	1.190	0.900	1.280
95% CI *	Upper	0.876	1.285	0.740	1.365	0.804	1.162	0.816	1.332	0.989	1.368
	Lower	0.573	1.017	0.593	0.913	0.678	1.046	0.695	1.076	0.861	1.205
Degree of freedom (*df*)		13	13	16	16	12	12	9	9	7	7
F-value		2.160	2.160	2.120	2.120	2.179	2.179	2.262	2.262	2.365	2.365
T-value		−0.609	−1.790	1.156	−1.257	1.380	0.199	1.053	0.117	0.221	0.079
p-value		0.552	0.097	0.265	0.226	0.193	0.846	0.319	0.909	0.831	0.939
Cohen's d (Effect size)		−0.321	−0.943	0.545	−0.577	0.770	0.111	0.638	0.071	0.148	0.053

* CI = Confidence Interval of this age group.

Figure 14 further shows the results of average Jitter and average TET for the 31–50 years old age group. We can find that Jitter and TET become longer as the volume of water increases. Moreover, when we increase the volume of water to 20 mL, the male participants display piecemeal swallowing situations. In this case, they need to swallow the 20 mL water in many times, therefore their Jitter and TET cannot be measured.

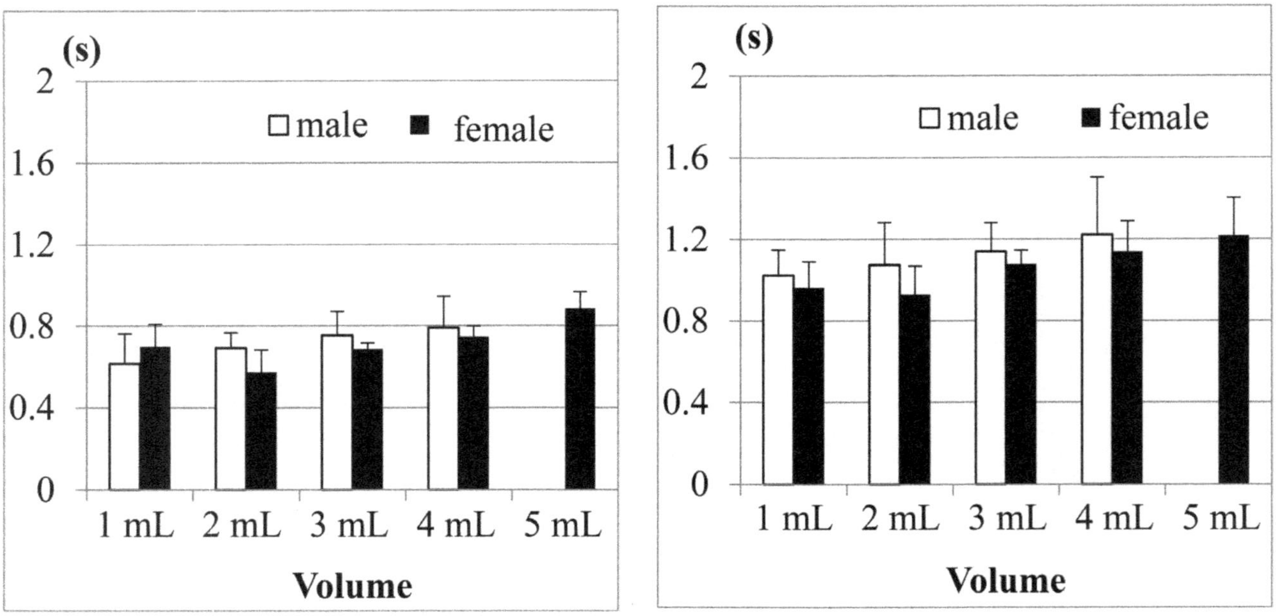

(a) (b)

Figure 14. Average duration of (**a**) Jitter and (**b**) TET for the age group 31–50 years old.

For this age group (31–50), we also find that the male participants in most measurements need longer Jitter and TET durations to finish the swallowing than the female participants. We did pairwise T-tests to compare the male and female groups. Table 3 shows the statistical analysis of average Jitter and average TET between the male and female groups aged from 31 to 50 years old for different water volumes. We can find that in the 3 mL and 5 mL trials, the average Jitter shows a statistical difference (*i.e.*, *T*-value > *F*-value), while the average TET does not. This result is consistent with the study in [24], where they found that the male participants performed slightly worse than the female participants in this age group.

Table 3. Statistical analysis of average Jitter and average TET between the male and female groups aged from 31 to 50 years old for different water volumes.

(α = 0.05)		1 mL		3 mL		5 mL		10 mL		20 mL	
		Jitter	**TET**	**Jitter**	**TET**	**Jitter**	**TET**	**Jitter**	**TET**	**Jitter**	**TET**
Male's mean		0.616	1.023	0.691	1.073	0.754	1.139	0.789	1.221	N/A	N/A
Female's mean		0.698	0.962	0.573	0.928	0.686	1.076	0.746	1.137	0.883	1.215
95% CI *	Upper	0.698	1.032	0.679	1.088	0.764	1.201	0.805	1.277	0.914	1.237
	Lower	0.573	0.905	0.550	0.921	0.682	1.054	0.701	1.042	0.852	1.074
Degree of freedom (*df*)		18	18	14	14	19	19	17	17	19	19
F-value		2.101	2.101	2.145	2.145	2.093	2.093	2.110	2.110	2.093	2.093
T-value **		−1.763	0.947	**5.147**	1.481	**2.424**	1.800	0.533	0.304	N/A	N/A
p-value		0.095	0.356	**0.001**	0.156	**0.025**	0.088	0.601	0.765	N/A	N/A
Cohen's d (Effect size)		−0.788	0.424	2.658	0.662	1.069	0.794	0.245	0.139	N/A	N/A

* CI = Confidence Interval; ** The bold number means that *T*-value > *F*-value.

Finally, the results of the 51–60 year old age group are shown in Figure 15. Compared with the previous two age groups, the participants aged from 51 to 60 obviously needed even longer durationd of Jitter and TET to finish swallowing in all measurements. Particularly, in this age group, we note that the female participants on average need longer Jitter and TET durationd than the male participants, which is quite different from the results in the two previous age groups.

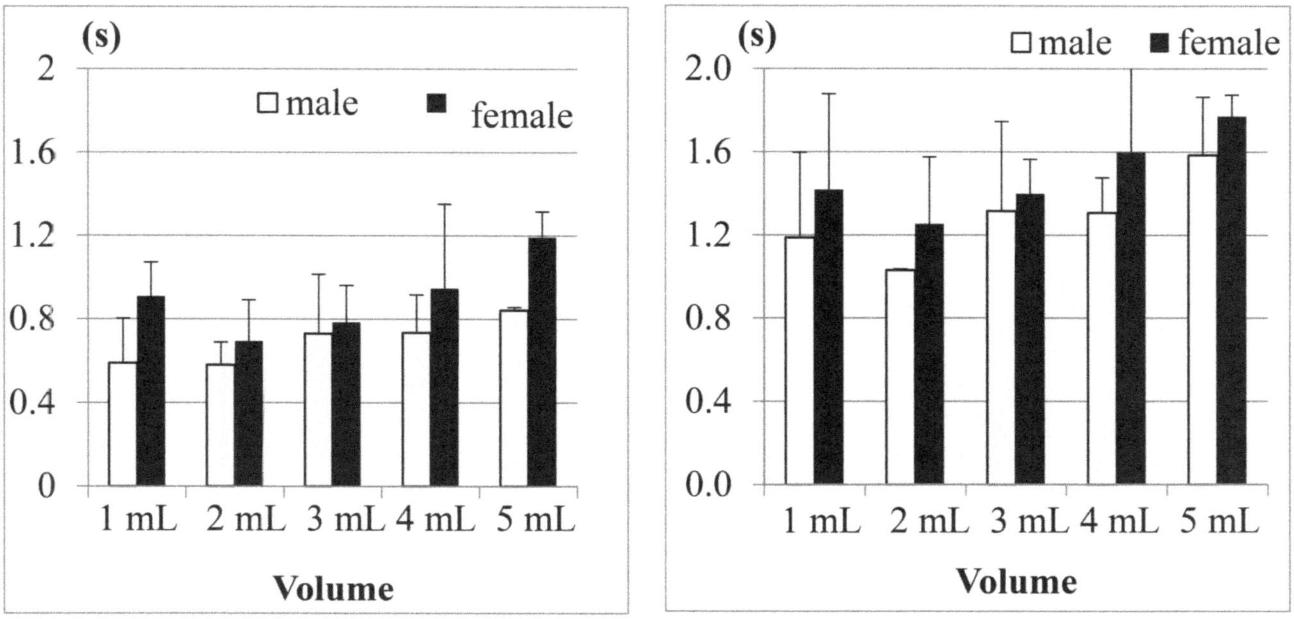

(**a**) (**b**)

Figure 15. Average duration of (**a**) Jitter and (**b**) TET for the age group 51–60 years old.

If we apply a pairwise T-test to compare the male and female participants in this age group (51–60), we find that the durations of Jitter in the 1, 10 and 20 mL measurements show more significant differences (*i.e.*, *T*-value < −*F*-value) than the durations of TET (see Table 4). From Figures 13–15, our testing suggests that the measurement of Jitter can have more distinctive sensitivity than the measurement of TET.

Table 4. Statistical analysis of average Jitter and average TET between the male and female groups aged from 51 to 60 years old for different water volumes

		1 mL		3 mL		5 mL		10 mL		20 mL	
		Jitter	**TET**	**Jitter**	**TET**	**Jitter**	**TET**	**Jitter**	**TET**	**Jitter**	**TET**
Male's mean		0.589	1.187	0.580	1.029	0.728	1.316	0.733	1.307	0.842	1.584
Female's mean		0.908	1.418	0.693	1.253	0.784	1.399	0.946	1.598	1.192	1.771
95% CI *	Upper	0.904	1.524	0.732	1.302	0.917	1.579	1.096	1.754	1.234	1.852
	Lower	0.594	1.081	0.538	1.001	0.603	1.135	0.660	1.278	0.917	1.566
Degree of freedom (*df*)		10	10	9	9	7	7	7	7	7	7
F-value		2.228	2.228	2.262	2.262	2.365	2.365	2.365	2.365	2.365	2.365
T-value **		**−3.004**	−1.216	−1.615	−1.889	-0.645	−0.592	**−2.725**	**−2.746**	**−4.099**	**−5.651**
p-value		**0.013**	0.252	0.141	0.091	0.539	0.572	**0.029**	**0.028**	**0.004**	**0.001**
Cohen's d (Effect size)		−1.734	−0.702	−0.978	−1.144	−0.456	−0.419	−1.828	−1.942	−2.898	−3.996

* CI = Confidence Interval; ** The bold number means that *T*-value < −*F*-value.

(2) Effects of the Age

The authors in the study [18,19] mentioned that age will have more impact on the swallowing ability; when the age increases, the duration of Jitter and TET will become longer. According to this point, we rearrange the data and compare the swallowing ability among different age groups.

Figure 16 shows the results of Jitter and TET for the male participants in three age groups (the data of the participants aged from 31 to 50 years old in the 20 mL measurement was not included due to the piecemeal swallowing).

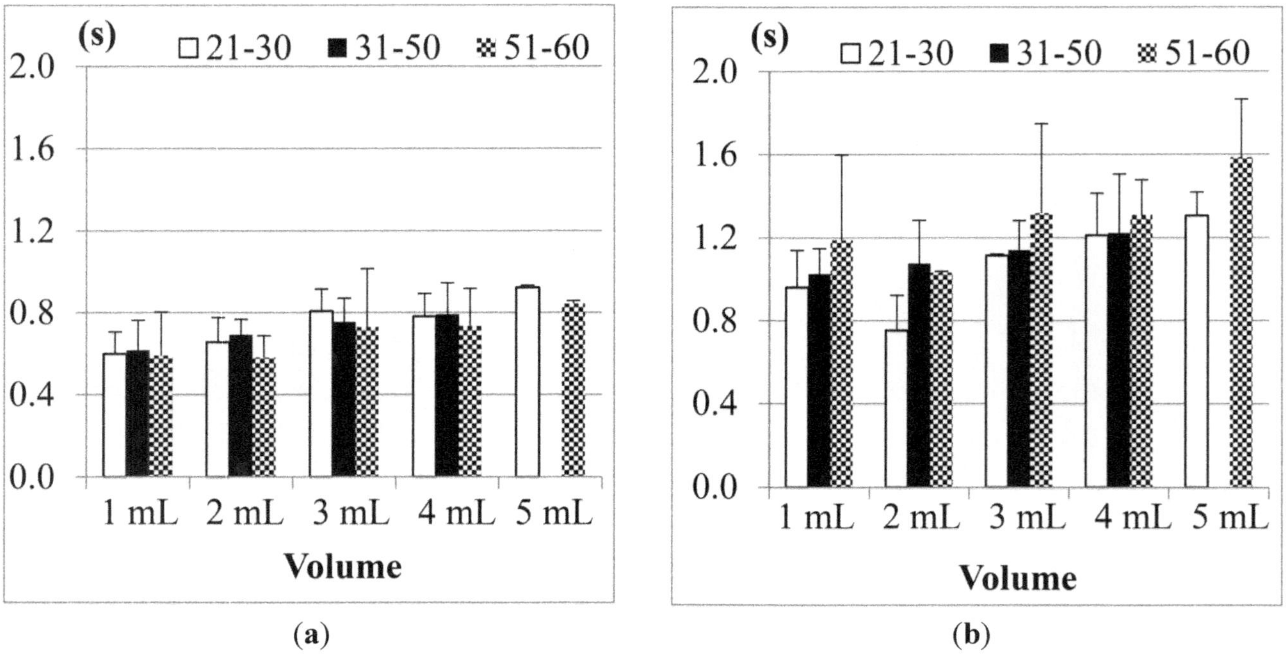

Figure 16. Average results of the male participants in different age groups for (**a**) Jitter and (**b**) TET.

We find that the participants aged from 51 to 60 years old, on average, need the longest TET (total swallowing time) but the shortest Jitter (the duration of the thyroid cartilage pushing the bolus into the esophagus) to finish the swallowing of water, compared with the other two age groups. This shows that the overall swallowing ability of the elder participants is not as good as the younger participants such that their swallowing needs longer time in the pharyngeal stage. On the other hand, the strength of the thyroid cartilage for the elder participants may be not so good, too. That is why the water bolus will fall into the esophagus in a shorter period, which could potentially bring higher risks of choking injuries.

Figure 17 showed the results of the female participants. As we can see, the female participants aged from 51 to 60 years old also have the longest TET but shorter Jitter, even worse than the male participants. It shows that both of their swallowing function of the thyroid cartilage could deteriorate at a certain rate at this age, but the female participants in this age group may have a higher risk of choking injuries than the male group in their daily eating.

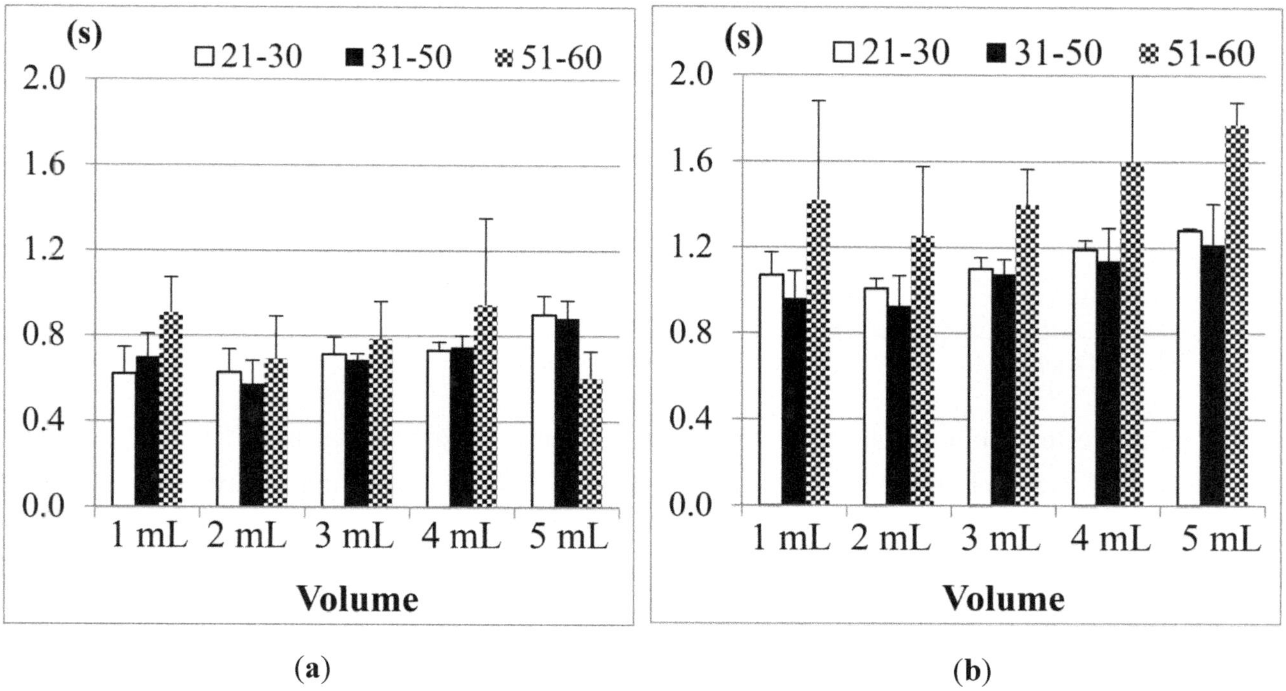

Figure 17. Average results of the female participants in different age groups for (**a**) Jitter and (**b**) TET.

We would like to know if the results shown in Figures 16 and 17 have statistical meaning. Therefore we use Analysis of Variance (ANOVA) testing to compare the three different age groups (*i.e.*, 21–30, 31–50, and 51–60). Table 5 shows the testing results for Jitter, and Table 6 for TET. In both tables we list the degree of freedom and the sum of squares for each trial. Also, F-values are calculated and compared with F-boundaries at the $\alpha = 0.05$ level. If (F-value > F-boundary) or (F-value < −F-boundary), the null hypothesis (*i.e.*, those three age groups did not have obvious difference) can be rejected. Finally the values of "eta squared" are listed in the last row for reference.

From Table 5, we find that the parameter "Jitter" does not show significant differences between the different age groups in most water volumes except 20 mL because their F-values were not large enough. From Table 6, however, the "TET" parameter shows totally different results. In those five different water volumes, we have up to four F-values (*i.e.*, 1, 5, 10, and 20 mL) larger than F-boundaries. It shows that these three age groups really have a statistical difference in the total swallowing time (TET).

Table 5. ANOVA testing of "Jitter" for three different age groups in different water volumes.

ANOVA of Jitter	1 mL	3 mL	5 mL	10 mL	20 mL
Between groups *df* *	2	2	2	2	2
Within groups *df* *	44	42	41	36	25
F-boundary ($\alpha = 0.05$)	3.20	3.22	3.22	3.26	3.39
F-value **	1.192	0.754	0.255	1.841	**15.818**
p-value	0.313	0.477	0.776	0.173	**0.001**
eta squared η^2	0.051	0.035	0.012	0.093	0.559

* Degree of freedom; ** The bold number means that (F-value > F-boundary).

Table 6. ANOVA testing of "TET" for three different age groups in different water volumes.

ANOVA of TET	1 mL	3 mL	5 mL	10 mL	20 mL
Between groups *df* *	2	2	2	2	2
Within groups *df* *	44	42	41	36	25
F-boundary ($\alpha = 0.05$)	3.20	3.22	3.22	3.26	3.39
F-value **	**7.086**	1.073	**5.864**	**6.166**	**7.854**
p-value	**0.002**	0.350	**0.005**	**0.004**	**0.002**
eta squared η^2	0.243	0.044	0.222	0.255	0.428

* Degree of freedom; ** The bold number means that (F-value > F-boundary).

Next, we are interested about which age group has an obvious mean difference of TET from the other two age groups. Thus we apply the Scheffé *post hoc* test to Table 6 for comparing any two age groups. Tables 7–10 show the *post hoc* test results of TET for the water volumes of 1, 5, 10 and 20 mL, respectively. In each table, if the 95% confidence interval includes 0, the mean difference ($\mu_i - \mu_j$) cannot be considered a significant result. As we can see, the 51–60 year-old age group has obvious mean differences (*i.e.*, the bold numbers) from the other two age groups.

Table 7. *Post hoc* test of TET for 1 mL water volume.

Post Hoc Test (1 mL)	Age group (i)	Age Group (j)	Mean Difference ($\mu_i - \mu_j$)	95% Confidence Interval	
				Lower	Upper
Scheffé's *	51–60	21–30	0.152	−0.077	0.381
	51–60	31–50	**0.334**	0.118	0.551
	21–30	31–50	0.183	−0.020	0.385

* The bold number means that the 95% confidence interval did not include 0.

Table 8. *Post hoc* test of TET for 5 mL water volume.

Post Hoc Test (5 mL)	Age Group (i)	Age Group (j)	Mean Difference ($\mu_i - \mu_j$)	95% Confidence Interval	
				Lower	Upper
Scheffé's *	51–60	21–30	**0.253**	0.061	0.445
	51–60	31–50	**0.229**	0.051	0.408
	21–30	31–50	−0.023	−0.178	0.131

* The bold number means that the 95% confidence interval did not include 0.

Table 9. *Post hoc* test of TET for 10 mL water volume.

Post Hoc Test (10 mL)	Age Group (i)	Age Group (j)	Mean difference ($\mu_i - \mu_j$)	95% Confidence Interval	
				Lower	Upper
Scheffé's *	51–60	21–30	**0.312**	0.035	0.589
	51–60	31–50	**0.357**	0.108	0.607
	21–30	31–50	0.045	−0.189	0.278

* The bold number means that the 95% confidence interval did not include 0.

Table 10. *Post hoc* test of TET for 20 mL water volume.

Post Hoc Test (20 mL)	Age Group (*i*)	Age Group (*j*)	Mean Difference ($\mu_i - \mu_j$)	95% Confidence Interval	
				Lower	Upper
	51–60	21–30	**0.423**	0.051	0.795
Scheffé's *	51–60	31–50	**0.554**	0.182	0.926
	21–30	31–50	0.131	−0.241	0.503

* The bold number means that the 95% confidence interval did not include 0.

Therefore we can conclude that the factor of age actually has more significant impact on the swallowing ability than the gender factor. The results also suggest that after 50 years of age, no matter whether men or women are considered, the swallowing ability could deteriorate very obviously.

4.1.3. Piecemeal Swallowing

Some researchers have pointed that a 20 mL bolus volume would be the limit for human swallowing [13,25]. In our measurements, some participants also displayed instances of piecemeal swallowing. Table 11 shows the frequency of piecemeal swallowing in the different age groups in our measurements. We can find that in total 57 measurements, or about 42.1%, are cases of piecemeal swallowing, especially for the male participants after 31 years of age. If this is the case, the 20 mL water bolus could be used as an indicator to detect the deterioration of the swallowing in aging. This is still an open problem. More evidence and experiments are required to verify this.

Table 11. Frequency of the Piecemeal Swallowing.

Age	Gender	Number of the Participants	Total Trials	Piecemeal Swallowing of 20 mL	The Frequency of Piecemeal Swallowing
21–30	Male	4	12	0	0%
21–30	Female	3	9	3	33%
31–50	Male	4	12	12	100%
31–50	Female	4	12	6	50%
51–60	Male	2	6	3	50%
51–60	Female	2	6	0	0%
Total		19	57	24	42.1%

4.1.4. Discussion

From the results described above, we conclude that the FSR throat belt can effectively measure and differentiate the swallowing parameters, Jitter and TET, between different age groups for different volumes of water. Compared with other two parameters sEMG and SAD (nasal air flow), we show that the parameters Jitter and TET have better sensitivity for evaluating the swallowing status. By our testing, we also show that a 20 mL volume would be a swallowing limitation. This fact is very important and useful in developing related healthcare applications [26,27]. Also, we show that differences indeed exist between male and female participants, and a deterioration of the swallowing ability will occur with increasing age.

Currently instrument-based approaches (e.g., VFSS) have advantages and limitations. Umay *et al.* recommended that bedside tests should be used mainly as initial screening tests [26]. Therefore the instrument-based approaches should be performed in patients who are at risk for swallowing disorders [26,27]. They are advantageous because they can detect aspiration [28,29], which aids in the clinical management planning of tube or oral feeding. Nonetheless, the disadvantages of radiation exposure and non-portability make it difficult to apply them in cases of acute stage disease, sickness, in large studies, and for repeated measurements within a short period in follow-up studies. In this paper we use a non-invasive assessment method that does not cause any stress to patients or pose a radiation risk. Furthermore, it should be easily carried or worn, and portable. Equally important, this non-invasive swallowing study tool should be capable of being combined with respiration monitoring for swallowing and respiration coordination studies.

Figure 18 shows the applications that can be developed by the system proposed in this paper. At first, this system can be applied to healthy subjects for building the indicators of normal swallowing patterns. Then we can use this monitoring system to measure the swallowing motions of dysphagia patients. The results can be evaluated by the physicians for eating strategy adjustment or for biofeedback training. All above steps can be repeated for regular tracking and monitoring.

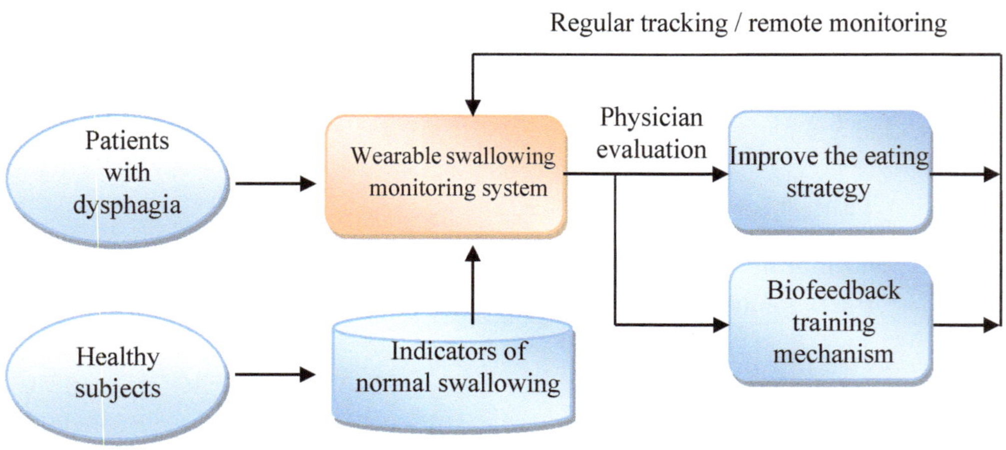

Figure 18. The applications of the wearable dysphagia monitoring system.

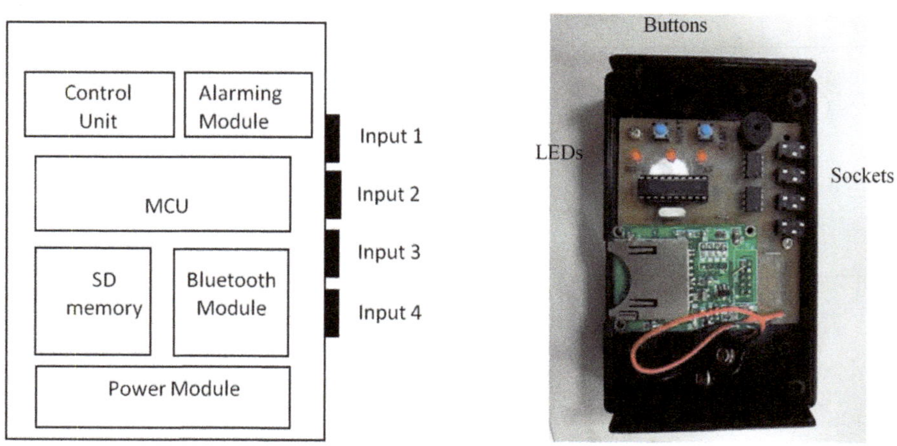

Figure 19. Design of the holter.

We can also connect the FSR throat belt with a holter to develop portable applications. The holter we implemented is shown in Figure 19. It includes four input signal sockets, two control buttons, three output LEDs and one signal processing circuit. To unify the connection between the holter and different kinds of sensors (*i.e.*, FSR, sEMG, nasal airflow *etc.*), we use 3.5 mm stereo connectors as the input port and refit the sensor output pins with those connectors. In our holter design, we reserved four input ports for future possible connections to other sensors for multivariate data measurements. The signal processing circuit is composed of a microcontroller (MCU), a Bluetooth module, a SD memory module and a power-supply module. The MCU we used is an AT89C2051 [30], which is a low-power but high performance chip. We use it to do the digital signal preprocessing operations. The data measured by the sensors can be collected on the SD memory card, or be transferred to a mobile phone via the Bluetooth module.

5. Conclusions

In our study, we use the FSR sensor to develop a throat belt to measure swallowing ability. We use different water bolus volumes to observe the swallowing ability differences between male and female participants in different age groups. The experiment results show that the FSR sensor can effectively measure the motions of the thyroid cartilage during swallowing.

We also use this belt to develop a portable holter and an automatic data analysis program. This program can analyze the onset time and the offset time from sEMG, SAD (nasal air flow), and throat belt signals in a synchronization presentation. The functions and the precision of the throat-belt are proven by real measurements. We show that this throat belt design would be an easy-to-use, objective, and non-invasive measurement or assessment tool for developing dysphagia-related services.

While this work focuses on measurements on the healthy participants, in the future, we plan to apply the proposed system to measure other important swallowing indicators for particular diseases. Take the work in [31] for example. They used the acoustic monitoring to measure the frequency of spontaneous swallowing for Parkinson's disease. The limitation of this approach, however, was the need for a quiet environment to record the sounds of swallowing. Using our FSR-based approach, on the other hand, one could perform such monitoring for is under investigation.

Acknowledgments

This work was supported by Ministry of Science and Technology, Taiwan, under grant number: MOST 103-2221-E-182-020-MY2, and a grant from Chang Gung University Taiwan. The authors also thank the reviewers for their valuable suggestions to this work.

Author Contributions

The contribution of the Wann-Yun Shieh was to design the architecture of the hardware and the software of this work, and to perform the functional verification. The contribution of the Chin-Man Wang was to design and monitor all the experiments, as well as the clinical evaluation of all participants. Both of them contributed equally to the writing of this paper. The contribution of Chia-Shuo Chang was to assist in the experiments and manage the collected data.

Conflicts of Interest

The authors declare no conflict of interest.

References

1. Greco, C.S.S.; Nunes, L.G.M.Q.; Melo, P.L. Instrumentation for Bedside Analysis of Swallowing Disorders. *Conf. Proc. IEEE Eng. Med. Biol. Soc.* **2010**, *2010*, 923–926.
2. Langmore, S.E.; Terpenning, M.S.; Schork, A.; Chen, Y.; Murray, J.T.; Lopatin, D.; Loesche, W.J. Predictors of Aspiration Pneumonia: How Important Is Dysphagia? *Dysphagia* **1998**, *13*, 69–81.
3. Ertekin, C.; Pehlivan, M.; Aydoğdu, I.; Ertaş, M.; Uludağ, B.; Celebi, G.; Colakoğlu, Z.; Sağduyu, A.; Yüceyar, N. An electrophysiological investigation of deglutition in man. *Muscle Nerve* **1995**, *18*, 1177–1186.
4. Prosiegel, M.; Heintze, M.; Sonntag, E.W.; Schenk, T.; Yassouridis, A. Kinematic Analysis of Laryngeal Movements in Patients with Neurogenic Dysphagia before and after Swallowing Rehabilitation. *Dysphagia* **2000**, *15*, 173–179.
5. Esteves, G.P.; Silva, E.P.; Nunes, L.; Greco, C.; Melo, P.L. Configurable Portable/Ambulatory Instrument for the Analysis of the Coordination between Respiration and Swallowing. *Conf. Proc. IEEE Eng. Med. Biol. Soc.* **2010**, 90–93.
6. Martin-Harris, B.; Brodsky, M.B.; Michel, Y.; Ford, C.L.; Walter, B.; Heffner, J. Breathing and swallowing dynamics across the adult lifespan. *Arch. Otolaryngol. Head Neck Surg.* **2005**, *131*, 762–700.
7. Martin-Harris, B.; Brodsky, M.B.; Michel, Y.; Price, C.C.; Walter, B. Temporal coordination of pharyngeal and laryngeal dynamics with breathing during swallowing: Single liquid swallows. *J. Appl. Physiol.* **2003**, *94*, 1735–1734.
8. Dozier, T.S.; Brodsky, M.B.; Michel, Y.; Walters, B.; Martin-Harris, B. Coordination of swallowing and respiration in normal sequential cup swallows. *Laryngoscope* **2006**, *116*, 1489–1493.
9. Logemann, J.A. *Manual for the Videofluorographic Study of Swallowing*, 2nd ed.; PRO-ED: Austin, TX, USA, 1993; pp. 157–162.
10. Logemann, J.A. *Evaluation and Treatment of Swallowing Disorders*, 1st ed.; College Hills: San Diego, CA, USA, 1983; p. 121.
11. Steven, B.; Lynn, M.; Lisitano, H.L.; Murray, J.T. Fiberoptic Endoscopic Evaluation of Swallowing (FEES) with and without Blue-Dyed Food. *Dysphagia* **2005**, *20*, 157–162.
12. Lee, J.; Sejdić, E.; Steele, C.M.; Chau, T. Effects of liquid stimuli on dual-axis swallowing accelerometry signals in a healthy population. *Biomed. Eng. Online* **2010**, *9*, 7.
13. Selçuk, B; Uysal, H.; Aydogdu, I.; Akyuz, M. Effect of temperature on electrophysiological parameters of swallowing. *J. Rehabil. Res. Dev.* **2007**, *44*, 373–380.

14. Aminian, K.; Rezakhanlou, K.; de Andres, E.; Fritsch, C.; Leyvraz, P.; Robert, P. Temporal feature estimation during walking using miniature accelerometers: An analysis of gait improvement after hip arthroplasty. *Med. Biol. Eng. Comput.* **1999**, *37*, 686–691.
15. Rana, N.K. Application of Force Sensing Resistor (FSR) in Design of Pressure Scanning System for Plantar Pressure Measurement. In Proceeding of the Second International Conference on Computer and Electrical Engineering, Dubai, United Arab Emirates, 28–30 December 2009; Volume 2, pp. 678–685.
16. Shieh, WY. The Development of a Portable and Non-invasive Swallowing Assessment Approach. In Proceeding of the International Symposium on InfoComm and Media Technology in Bio-Medical and Healthcare Application, Shanghai, China, 21–24 December 2014; pp. 41–43.
17. Ball, S.; Idel, O.; Cotton, S.M.; Perry, A. Comparison of Two Methods for Measuring Tongue Pressure during Swallowing in People with Head and Neck Cancer. *Dysphagia* **2006**, *1*, 28–37.
18. Utanohara, Y.; Hayashi, R.; Yoshikawa, M.; Yoshida, I.; Tsuga, K.; Akagawa, Y. Standard Values of Maximum Tongue Pressure Taken Using Newly Developed Disposable Tongue Pressure Measurement Device. *Dysphagia* **2008**, *23*, 286–290.
19. Ono, T.; Hori, K.; Masuda, Y.; Hayashi, T. Recent Advances in Sensing Oropharyngeal Swallowing Function in Japan. *Sensors* **2010**, *10*, 176–202.
20. BIOPAC Systems. Available online: http://www.biopac.com/ (accessed on 5 May 2015).
21. Wang, C.M.; Chen, J.Y.; Chuang, C.C.; Tseng, W.J.; Wong, A.M.K.; Pei, Y.C. Aging-related changes in swallowing and in the coordination of swallowing and respiration determined by novel non-invasive measurement techniques. *Geriatr. Gerontol. Int.* **2014**, doi:10.1111/ggi.12343.
22. Łeski, J.M.; Henzel, N. ECG Baseline Wander and Powerline Interference Reduction Using Nonlinear Filter Bank. *Signal Process.* **2005**, *85*, 781–793.
23. Wang, C.M.; Chuang, C.C.; Chen, C.P.; Tseng, W.C.; Chen, J.Y. Non-invasive Swallowing Test for Young Healthy Adults: Finding the Best Location to Monitor Thyroid Cartilage Movements. *J. Med. Biol. Eng.* **2014**, *34*, 393–398.
24. Susan, G.; Hiss, M.S.; Treole, K.; Stuart, A. Effects of Age, Gender, Bolus Volume, and Trial on Swallowing Apnea Duration and Swallow/Respiratory Phase Relationships of Normal Adults. *Dysphagia* **2001**, *16*, 128–135.
25. Aydogdu, I.; Kiylioglu, N.; Tarlaci, S.; Tanriverdi, Z.; Alpaydin, S.; Acarer, A. Diagnostic Value of "Dysphagia Limit" for Neurogenic Dysphagia: 17years of Experience in 1278 Adults. *Clin. Neurophysiol.* **2014**, *126*, 634–643.
26. Umay, E.K.; Unlu, E.; Saylam, G.K.; Cakci, A.; Korkmaz, H. Evaluation of Dysphagia in Early Stroke Patients by Bedside, Endoscopic, and Electrophysiological Methods. *Dysphagia* **2013**, *28*, 395–403.
27. Rofes, L.; Arreola, V.; Mukherjee, R.; Clave, P. Sensitivity and Specificity of the Eating Assessment Tool and the Volume-Viscosity Swallow Test for Clinical Evaluation of Oropharyngeal Dysphagia. *Neurogastroenterol. Motil.* **2014**, *26*, 1256–1265.

28. Dziewas, R.; Warnecke, T.; Olenberg, S.; Teismann, I.; Zimmermann, J.; Kramer, C. Towards a basic endoscopic assessment of swallowing in acute stroke—Development and evaluation of a simple dysphagia score. *Cerebrovasc. Dis.* **2008**, *26*, 41–47.
29. Stoeckli, S.J.; Huisman, T.A.; Seifert, B.; Martin-Harris, B.J. Interrater reliability of videofluoroscopic swallow evaluation. *Dysphagia* **2003**, *18*, 53–57.
30. Atmel AT89C2051. Available online: http://www.atmel.com/devices/at89c2051.aspx (accessed on 5 May 2015).
31. Golabbakhsh, M.; Rajaei, A.; Derakhshan, M.; Sadri, S.; Taheri, M.; Adibi, P. Automated Acoustic Analysis in Detection of Spontaneous Swallows in Parkinson's Disease. *Dysphagia* **2014**, *29*, 572–577.

7

Amorphous Silicon p-i-n Structure Acting as Light and Temperature Sensor

Giampiero de Cesare [1,†,*], Augusto Nascetti [2,†] and Domenico Caputo [1,†]

[1] Department of Information Engineering, Electronics and Telecommunications, "La Sapienza" University of Rome, via Eudossiana 18, 00184 Rome, Italy; E-Mail: caputo@diet.uniroma1.it

[2] Department of Astronautics, Electrical and Energetic Engineering, "La Sapienza" University of Rome, via Eudossiana 18, 00184 Rome, Italy; E-Mail: augusto.nascetti@uniroma1.it

[†] These authors contributed equally to this work.

* Author to whom correspondence should be addressed; E-Mail: decesare@diet.uniroma1.it

Academic Editor: Vittorio M.N. Passaro

Abstract: In this work, we propose a multi-parametric sensor able to measure both temperature and radiation intensity, suitable to increase the level of integration and miniaturization in Lab-on-Chip applications. The device is based on amorphous silicon p-doped/intrinsic/n-doped thin film junction. The device is first characterized as radiation and temperature sensor independently. We found a maximum value of responsivity equal to 350 mA/W at 510 nm and temperature sensitivity equal to 3.2 mV/K. We then investigated the effects of the temperature variation on light intensity measurement and of the light intensity variation on the accuracy of the temperature measurement. We found that the temperature variation induces an error lower than 0.55 pW/K in the light intensity measurement at 550 nm when the diode is biased in short circuit condition, while an error below 1 K/μW results in the temperature measurement when a forward bias current higher than 25 μA/cm^2 is applied.

Keywords: temperature sensors; photosensors; amorphous silicon devices; Lab-on-Chip

---

1. Introduction

Lithography-based microtechnology, initially used to realize integrated semiconductor structures for microelectronic chips, was soon applied to fabricate microsystem devices suitable for several applications. In particular, the introduction of the concept of the micro Total Analysis Systems (µTAS) [1] and the development of the Micro Electro Mechanical Systems (MEMS) led to the development of Lab-on-Chip (LOC) systems as a powerful tool for biomolecular analysis in point-of-care applications [2–5]. LOC device is an example of a system where the high miniaturization level allows accomplishing several laboratory functions, usually done at lab-scale, with a fast response time, low sample consumption and on-site operation [6,7]. Although the application of LOCs is still novel and modest, a growing interest of companies and research groups is observed in different fields, such as chemical analysis, environmental monitoring, and medical diagnostics. However, only few LOC are currently commercially available and in many cases laboratory equipment is still required for their operation. Thus, current research efforts are directed towards the integration, on the same substrate (that can be either glass, plastic or crystalline silicon), of different physical and optical sensors, together with microfluidic devices, in order to produce a single analysis tool, where all the steps of the analysis, such as sample preparation, sample handling and analytical detection, can be carried out [8–10].

Sample preparation is usually implemented by heating the substrate for thermal treatments of the biomolecules and/or to improve the surface functionalization [11,12]. Monitoring and control of temperature is often performed with thin film sensors due to their high degree of integration [13].

Detection is usually performed using off-chip detection systems able to measure the fluorescence emitted by fluorescent dyes attached to the target molecules [14]. Recently, different groups have integrated electrical and/or optical sensors with the microfluidics, with the aim of achieving on-chip detection for improving the system sensitivity and compactness. In particular, optical detection of biomolecules based on organic [15–17] and inorganic [18] thin film photosensors has been developed.

One of the most promising materials to this aim is hydrogenated amorphous silicon (a-Si:H) and its alloy. The low deposition temperature (below 250 °C) and its physical characteristics prompt the use of this material in different devices such as solar cells [19], electronic switching [20], strain sensors [21,22] and photosensors [23]. The use of thin film a-Si:H photosensors for the detection of biomolecules has already been developed by different research groups [24–29], and in particular by the authors in both labeled and label free techniques [30–32]. Furthermore, the same a-Si:H structure has recently been used as temperature sensor in a lab-on-glass system for the molecular amplification by Polymerase Chain Reaction (PCR) technique, in order to achieve a point-monitoring of the temperature distribution within the device area [33–35].

In this work, we propose the use of a single a-Si:H diode as multi-parametric sensor to measure both temperature and radiation intensity, in order to increase the level of integration and the miniaturization of LOC systems. In particular, we present a detailed investigation of the requirements that the structure has to satisfy, taking into account that the temperature variation affects the photo-response and that, on the other hand, light intensity variation can induce a reduction of accuracy in the temperature measurement.

The paper is organized as follows: details of the device fabrication process are reported in the Section 2 "Materials and Methods", in Section 3 the structure and the fabrication process of the a-Si:H device are described, together with the characterization of the structure as radiation and temperature sensor;

in Section 4, experimental results are described and discussed in details with focus on the reciprocal influence of the two measured parameters; and in Section 5, conclusions are drawn..

2. Material and Methods

Fabrication Process of the Device

We have fabricated several 2 × 2 mm^2 diodes arranged in array structure by the use of standard microelectronic technologies and four photolithographic steps for the geometry patterning.

Below are described details of the technological steps used for the fabrication of the diode array.

1. Cleaning of the glass substrate:
   a. ultrasonic cleaning in Detergent 8 (by Alconox) diluted at 5% in deionized water (DI) for 30 min;
   b. ultrasonic rinse in DI at 50 °C for 15 min;
   c. ultrasonic cleaning in Liquinox (by Alconox) diluted at 2% in DI for 30 min;
   d. ultrasonic rinse in DI at 50 °C for 15 min;
   e. drying in nitrogen flow; and
   f. immediately positioning of the glass substrate inside the RF magnetron sputtering system.
2. Sputter deposition (in Material Research Corporation system) of 200 nm thick layers from a 6-in-diameter Indium Tin Oxide (ITO) target (with 90% In_2O_3–10% SnO_2 in weight composition), at 200 W of RF power, 2.7 mTorr of pressure process, 25 sccm of argon flow and 120 °C substrate temperature (Figure 1a).
3. First lithographic step:
   a. spin coating of photoresist (AZ1518 from Shipley);
   b. soft bake at 100 °C for 1 min;
   c. UV exposure through a lithographic mask at 320 mJ/cm^2 for 45 s; and
   d. development of the photoresist in AZ351B diluted at 25% in DI for 45 s.
4. Patterning of ITO layer (with mask #1 in Figure 1b) by Sputter Etching at 180 W, 50 sccm Argon flow and 90 mTorr.

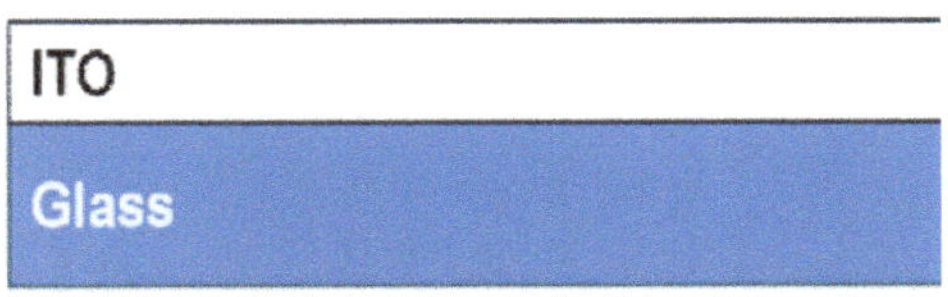

a)

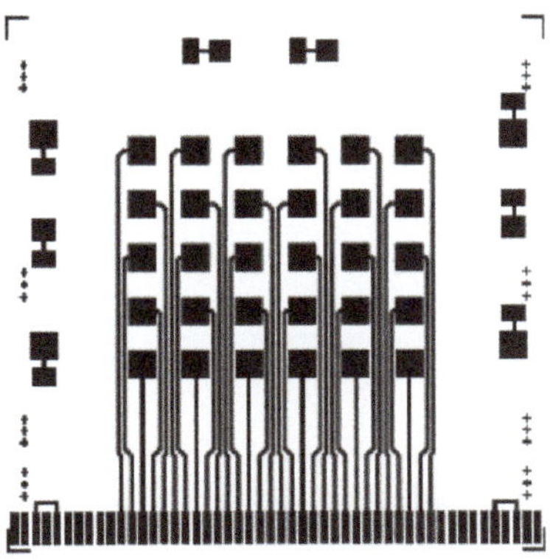

b) MASK #1

Figure 1. (**a**) Structure of the sensor after step 4 of the fabrication process. (**b**) Photolithographic mask used for the patterning of ITO bottom electrode of the device.

5. Covering the ITO markers with kapton tape to allow the subsequent alignments.

6. Sequential deposition, by Plasma Enhanced Chemical Vapor Deposition (PECVD) in a three UHV chamber system, of the hydrogenated p-type a-SiC, intrinsic and n-type a-Si layers, that constitute the stacked structure of the diode. The following Table 1 reports the deposition parameters of the PECVD process.

Table 1. PECVD parameters used in the deposition of the a-Si:H and a-SiC:H layers. The gases are: SiH_4 pure silane, PH_3 silane diluted (5%), B_2H_6 helium diluted (5%), CH_4 pure methane; P_D is the process pressure; P_{RF} is the power density of the plasma discharge; T_D is the substrate temperature; t_D is the deposition time. The thicknesses reported above have been estimated from the growth rate of the different materials.

Layer type/material	SiH_4 (sccm)	PH_3 (sccm)	B_2H_6 (sccm)	CH_4 (sccm)	P_D (Torr)	P_{RF} (mW/cm^2)	T_D (°C)	t_D (sec)	Thickness (nm)
p/a-SiC:H	40		5	60	0.7	25	150	60	10
i/a-Si:H	40				0.68	25	180	2100	400
n/a-Si:H	40	10			0.3	25	200	180	50

7. Deposition of a three metal layer stack of Cr/Al/Cr with thickness of 30/200/30 nm, respectively, in an Ultra Vacuum system (thermal evaporator by Balzers) with measured growth rate of 0.3 nm for Cromium and 1 nm for Aluminum.
8. Peeling of the kapton tape, for the subsequent alignments.
9. Lithographic steps (see point 3) with Mask #2 in Figure 2b.
10. Patterning of the Cr layer by wet etching in a solution of 30 g $Ce(NH_4)_2(NO_3)_6$, 9 mL CH_3COOH and 200 mL DI for 1 min.
11. Patterning of the Aluminum layer by wet etching in a solution of 80 mL H_3PO_4 (85%), 5 mL HNO_3 and 10 mL DI.
12. Patterning of the Cr layer by wet etching (see point 10).
13. Patterning of the a-Si:H n-i-p structure by dry etching in a Reactive Ion Etching system (by IONVAC) (Figure 2a) with the following parameters:
    a. Oxigen (O_2) flow O_2: 10 sccm;
    b. Carbon tetrafluoride (CF_4) flow: 100 sccm; and
    c. Radio Frequency Power Density: 300 mW/cm^2.

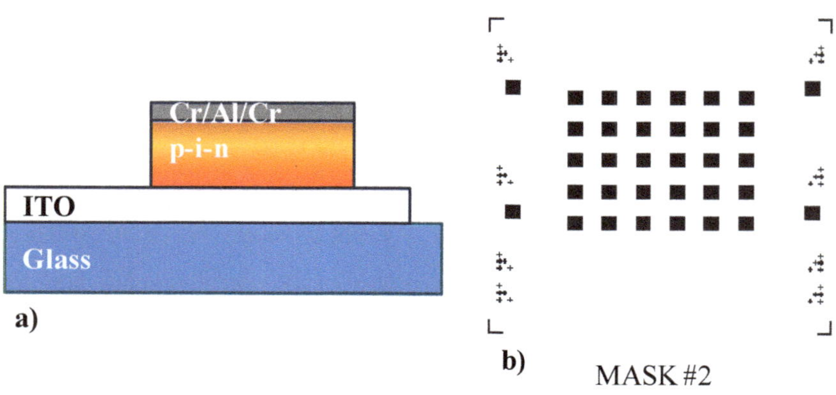

Figure 2. (**a**) Structure of the sensor after the step 13 of the fabrication process. (**b**) Photolithographic mask used for the patterning of the mesa structure of the device.

14. Deposition by spin coating of a 5 μm-thick SU8 (from Micro-Chem, MA, USA) and its pattering through Mask #3 in Figure 3b for opening via holes over the diodes (Figure 3a). The deposition of the SU-8 film is implemented through the following steps:
    a. spin coating at 500 rpm for 5 s followed by another run at 3000 rpm for 30 s;
    b. bake at 65 °C for 1 min followed by another bake at 95 °C for 2 min;
    c. UV-light exposition at 250 mJ/cm^2;
    d. bake at 65 °C for 1 min followed by another bake at 95 °C for 1 min;
    e. developing in SU-8 remover;
    f. rinse in isopropyl alcool; and
    g. hard bake at 150 °C for 30 min.

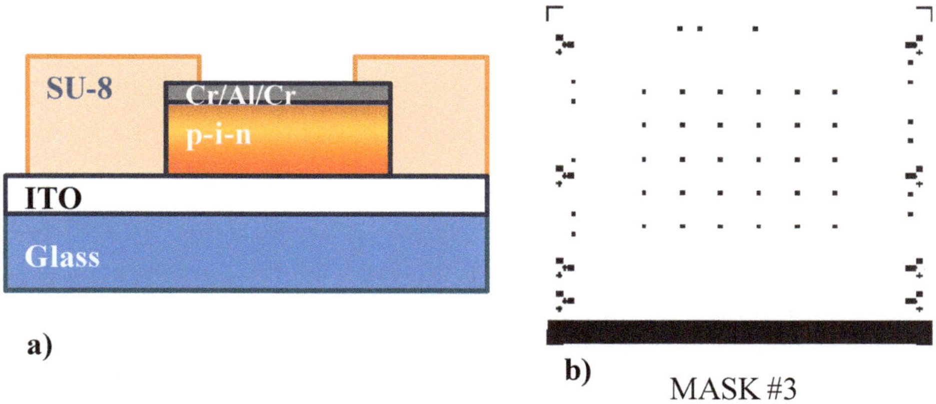

Figure 3. (**a**) Structure of the sensor after step 14 of the fabrication process. (**b**) Photolithographic mask used for the patterning of the via on the SU8 insulation layer.

15. Deposition of a three metal layer stack of Cr/Al/Cr with thickness of 30/200/30 nm, respectively, in an Ultra Vacuum system (see point 7).
16. Lithographic steps (see point 3) with Mask #4 in Figure 4b.
17. Patterning of the Cr/Al/Cr layers (see points 10 an 11).

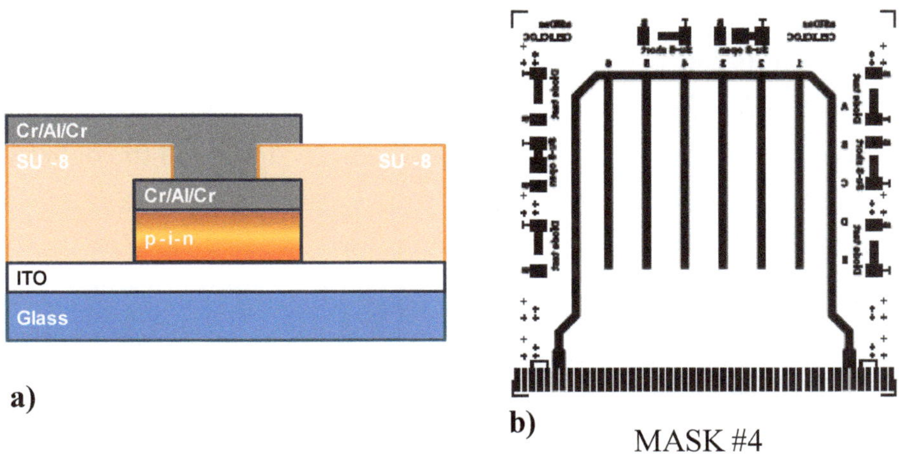

Figure 4. (**a**) Structure of the sensor after step 17 of the fabrication process; (**b**) Photolithographic mask used for the patterning of the metallic lines for the electrical connection of the top electrode of the device to the edge of the glass substrate.

18. Deposition by spin coating of a 5 μm-thick SU-8 as passivation layer and hard baking at 150 °C for 30 min.

3. Structure and Characterization of the Device

3.1. Sensor Structure

The basic structure of the device is a p-type doped amorphous silicon carbide (a-SiC:H)/intrinsic a-Si:H/n-type doped a-Si:H heterojunction. The diode is deposited by Plasma Enhanced Chemical Vapor Deposition (PECVD) on a glass substrate covered with a Indium Tin Oxide (ITO) transparent conductive layer [36]. This layer acts as bottom electrode of the device, and for the light sensor application, as window layer for the light impinging on the photodetector through the glass. A Cr/Al/Cr stacked structure vacuum evaporated on the a-Si:H diode behaves as the top contact of the structure. An SU-8 insulation layer is deposited over the diode in order to avoid the short circuit between top and bottom electrodes and to reduce the current component induced by or the exposed diode perimeter [37].

We have fabricated several 2 × 2 mm^2 diodes, arranged in array structure, by the use of standard microelectronic technologies: Physical and Chemical Vapor Deposition of thin film, Dry and Wet etching of different materials, and four photolithographic steps for the geometry patterning. All the details of the device fabrication process are reported in Section 5 "Material and Methods".

The device current-voltage characteristics were evaluated using a Keithley 236 Source Measure Unit (SMU). We achieved a current of 10^{-11} A/cm^2 at small (10 mV) reverse voltage, with an excellent reproducibility among the samples. This value determines a noise current contribution in the order of 2 $fA/\sqrt{Hz}$, which is below the minimum detectable signal in our experimental set-up.

3.2. Characterization as Radiation Sensor

Due to the a-Si:H optical properties, the p-i-n junction is widely used as thin film devices for light detections. The active layer of the sensor is the intrinsic one, where the holes/electrons, photo-generated by the absorbed light, are swept toward the doped-regions by the electric field existing in the intrinsic region. The doped layers provide the built-in potential of the junction, but do not contribute to the photocurrent, because the high density of charged dangling bond defects, induced by doping in amorphous silicon, strongly traps the minority photo-generated carriers [37].

The spectral responsivity of the p-i-n photodiode can be designed through the optical absorption of the three layers, by tuning their thicknesses and energy gaps. In our devices, the energy gap of the p, i and n-type materials, measured on single film deposited on Corning glass substrates, are equal to 1.92, 1.79 and 1.71 eV, respectively, while the thicknesses are equal to 10, 400 and 50 nm, respectively.

The structure has been designed with the aim to maximize the responsivity at wavelengths between 400 and 600 nm, which ensures the suitability of our photosensors for biochemical applications, where chemiluminescent or fluorescent signals have to be detected [38–40].

Figure 5a reports the quantum efficiency (QY) curve of one photodiode, measured in short circuit condition. As reported in [41], this operation mode minimizes the peripheral leakage current and therefore this is the bias condition we have chosen when the diode acts as radiation sensor.

Characterization has been performed on a double-arm set-up, including a tungsten light source, a monochromator Spex 340E, a beam splitter and a calibrated UV-enhanced crystalline silicon diode.

From this characterization and from the following equation:

$$R = \frac{QY \cdot \lambda \cdot q}{hc} \tag{1}$$

where λ is the wavelength, q the electron charge, h the Planck constant and c the light speed, we derive that the photosensor responsivity (R) reported in Figure 5b. R is around 280 mA/W at 450 and 600 nm and shows a maximum equal to 350 mA/W at 510 nm.

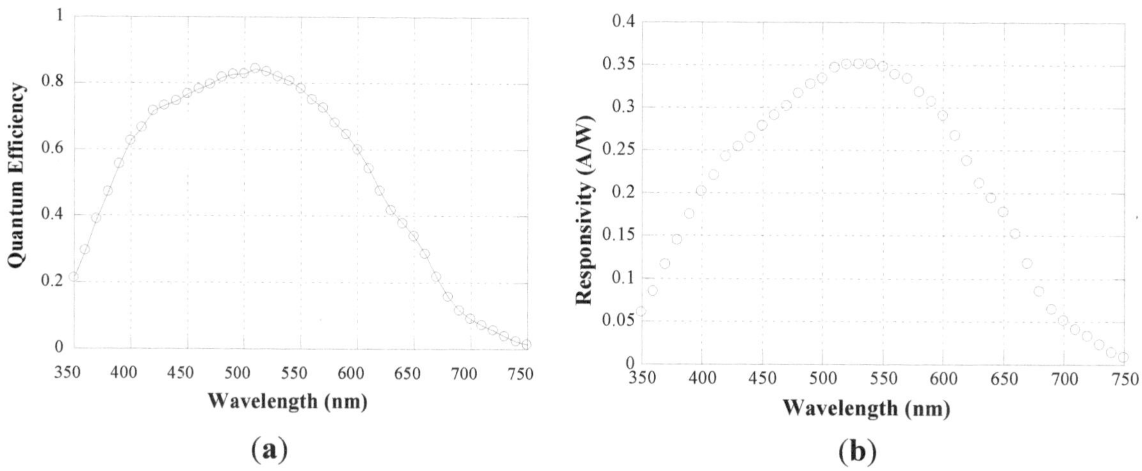

Figure 5. Quantum efficiency (**a**) and responsivity (**b**) curves of p-i-n a-Si:H photodiode measured in short circuit condition.

3.3. Characterization as Temperature Sensor

The same a-SiC:H/a-Si:H p-i-n diodes have been characterized as temperature sensors measuring their current–voltage characteristics as a function of temperature in the range 30 °C–90 °C, both under forward and reverse bias voltage conditions. Measurements have been performed in a probe station with a temperature-controlled chuck, by using a thermocouple as temperature reference. As expected [42], at constant voltage, in reverse bias condition, we observe an exponential dependence of the reverse saturation current with temperature, while at constant current, in forward bias condition, a linear decrease of the voltage across the diode with temperature results. These results are summarized in Figure 6. In particular, Figure 6a reports the diode current as a function of temperature when the diode is biased at 0.2, 0.6 or 1 V reverse voltage, while Figure 6b shows the voltage drop across the diode as a function of temperature when a forward current of 10 or 20 nA is flowing in the device.

In forward bias condition, with a bias current higher than 10 nA, the achieved temperature sensitivity (S_T) is equal to 3.2 mV/K, greater than the one of a crystalline silicon diode [43], showing that the a-Si:H diode can be considered a very promising candidate for detecting very small temperature variations in LOC applications.

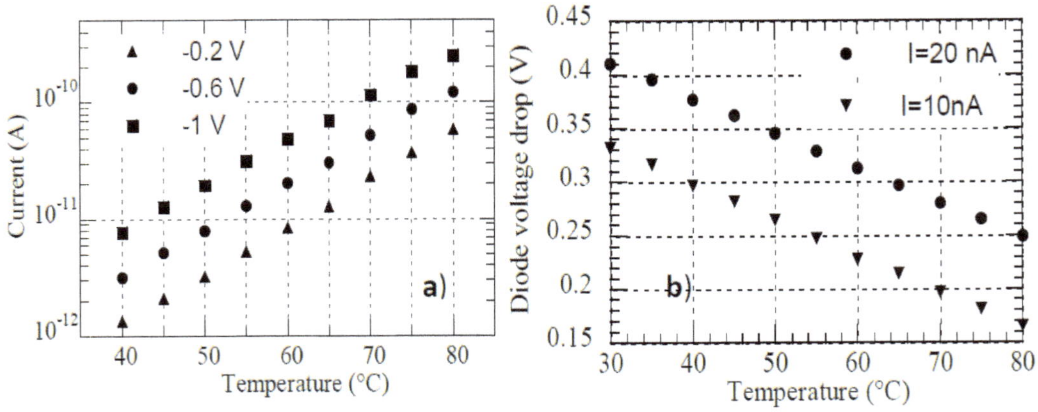

Figure 6. (**a**) Current flowing in the device diode as a function of temperature at reverse voltage bias equal to −0.2 V (triangle symbols), −0.6 V (circular symbols) and −1 V (square symbols); (**b**) Voltage drop across the diode as a function of temperature when a forward current of 10 nA (triangle symbols) or 20 nA (circular symbols) is flowing in the device.

4. Results and Discussion

The use of the same device as both light and temperature sensor has to consider the mutual influence of these two physical parameters. Indeed, temperature variation affects the current flowing through an irradiated diode, and, on the other hand, light intensity variation can induce a reduction of accuracy in the temperature measurement. In the following, we analyze in some detail the effect of one parameter (light intensity or temperature) on the measurement of the other parameter (temperature or light intensity, respectively).

4.1. Temperature Effect on Light Measurement

To investigate the effect of temperature on the sensor photo-response, we measured the current flowing in the a-Si:H diode as a function of temperature. In particular, Figure 7 reports the diode current, measured through a charge sensitive circuit with the device in dark condition and biased to the virtual ground of the amplifier, in the temperature range 25–80 °C.

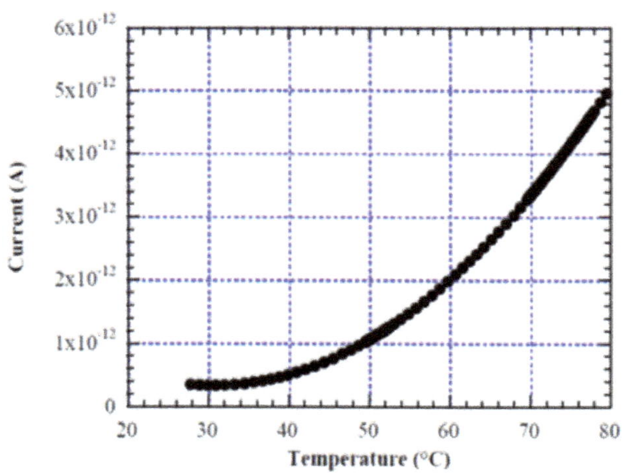

Figure 7. Diode dark current as function of temperature, with the device biased to the virtual ground of the amplifier.

We see an increase of the dark current (I_D) from 0.25 pA at room temperature to 5 pA at 80 °C. The slope of this curve increases exponentially with temperature, with a maximum of 195 fA/K.

When the device is used as light sensor the output signal is a current proportional to the light intensity. In this case, a current variation (ΔI_D) due to a temperature variation (ΔT) induces an error in the measure of the light power (ΔP_T), defined as:

$$\Delta P_T = (\Delta I_D/\Delta T)/R_\lambda \quad [\mathrm{W/K}] \tag{2}$$

For example, when the temperature presents small variations around 80 °C, applying Equation (2) and considering R_λ values at 450 and 600 nm, the error in the light power measurement is lower than 0.7 pW/K, while at 510 nm wavelength, corresponding to the maximum of R_λ, the error does not exceeds 0.55 pW/K.

Under large temperature variation (between T_1 and T_2), the total error ($\Delta P_{T,tot}$) in the light power measurement is given by:

$$\Delta P_{T,tot} = ((I_D(T_2) - I_D(T_1))/R_\lambda \quad [\mathrm{W}] \tag{3}$$

This behavior is illustrated in Figure 8. The two current values, measured under the same monochromatic (465 nm).

Light pulse at 23 and 70 °C, have a difference of 2.1 pA. This value represents the influence of the temperature on the measurement of light intensity. Referring to Equation (3), with T_1 = 23 °C and T_2 = 70 °C, and considering a photosensor responsivity (R) around 290 mA/W at 465 nm, we calculate a total error in the light intensity measurement equal to 7.25 pW.

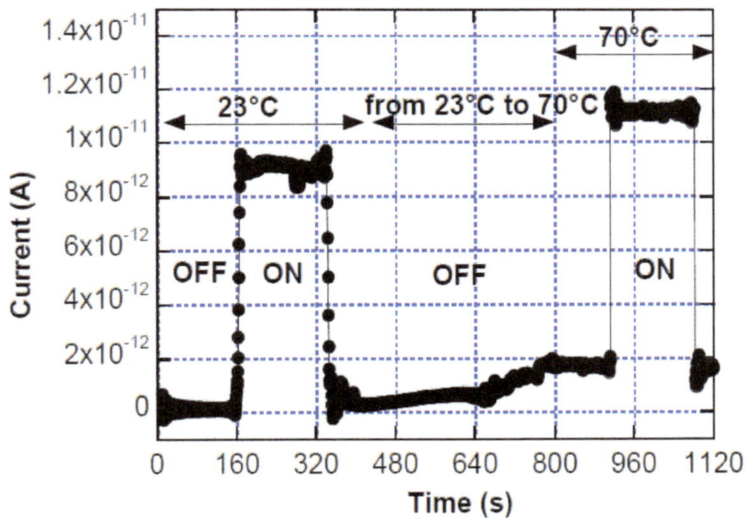

Figure 8. Diode current under two equal light pulses at temperature equal to 23 and 70 °C. ON and OFF refer to turning on and off a light source impinging on the device.

4.2. Effect of Light Intensity on Temperature Measurement

When the sensor is used to measure the temperature, with the diode biased by a constant forward current, the output signal device is a voltage inversely proportional to the measured temperature. A voltage variation (ΔV) induced by a light power variation (ΔP) corresponds to an error in the measurement of the temperature (ΔT_L), defined as:

$$\Delta T_L = (\Delta V/\Delta P)/S_T \quad [\mathrm{K/W}] \tag{4}$$

This is illustrated in Figure 9, which reports the current-voltage characteristics measured, at stabilized temperature (30 °C), in dark conditions and under 0.5 μW monochromatic illumination at 465 nm. The photocurrent measured in reverse bias is about 10^{-7} A, four orders of magnitude higher than the dark current (10^{-11} A), while in forward bias (above 0.5 V), the two curves are almost superimposed. In particular, we found that, at a bias current of 1 μA, the voltage difference between the two curves, due the light intensity, is equal to 0.85 mV. Taking into account the thermal sensitivity (S_T = 3.2 mV/K) of the a-Si:H diode, we derive that the error in the temperature measurement is equal to 0.5 K at 0.5 μW light intensity. This error can be considered negligible in biological analysis [38], where 0.5 μW is well above the optical power to be detected.

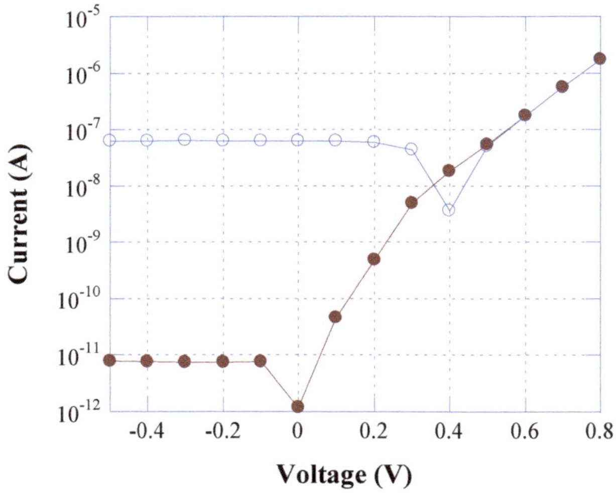

Figure 9. Current-voltage characteristics measured at room temperature in dark conditions (solid symbols) and under monochromatic illumination at 465 nm with light intensity equal to 0.5 μW (open symbols).

5. Conclusions

We have proposed the use of an amorphous silicon p-i-n junction as both temperature and radiation sensor. In particular, we have analyzed the mutual effect of the temperature on light intensity measurements and of the light intensity variation on the accuracy in the temperature measurement. These effects induce very low errors in both light and temperature measurements if the diode is biased in short circuit or in high injection conditions, respectively. These positive characteristics of the structure make the device a very promising candidate as thin film sensor in integrated LOC systems, where optical detection of biomolecules is often required during a heat treatment of the analyte.

Acknowledgments

Authors would like to thank the financial support of the Center for Life Nano Science@Sapienza, Istituto Italiano di Tecnologia (Rome, Italy) and the University Project Research 2013—prot. C26A13HKFB “Lab-on-Chip system for mycotoxin detection in food commodities”.

Author Contributions

The manuscript was written through contributions of all authors. In particular, contributions of Giampiero de Cesare focus on the design and the fabrication of the proposed device and on the data analysis. Augusto Nascetti performed the characterizations of the device as radiation and temperature sensor. Domenico Caputo performed the experiments and analyzed the data.

Conflicts of Interest

The authors declare no conflict of interest.

References

1. Manz, A.; Graber, N.; Widmer, H.M. Miniaturized total Chemical Analysis systems: A Novel Concept for Chemical Sensing. *Sens. Actuators B Chem.* **1990**, *1*, 244–248.
2. Rabner, A.; Shacham, Y. A Concept for a Sensitive Micro Total Analysis System for High Throughput Fluorescence Imaging. *Sensors* **2006**, *6*, 341–349.
3. Caputo, D.; de Cesare, G.; Scipinotti, R.; Mirasoli, M.; Roda, A.; Zangheri, M.; Nascetti, A. Chemiluminescence-Based Micro-Total-Analysis System with Amorphous Silicon Photodiodes. *Lect. Notes Electr. Eng.* **2014**, *268*, 207–211.
4. Guiducci, C.; Stagni, C.; Fischetti, A.; Mastromatteo, U.; Benini, L.; Riccò, B. Microelectrodes on a Silicon Chip for Label-Free Capacitive DNA Sensing. *IEEE Sensors J.* **2006**, *6*, 1084–1093.
5. Waleed Shinwari, M.; Zhitomirsk, D.; Deen, I.A.; Selvaganapathy, P.R.; Jamal Deen, M.; Landheer, D. Microfabricated Reference Electrodes and their Biosensing Applications. *Sensors* **2010**, 10, 1679–1715.
6. Miled, M.A.; Sawan, M. Dielectrophoresis-Based Integrated Lab-on-Chip for Nano and Micro-Particles Manipulation and Capacitive Detection. *IEEE Trans. Biomed. Circuits Syst.* **2012**, *6*, 120–132.
7. Lee, T.M.H.; Carles, M.C.; Hsing, I.M. Microfabricated PCR-electrochemical device for simultaneous DNA amplfication and detection. *Lab Chip* **2003**, *3*, 100–105.
8. Medoro, G.; Nastruzzi, C.; Guerrieri, R.; Gambari, R.; Manaresi, N. Lab on a Chip for Live-Cell Manipulation. *IEEE Des. Test* **2007**, *24*, 26–36.
9. Caputo, D.; de Cesare, G.; Dolci, L.S.; Mirasoli, M.; Nascetti, A.; Roda, A.; Scipinotti, R. Microfluidic chip with integrated a-Si:H photodiodes for chemiluminescence-based bioassays. *IEEE Sens. J.* **2013**, *13*, 2595–2602.
10. Dutse, S.W.; Yusof, N.A. Microfluidics-Based Lab-on-Chip Systems in DNA-Based Biosensing: An Overview. *Sensors* **2011**, *11*, 5754–5768.
11. Bahşi, Z.B.; Büyükaksoy, A.; Ölmezcan, S.M.; Şimşek, F.; Aslan, M.H.; Oral, A.Y. A Novel Label-Free Optical Biosensor Using Synthetic Oligonucleotides from E.Coli O157:H7: Elementary Sensitivity Tests. *Sensors* **2009**, *9*, 4890–4900.
12. Jha, S.K.; Joo, G.-S.; Ra, G.-S.; Lee, H.H.; Kim, Y.-S. Development of PCR Microchip for Early Cancer Risk Prediction. *IEEE Sens. J.* **2011**, *11*, 2065–2070.
13. Wu, J.; Cao, W.; Wen, W.; Chang, D.C.; Sheng, P. Polydimethylsiloxane microfluidic chip with integrated microheater and thermal sensor. *Biomicrofluidics* **2009**, *3*, doi:10.1063/1.3058587.

14. Matos Pires, N.M.; Dong, T.; Hanke, U.; Hoivik, N. Recent Developments in Optical Detection Technologies in Lab-on-a-Chip Devices for Biosensing Applications. *Sensors* **2014**, *14*, 15458–15479.
15. Hofmann, O.; Miller, P.; Sullivan, P.; Jones, T.S.; deMello, J.C.; Bradley, D.D.C.; deMello, A.J. Thin-film organic photodiodes as integrated detectors for microscale chemiluminescence assays. *Sens. Actuators B Chem.* **2005**, *106*, 878–884.
16. Wang, X.; Amatatogchai, M.; Nacaprocha, D.; Hofmann, O.; de Mello, J.C.; Bradley, D.D.C.; de Mello, A.J. Thin-film organic photodiodes for integrated on-chip chemilunescence detection – application to antioxidant capacity screening. *Sens. Actuators B Chem.* **2009**, *140*, 643–648.
17. Matos Pires, N.M.; Dong, T.; Hanke, U.; Hoivik, N. Integrated optical microfluidic biosensor using a polycarbazole photodetector for point-of-care detection of hormonal compounds. *J. Biomed. Opt.* **2003**, *18*, doi:10.1117/1.JBO.18.9.097001.
18. Liu, C.-H.; Chang, Y.-C.; Norris, T.B.; Zhong, Z. Graphene photodetectors with ultra broadband and high responsivity at room temperature. *Nat. Nanotechnol.* **2014**, *9*, 273–278.
19. Carlson, D.E. Amorphous silicon solar cells. *IEEE Trans. Electron Devices* **1977**, *ED-24*, 449–453.
20. Caputo, D.; de Cesare, G. New a-Si:H Two Terminal Switching Device for Active Display. *J. Non-Cryst. Solids* **1996**, *198–200*, 1134–1137.
21. Ibaraki, N. a-Si TFT technologies for AM-LCDs. *MRS Proc.* **1994**, doi:10.1557/PROC-336-749.
22. De Cesare, G.; Gavesi, M.; Palma, F.; Riccò, B. A novel a-si:H mechanical stress sensor. *Thin Solid Films* **2003**, *427*, 191–195.
23. Caputo, D.; de Cesare, G.; Nascetti, A.; Tucci, M. Detailed study of amorphous silicon ultraviolet sensor with chromium silicide window layer. *IEEE Trans. Electr. Dev.* **2008**, *55*, 452–456.
24. Schaefer, H.; Seibel, K.; Walder, M.; Schoeler, L.; Pletzer, T.; Waidelich, M.; Ihmels, H.; Ehrhardt, D.; Boehm, M. Monolithic integrated optical detection for microfluidic systems using thin film photodiodes based on amorphous silicon. In Proceedings of the 18th IEEE International Conference on Micro Electro Mechanical Systems, Miami, FL, USA, 30 January–3 February 2005; pp. 758–761.
25. Costantini, F.; Nascetti, A.; Scipinotti, R.; Domenici, F.; Sennato, S.; Gazza, L.; Bordi, F.; Pogna, N.; Manetti, C.; Caputo, D.; *et al.* On-chip detection of multiple serum antibodies against epitopes of celiac disease by an array of amorphous silicon sensors. *RSC Adv.* **2014**, *4*, 2073–2080.
26. Kamei, T.; Toriello, N.M.; Lagally, E.T.; Blazej, R.G.; Scherer, J.R.; Street, R.A.; Mathies, R.A. Microfluidic Genetic Analysis with an Integrated a-Si:H Detector. *J. Biomed. Microdevices* **2005**, *7*, 147–152.
27. Martins, R.; Baptista, P.; Raniero, L.; Doria, G.; Silva, L.; Franco, R.; Fortunato, E. Amorphous/nanocrystalline silicon biosensor for the specific identification of unamplified nucleic acid sequences using gold nanoparticle probes. *Appl. Phys. Lett.* **2007**, *90*, 023903–1/023903–3.
28. Pimentel, A.C.; Pereira, A.T.; Chu, V.; Prazeres, D.M.F.; Conde, J.P. Detection of chemiluminescence using an amorphous silicon photodiode. *IEEE Sens. J.* **2007**, *7*, 415–416.
29. Joskowiak, A.; Chu, V.; Prazeres, D.M.F.; Conde, J.P. Amorphous Silicon Photosensors for Detection of Intrinsic Cell Fluorophores. *MRS Proc.* **2011**, *1321*, 435–439.
30. Caputo, D.; de Cesare, G.; Fanelli, C.; Nascetti, A.; Ricelli, A.; Scipinotti, R. Amorphous silicon photosensors for detection of Ochratoxin A in wine. *IEEE Sens. J.* **2012**, *12*, 2674–2679.
31. Caputo, D.; de Cesare, G.; Nascetti, A.; Negri, R. Spectral tuned amorphous silicon p-i-n for DNA detection. *J. Non-Cryst. Solids* **2006**, *352*, 2004–2006.

32. Caputo, D.; de Cesare, G.; Fanelli, C.; Nascetti, A.; Ricelli, A.; Scipinotti, R. Innovative detection system of ochratoxin A by thin film photodiodes. *Sensors* **2007**, *7*, 1317–1322.
33. Gibbs, R.A. DNA Amplification by the Polymerase Chain Reaction. *Anal. Chem.* **1990**, *62*, 1202–1214.
34. Zhang, C.; Xu, J.; Ma, W.; Zheng, W. PCR microfluidic devices for DNA amplification. *Biotechnol. Adv.* **2006**, *24*, 243–284.
35. Caputo, D.; de Cesare, G.; Nardini, M.; Nascetti, A.; Scipinotti, R. Monitoring of temperature distribution in a thin film heater by an array of a-Si:H temperature sensors. *IEEE Sens. J.* **2012**, *12*, 1209–1213.
36. De Cesare, G.; Caputo, D.; Tucci, M. Electrical properties of ITO/crystalline silicon contact at different deposition temperatures. *IEEE Electron Devices Lett.* **2012**, *33*, 327–329.
37. Street, R.A. Large Area Image Sensor Array. In *Technology and Applications of Amorphous Silicon*; Springer-Verlag: Berlin, Germany, 2000; Chapter 4, p. 151.
38. Ibañez, G.A.; Escandar, G.M. Luminescence Sensors Applied to Water Analysis of Organic Pollutants—An Update. *Sensors* **2011**, *11*, 11081–11102.
39. Caputo, D.; de Cesare, G.; Nascetti, A.; Negri, R.; Scipinotti, R. Amorphous Silicon Sensors for Single and Multicolor Detection of Biomolecules. *IEEE Sens. J.* **2007**, *7*, 1274–1280.
40. Caputo, D.; de Cesare, G.; Scipinotti, R.; Stasio, N.; Costantini, F.; Manetti, C. On Chip Diagnosis of Celiac Disease by an Amorphous Silicon Chemiluminescence Detector. *Lect. Notes Electr. Eng.* **2014**, *268*, 183–187.
41. Schiff, E.A.; Street. R.A.; Weisfield, W. Reverse bias currents in amorphous silicon nip sensors. *J. Non-Cryst. Solids* **1996**, *198–200*, 1155–1159.
42. Sze, S.M. p-n Junction Diode. In *Physics of Semiconductor Device*, 2nd ed.; John Wiley and Sons: Hoboken, NJ, USA, 1981; Chapter 2, pp. 84–94.
43. Forward Bias Characteristics of 1N400x Family Diodes. Available online: http://www.cliftonlaboratories.com/1n400x_diode_family_forward_voltage.htm (accessed on 25 May 2015).

8

Microfabrication and Integration of a Sol-Gel PZT Folded Spring Energy Harvester

Jonathan Lueke [1], Ahmed Badr [2], Edmond Lou [2] and Walied A. Moussa [1,*]

[1] Department of Mechanical Engineering, University of Alberta, University of Alberta, Edmonton, AB T6G 2G8, Canada; E-Mail: lueke@ualberta.ca

[2] Department of Electrical and Computer, University of Alberta, University of Alberta, Edmonton, AB T6G 2V4, Canada; E-Mails: abadr@ualberta.ca (A.B); elou@ualberta.ca (E.L.)

* Author to whom correspondence should be addressed; E-Mail: Walied.Moussa@ualberta.ca

Academic Editor: Yasufumi Enami

Abstract: This paper presents the methodology and challenges experienced in the microfabrication, packaging, and integration of a fixed-fixed folded spring piezoelectric energy harvester. A variety of challenges were overcome in the fabrication of the energy harvesters, such as the diagnosis and rectification of sol-gel PZT film quality and adhesion issues. A packaging and integration methodology was developed to allow for the characterizing the harvesters under a base vibration. The conditioning circuitry developed allowed for a complete energy harvesting system, consisting a harvester, a voltage doubler, a voltage regulator and a NiMH battery. A feasibility study was undertaken with the designed conditioning circuitry to determine the effect of the input parameters on the overall performance of the circuit. It was found that the maximum efficiency does not correlate to the maximum charging current supplied to the battery. The efficiency and charging current must be balanced to achieve a high output and a reasonable output current. The development of the complete energy harvesting system allows for the direct integration of the energy harvesting technology into existing power management schemes for wireless sensing.

Keywords: PZT sol-gel; piezoelectric energy harvesting; conditioning circuitry; microfabrication; MEMS; packaging; integration; feasibility study

---

1. Introduction

The focus of the majority of Microelectromechanical (MEMS)-based energy harvesting research is the optimization of the power output of a specific energy harvester. Regardless of the harvesting methodology or design [1–4], significant challenges exist in the design, fabrication, and implementation of the energy harvester. This paper will focus on the challenges encountered in the microfabrication; packaging and integration of a fixed-fixed folded spring vibration-based piezoelectric energy harvesting system [5]. The focus of this specific design of the folded spring energy harvester was to controllably and reliably reduce the natural/operational frequency of the harvester while remaining insensitive to out of plane static deformations due to residual stresses [5–9]. The folded spring allows for the effective length of the spring to be increased while allowing for free expansion, contraction, and rotation of individual beam elements of the spring to dissipate residual stresses.

In general, there are two methodologies available to fabricate the released structures required for the piezoelectric energy harvesters. The first method relies upon surface micromachining techniques to undercut the released structures [4,10–14]. Typically, this method uses a sacrificial material and an etch stop, such as a Silicon-On-Insulator (SOI) wafer, to release the structural members of the device. This fabrication methodology is capable of producing thin structural beams with high accuracy, which is desirable for low frequency energy harvesting applications. The second method relies on bulk micromachining to define the geometry and release the structure of the harvesters [4,15–18]. Typically, this method uses Deep Reactive Ion Etching (DRIE) to remove the structural silicon from both sides of the wafer to define the structural beams and proof masses required by the energy harvesters. Although the bulk micromachining methodology may introduce difficulties in fabricating thin structural members, integrated proof masses are difficult to achieve with surface micromachining techniques. A release methodology was developed for the energy harvesters designed in this research [5,9], based upon the bulk micromachining techniques, to allow for the required large proof masses and thin structural beams. Several difficulties encountered with this chosen method of releasing the energy harvester structure will be discussed at length in the following sections. Throughout the remaining microfabrication process flow, several obstacles were encountered, including difficulties caused by titanium residues from lower electrode etching, uneven and non-continuous films deposited by a simultaneous sol-gel and lift-off process, and wafer handling issues caused through the backside etching required to release the harvesters from the silicon wafer [9,19].

Additionally, the harvester operates under a base vibration, wafer-level testing was not possible. Therefore, full packaging of the harvesters were required, leading to several challenges dealing with adhesion of wirebonds and the overall design and implementation of suitable packaging [9,19]. Finally, significant challenges were encountered in designing and fabricating suitable conditioning circuitry including maximizing the efficiency and charging current delivered to the storage medium. Additionally, the relationship between the efficiency and battery charging current of the conditioning circuitry in response to variations in the input voltage, input frequency, and battery voltage level will be examined with a feasibility study. In the following sections, each of the microfabrication, packaging and integration and conditioning circuitry-based challenges will be examined in detail.

2. Microfabrication of PZT Sol-Gel Folded Spring Energy Harvesters

2.1. Energy Harvester Microfabrication Process Flow Overview

The process flow developed in [5,9] to microfabricate PZT sol-gel folded spring energy harvesters has undergone several iterations of development in order to overcome a variety of fabrication-based challenges. In response to these fabrication obstacles alternate fabrication processes were developed for multiple fabrication steps, including the lower electrode deposition, PZT sol-gel deposition, PZT patterning, and the definition and release of the energy harvesters. The entire fabrication flow is shown, with the optimum arrangement of the fabrication flow highlighted, in Figure 1 [9].

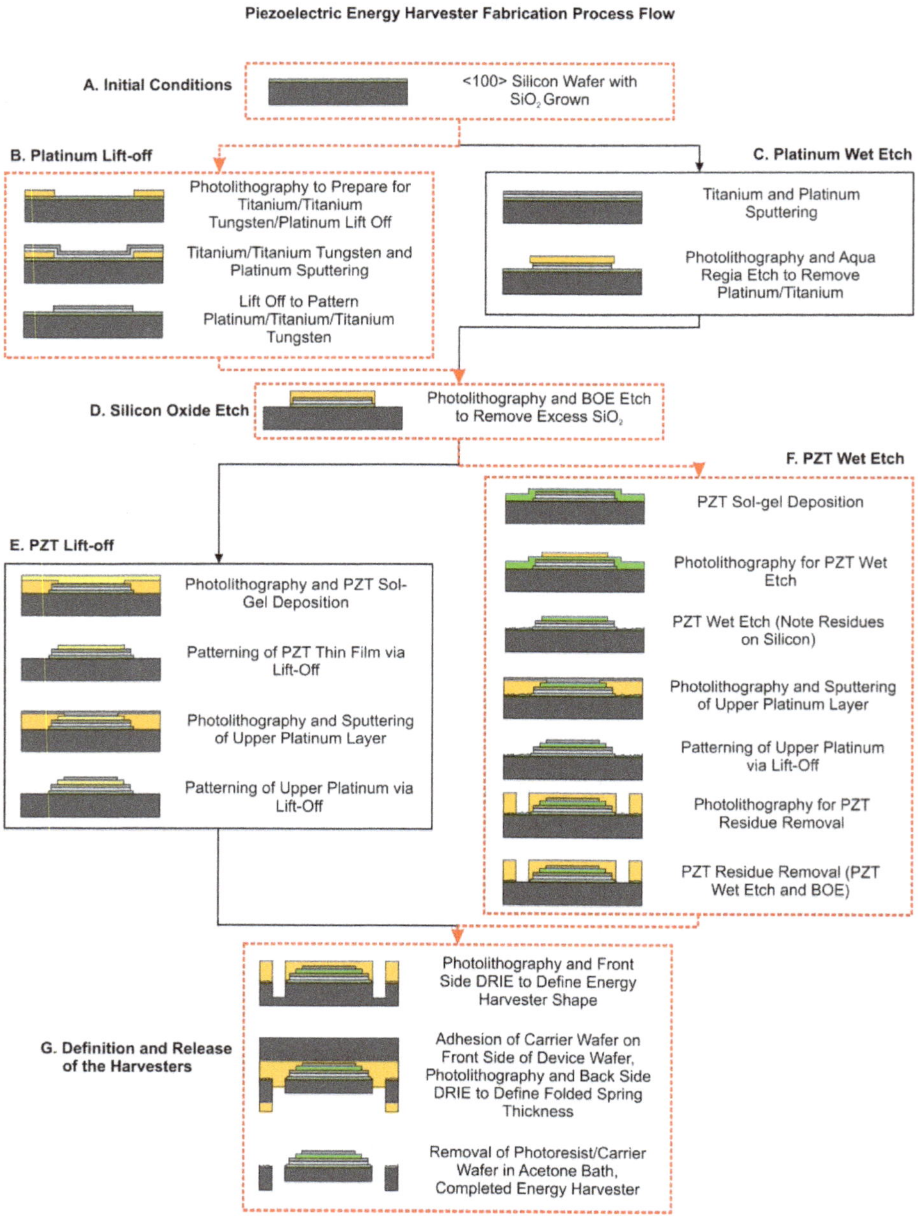

Figure 1. The PZT sol-gel folded spring energy harvester microfabrication process flow. The optimum arrangement of fabrication processes is highlighted in red [9].

The fabrication flow shown in Figure 1 is capable of producing piezoelectric energy harvesters of the same general cross section, of any configuration and beam thickness, using any sol-gel based piezoelectric film with suitable adhesion layers. The microfabrication fabrication process begins with a double side polished prime <100> silicon wafer. The thickness and uniformity of wafer chosen for this process will have a significant effect on the natural frequency of the harvesters, the volume of proof masses, and variation in folded spring thickness. Therefore, it is important to have wafers with as uniform thickness as possible. A layer of silicon oxide was then deposited with dry oxidation, as an electrical isolation barrier, diffusion barrier, and adhesion layer for the subsequent thin films. The deposition of lower electrodes can occur in two methods. The first method, as shown in "B. Platinum Liftoff" in Figure 1, uses a negative photoresist as a sacrificial material to lift off the required lower electrode layers. This option allows for the use of any lower electrode material, assuming adhesion to the underlying silicon oxide layer. The second procedure, as shown in "C. Platinum Wet Etch", uses an Aqua Regia Etch to pattern the titanium and platinum layers. Regardless of procedure used to pattern the lower electrodes of the piezoelectric stack, a buffered oxide etch is required to pattern the exposed silicon oxide layer. There are two methods available to deposit the required sol-gel PZT layer and the upper electrodes of the piezoelectric stack. The first, as shown in "E. Lift-off" in Figure 1, involves the sequential lift off of the PZT material and upper platinum electrode. The second, as shown in "F. PZT Wet Etch" in Figure 1, involves the sol-gel deposition and wet etch of the PZT and subsequent lift off of the upper platinum electrode. The microfabrication of the energy harvester is completed with a combination of two DRIE. The first etch, from the front of the wafer, defines the planar geometry of the energy harvester. The second etch, from the backside of the wafer, releases the structure of the energy harvesters and defines the overall thickness of the folded springs. The second DRIE etch also allows for the use of cleaving trenches to separate individual released energy harvesters from the wafer without the use of potentially harmful dicing.

Three major challenges were encountered in the development of this microfabrication process—overetching of the titanium films leading to the modification of the lower electrode deposition process from etching to lift off; the poor coverage, limited thickness and peeling of the PZT sol-gel film leading to an alternate PZT material and wet PZT etch; and the release and conservation of the rectangular cross section of the folded spring leading to a two-step DRIE release procedure. In each case, the resolution to the encountered obstacle resulted in a modification to the process flow in Figure 1.

2.2. Lower Electrode Deposition Process Challenges

The first microfabrication-based challenge encountered involved the lift off of titanium films due to uneven etching and deposition of the platinum and titanium films pre-deposited on the wafer. The initial iteration of the microfabrication process included an aqua regia wet etch to take advantage of the high etch rate of the etch, upwards of 13 nm/min. Typically, the isotropy of the aqua regia wet etch is not a drawback, however, when coupled with the non-uniformity of titanium/platinum sputtering, on the order of a few tens of nanometers across the wafer, the situation seen in Figure 2 can arise.

As highlighted in Figure 2, the nonuniformity in deposition across the wafer will later manifest as both underetched and overetched films. In areas of locally thicker titanium films, residual films will occur while the remainder of the wafer is properly etched. It is possible to remove the residual titanium, however, the correctly etched features will be overetched and removed. Alternatively, if the residual titanium film is not removed, the lower electrodes would be irreversibly damaged.

The etch rate of titanium in Buffered Oxide Etch (BOE) is extremely high [20], therefore any exposure to the BOE etchant in any subsequent step would cause the lower electrodes of the harvester to lift off from the wafer, as shown in Figure 2.

Figure 2. Platinum and titanium films etched by aqua regia. Residual titanium film from underetching is highlighted in the left figure, causing the overetch seen in the right figure in the following buffered oxide etch [9].

To rectify this issue, a lift off procedure was developed, using a negative photoresist as a sacrificial layer. The goal of this sacrificial layer is to prevent the adhesion of the sputtered metal to the wafer. After the metal is deposited, the photoresist is dissolved, and the unwanted material is removed from the wafer, as shown in Figure 3.

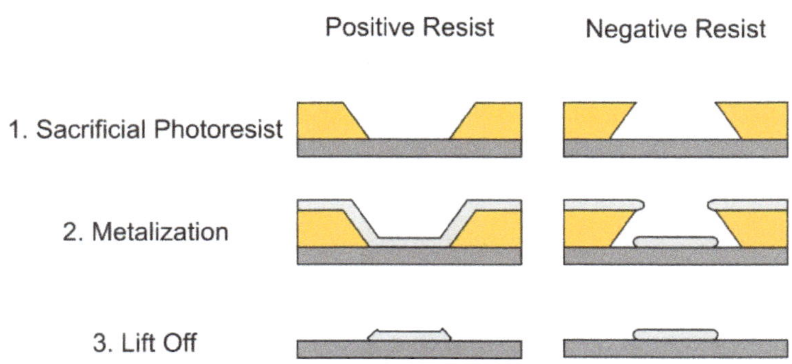

Figure 3. A schematic outlining the stages of the lift off process with both positive and negative photoresists. Sidewall angles are exaggerated for illustration [9].

For the lift off application, the polarity of the resist is not specifically important—the cross sectional profile of the resultant photoresist pattern is important. When a positive photoresist is used for lift-off, the cross sectional profile that is achieved is shown above in Figure 3. Although the sidewall angle is not as exaggerated as shown in Figure 3, the sidewalls will be coated with the sputtered metal, creating a continuous metal film. In order to complete the lift off procedure with a positive photoresist, ultrasonic cleaning and other harsh means are required to remove the unwanted metal. Alternatively, the sidewall profile of a negative photoresist aids in the liftoff patterning of the metallic film, preventing the metal layer from being deposited on the sidewalls of the photoresist. This allows for the lift off to occur easily and much more accurately. Additionally, the use of a robust lift off procedure allows for a variety of adhesion materials to be used to both overcome additional fabrication issues encountered later in the fabrication process and interchange piezoelectric materials in this process flow if required.

2.3. PZT Sol-Gel Deposition and Patterning Challenges

The second microfabrication challenge encountered in the fabrication of the harvesters was the poor adhesion and coverage of the PZT film. The initial development of the harvester, as shown as Section "E. PZT lift-off" in Figure 1, utilized a lift-off to pattern the sol-gel PZT film. This method worked well to develop the understanding of the sol-gel deposition procedure; however, there were significant film quality issues encountered. Due to the anneal/bake requirements of the sol-gel PZT and the thermal budget of the sacrificial photoresist, the lift-off and deposition must occur simultaneously. A sacrificial layer of photoresist is deposited and patterned to prevent the PZT film from adhering to the wafer in areas that the PZT film is not required. After the photoresist is patterned, the PZT sol-gel was then spun on to the wafer. Approximately 5 mL of the PZT sol-gel precursor is spread on the wafer at 250 RPM for 20 s, and then spun at 1000 RPM for 30 s. Once the spinning of the PZT precursor is complete, the wafer must be baked at 100 °C for 15 min to evaporate the sol-gel precursor solvent. When the baking is complete, a partial anneal is required in order to solidify the film sufficiently to allow for lift off to occur. The anneal is ramped to 150 °C and held for 15 min. After the partial anneal was complete, the temperature was ramped down to ambient. The wafer was immersed in acetone to remove the sacrificial photoresist and perform the PZT lift off. The PZT sol-gel film is partially soluble in acetone; therefore, it was necessary to limit the total exposure to limit the volumetric loss of the film. The optimum lift off occurred with a 5 min soak in acetone, followed by a 2 min, mid power, ultrasonic bath. After the lift off process was complete, the wafer was annealed to crystallize the PZT film. To prevent cracking from thermal shock, the crystallization anneal was also ramped to 475 °C, as required for this specific sol-gel material. The PZT film was annealed for one hour, and then ramped down to ambient over several hours. If the annealing furnace was opened prematurely, the thermal shock of the ambient air rushing into the furnace would damage the PZT film. This method of PZT deposition was capable of producing films as thick as 3 μm with repeated depositions.

During the lift-off, it was found that the semi-solid PZT film would locally dissolve from the wafer when exposed to acetone. This caused a net volume loss, which ultimately led to a contraction of the PZT film during annealing, causing the voids and cracks that are seen in Figure 4.

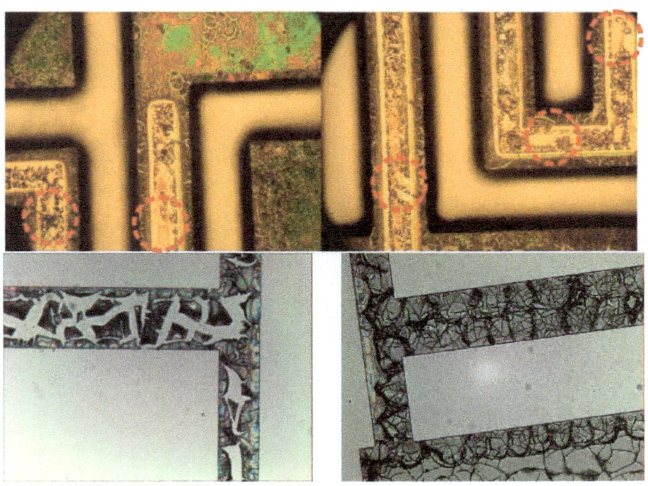

Figure 4. Examples of poor quality PZT film patterned by the lift off procedure. Circled areas denote short circuits and voided film.

To fill these voids shown in Figure 4, additional layers of PZT are required. However, to add additional layers of PZT, the entire lift-off process would be repeated. Although a fairy thick film is possible with this initial PZT sol-gel material and deposition procedure, the surface roughness and unevenness of the film makes deposition and continuity of the upper electrodes difficult. When it was apparent that an alternative PZT sol-gel precursor could allow for better coverage and uniformity, a second deposition/patterning technique was developed, as shown in "F. PZT Wet Etch" in Figure 1. The second PZT material, from Mitsubishi Materials Corporation (Schaumburg, IL, USA), required a much higher crystallization temperature of 700 °C for crystallization. Therefore, the lift-off process that was developed in the previous section would not be suitable for patterning. To pattern the second PZT material, a wet etch was selected from literature and then adapted for use in our fabrication facility [9,21].

The second PZT sol-gel deposition procedure was similar to the first deposition procedure, with some minor material dependent alterations. In this case, a repeated spin-bake process was required to build up the thickness of the film. The PZT sol-gel was spun on the wafer at 500 RPM for 5 s and then spread on the wafer at 3000 RPM for 30 s. Once the PZT sol-gel was spun, the film was baked on a hot plate at approximately 350 °C to solidify the film. At this point, this spin-bake process must be repeated three times to build up 0.24 μm of PZT, in steps of 0.08 μm. Once the spin-bake cycle was completed, the film was annealed to crystallize the PZT. The anneal takes place at 700 °C for 15 min. Once annealing is complete, the entire deposition process must be repeated to build up the film thickness in steps of 0.24 μm. In this work, PZT thicknesses of up to 0.96 μm have been achieved. Through X-Ray diffraction (XRD), the proper crystal structure of the PZT film has been confirmed, as shown in Figure 5.

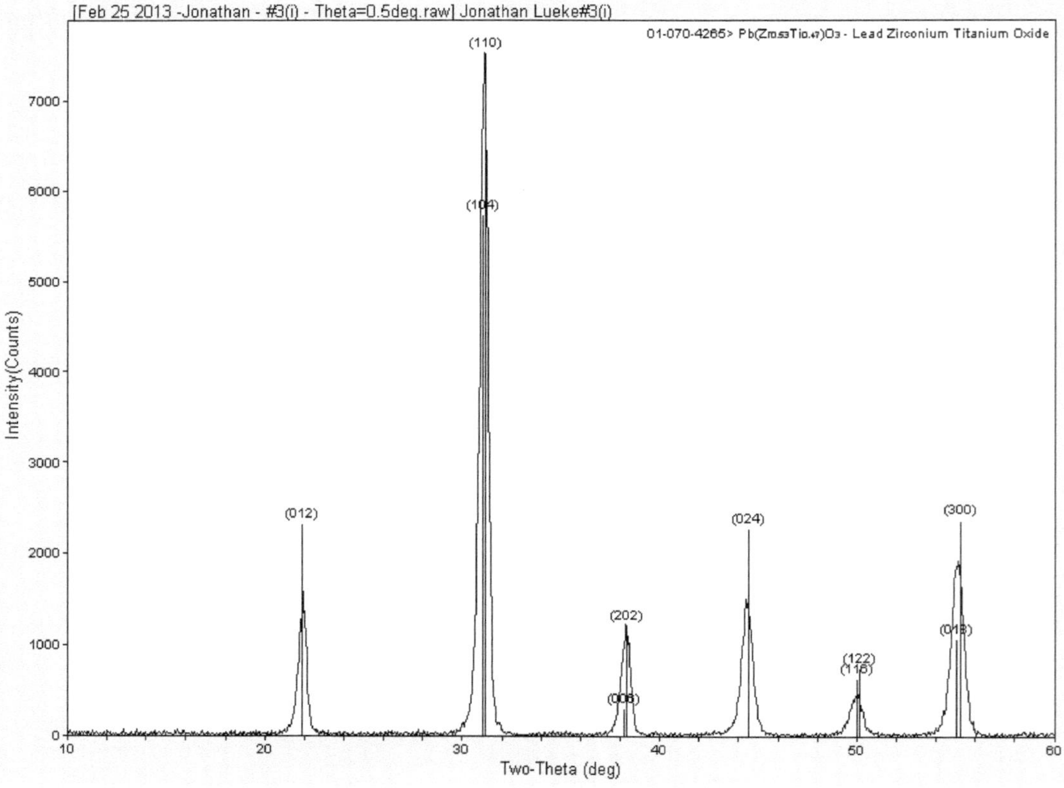

Figure 5. X-Ray Diffraction of the Mitsubishi Materials PZT material. This XRD is for a sample of 0.24 μm [9].

As shown in Figure 5, the expected peak locations for the crystal planes of the PZT film are shown as labelled vertical lines. The peaks of the XRD measurement correspond with the expected crystal planes, with the maximum peak aligned with the (110) peak of the PZT standard, suggesting that the deposited PZT film is aligned to the (110) crystal plane.

On areas of exposed silicon and silicon oxide, the PZT film tended to crack and produce discontinuous films. Ultimately, this is more of a cosmetic problem and does not cause any performance-related issues. An example of an annealed PZT film using this process can be seen in Figure 6.

As shown in Figure 6, the PZT sol-gel film forms correctly and continuously on the adhesion layers of platinum and titanium tungsten. On materials that the PZT is not expected to adhere well to, such as exposed silicon and silicon oxide, the film is rough and shows small cracks and discontinuities. This behavior has no impact on the energy harvesters, since the film in question is far from the physical locations of the harvesters and will be removed in a subsequent wet etch step.

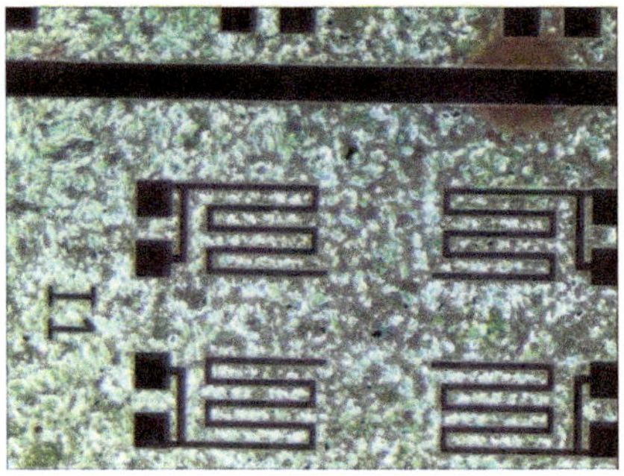

Figure 6. Correctly formed PZT thin film, continuous on the exposed platinum, cracked/discontinuous on the exposed silicon/silicon oxide [9].

An additional PZT-based challenge was encountered through the adaption of the PZT sol-gel deposition process. The crystallization anneal for this PZT film occurs at the same temperature titanium based silicides begin to form, ~700 °C [22]. Therefore, as the wafer is annealed, the titanium diffuses from the titanium layer into the silicon oxide, reducing the thickness and degrading the adhesion of the titanium layer. Given enough time in the annealing furnace, the titanium layer is sufficiently removed to cause the PZT to peel off the wafer due to residual stresses and poor adhesion between the remaining platinum and silicon oxide. To overcome this adhesion problem, the titanium layer was replaced with a titanium tungsten. The larger tungsten atoms act as a diffusion barrier, not allowing the titanium to freely migrate out of the film during the annealing processes. This prevents most of the titanium loss, maintaining the thickness and adhesion of the films, allowing for the deposition of continuous PZT films without major cracks/voids through multiple applications of the PZT sol-gel.

As was discussed previously, lift-off of this specific PZT material was not possible. Additionally, due to cross contamination concerns, dry plasma etching of PZT was not available. Therefore, a wet etch technique was chosen from literature, and adapted for use in our fabrication facility [21]. The wet etch adapted from literature is composed of three separate wet etches [21], the intermediate stages and final etch result are shown in Figure 7.

The initial unetched state of the unetched PZT film is shown in the left of Figure 7. First, the PZT is etched in a solution of Buffered Oxide Etch, hydrochloric acid, ammonium chloride, and water (1BOE:2HCl:4NH_4Cl:4H_2O) to convert the open PZT material to a white residue of PbClF [21]. In comparison to other available PZT wet etch procedures, the addition of the salt ammonium chloride (NH_4Cl) helps reduce the problematic undercut caused by the BOE etch, slowing the etch rate slightly, minimizing the lateral etch [21]. Second, the PbClF residue is etched with a solution of nitric acid and water ($2HNO_3{:}1H_2O$) producing a residue of water soluble $PbCl_2$ [21]. The middle of Figure 7 shows the remaining water soluble $PbCl_2$. Lastly, a prolonged soak in deionized water is used to remove the water soluble $PbCl_2$, resulting in the etched film shown in the right of Figure 7. This etch process is specifically designed to etch the PZT that is present on the platinum electrodes only—not the film deposited on exposed silicon and silicon oxide. Therefore, to prevent cross contamination of lead in equipment used in subsequent plasma etches, an additional PZT wet etched followed by an additional buffered oxide etch is required to remove the residual PZT film.

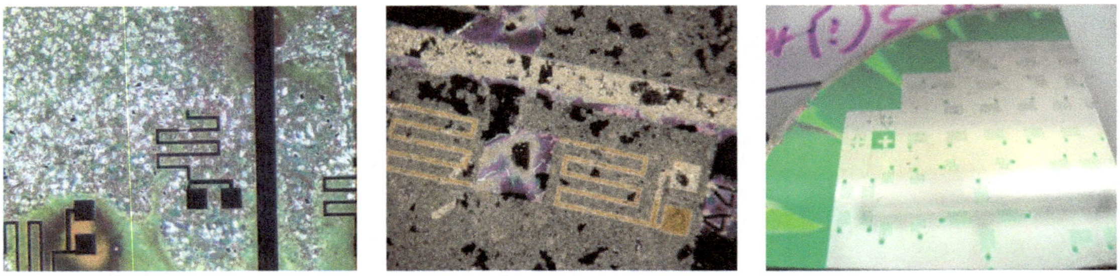

Figure 7. The PZT film during the etch process. The left figure shows the unetched PZT film. The middle figure shows the intermediate residue left after the second etch process. The right one shows the completed etch. PZT deposited on platinum appears dark green [9].

As shown in Figure 8, the etch area for this process is far from the piezoelectric stack which is distinguishable by the continuous platinum and PZT layers.

Figure 8. Mid-process photos of the second PZT etch needed to remove the residual PZT film from the silicon/silicon oxide. A color change in the PZT residual film denotes a height change from etching [9].

2.4. Challenges Involved with Definition and Release of the Energy Harvesters

The last microfabrication challenge encountered in the development of the harvesters involved the release of the harvesters and preservation of the cross section of the structural members of the harvesters. As shown as Section "G. Definition and Release of the Harvesters" in Figure 1, two sequential deep reactive ion etches (DRIE) were required to define the geometry of the folded springs and release the harvesters. The process uniformity of the DRIE process coupled with the thickness variation of the silicon wafer selected for fabrication can cause a significant difficulty in uniformity and accuracy of the etch depth. The DRIE non-uniformity is a function of the overall depth, therefore the total error in uniformity and etch depth increases cumulatively with etch depth [23–27]. This variability ultimately causes individual devices, of the same design on the same wafer, to have differing natural frequencies as discussed in [5]. There have been some attempts to tune the parameters of the DRIE process, such as RF power, chemistries used in the etch, cooling backpressure, dummy structures, to achieve a more consistent and accurate etch [23–28]. The methodology used to define the planar geometry and release the energy harvesters is based upon overlapping the etched volumes created from two sequential etch steps, as shown in Figure 9.

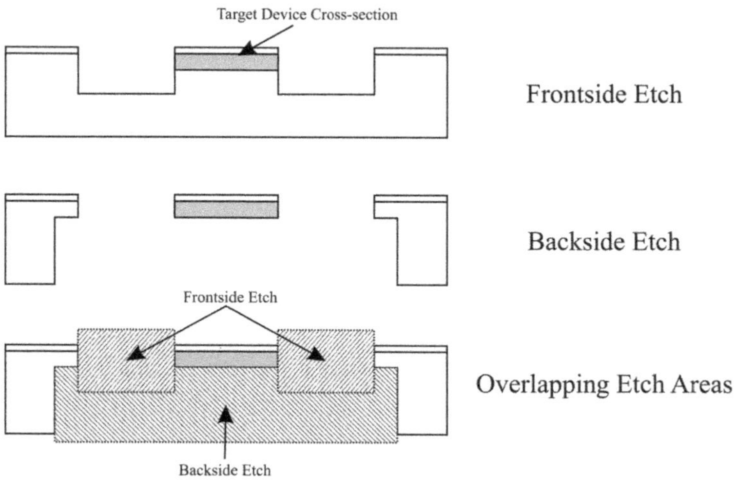

Figure 9. Overlapping Etch Areas releasing and defining the cross section of the harvesters [9,19].

The two-step methodology, shown in Figure 9, begins with a deep front side DRIE (at least 100 μm) to define the planar geometry, followed by a deep backside DRIE to define the thickness of the folded springs. The depth of the front side etch is significantly larger than the desired thickness of the energy harvester to ensure that the folded springs of the harvester have near vertical sidewalls. The deep backside DRIE then completes the near-rectangular cross section of the folded silicon beam. As shown in [5], the undercut of this etch can significantly alter the stiffness of the beam, deviating the natural frequency of the system from the expected design. The methodology used to release the energy harvesters through the combination of etch steps adds a major wafer handling difficulty. Since deep wells are etched through the wafer along crystal planes in the final etch process, the device wafer becomes progressively more delicate. Typically, the DRIE process requires the wafer to be gas cooled to maintain a constant etch rate. The slight pressure differential between the near-vacuum required for the plasma and the few millitorr of helium required to cool the wafer will cause the increasingly delicate wafer to burst during the final stages of the etch process. If the wafer survives the final DRIE etch, the weakened wafer can spontaneously cleave along etched areas and/or major crystallographic planes. Due to this problem, it was necessary to stabilize the device wafer with a carrier wafer previous to the final etch step.

To prevent damage and autocleaving during the final DRIE release etch, a carrier wafer was adhered to the devices wafer using photoresist as an adhesive. Although there are multiple methods to adhere the device and carrier wafer, including photoresist, Crystalbond adhesive, and double sided thermal release tape, it was found that photoresist was the only method available capable of sufficient heat transfer to cool the etched surface of the device wafer [9,29]. As the wafer is etched, both the mechanical etching and the plasma heat the top surface of the wafer. Silicon, having a high thermal conductivity, transfers the heat almost instantaneously. In typical situations, the backcooling of the DRIE process is tuned to offset the heating of the etch such that the photoresist mask does not exceed its maximum allowable temperature. If the resist is heated above its maximum temperature, the resist will degrade, either reflowing, losing the pattern of the mask completely, or irreversibly adhering to the surface of the wafer. This typically causes poor quality etching, non-uniformities, etch artifacts, global wafer etching, and limited etch depth. The photoresist-based method was the only one to perform suitably in both adhesion and heat transfer, therefore, it will be used as the main method of adhering the carrier wafer.

The use of a carrier wafer allows for—the separation of individual harvesters for packaging and testing without damaging the harvesters. The most common method of separating MEMS devices, dicing, is not well suited for this application. Dicing the wafer could adversely affect the overall yield of the harvesters by damaging the flexible, released structures. In the dicing process, a high speed rotary diamond blade, lubricated by a water jet, is used to cut the wafer into smaller sections. Both the vibration caused by the cut and the force of the water jet can cause serious damage to the individual energy harvesters during dicing. Cleaving can be used as an alternative for dicing to separate individual energy harvesters [19,30–33]. For this specific application, the use of a carrier wafer in the final DRIE etch step in the microfabrication process shown in Figure 1, allowed for the addition of deep cleaving trenches to the final backside etch pattern. The trenches integrated into the backside well pattern, as shown in Figure 10, followed major crystallographic planes of the silicon and were etched to the same thickness as the folded springs.

Figure 10. Cleaving trenches and backside wells etched on the backside of a testing wafer [9].

To remove the carrier wafer, both wafers are immersed in acetone and soaked for at least 24 h. The immersion in acetone completely dissolves all photoresist present. Additionally, the low surface tension of the acetone prevents any mechanical damage due to stiction during the separation of the wafers. Once the wafers are separated, the device wafer must be removed from the acetone bath and placed in a deionized water bath. The bottom of the bath should be lined with a soft material, such as cleanroom wipes, in order to provide a semi-flexible surface for cleaving. The cleaving must take place under water so that the energy released by the cleaving is damped sufficiently to prevent device damage. Once the wafer is submerged, the wafer is cleaved into sequentially smaller pieces until each harvester is separated. Light pressure applied by a sharp object will provide sufficient force to cleave the wafer along the etched cleaving trenches. This cleaving procedure was the precursor to the work published in [33].

3. Packaging and Integration of the Energy Harvesters

Although are several packaging schemes available for MEMS-based devices, there are limited complete packaging solutions for vibration-based energy harvesting. The energy harvesters operate under a base excitation; therefore it was not possible to directly connect to the energy harvesters in a typical probing station. The packaging scheme must allow for unhindered movement of the harvester to prevent damage and electrical losses. Additionally, suitable electrical connections must be made to the harvester to capture the voltage created through the vibration of the structure. Finally, the packaged energy harvester must be integrated into conditioning circuitry capable of efficiently rectifying and storing the harvested power.

3.1. Overview of the Packaging Methodology for the Energy Harvesters

It was chosen to package the harvester onto printed circuit boards to enable *in-situ* characterization. Cyanoacrylate glue is used to mechanically adhere the piezoelectric energy harvester onto the printed circuit board (PCB). The cyanoacrylate glue is biocompatible, and strongly adheres the silicon base of the energy harvester to the PCB. As shown in Figure 11, in order to allow free vibration, a hole is punched in the PCB where the proof mass of the harvester is expected to be located. This ensures that the energy harvester will not contact the circuit board during operation, preventing harvesting losses and potential damage.

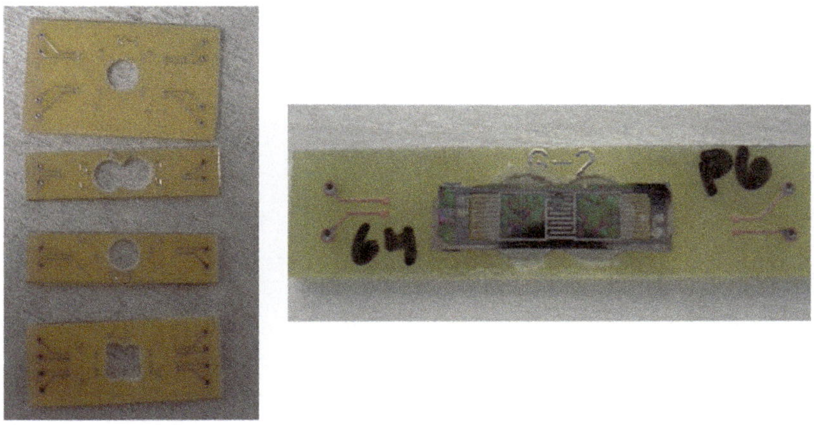

Figure 11. Solid PCBs developed for harvester packaging. (**Left**) separated and prepared individual PCBs; (**Right**) a mechanically packaged harvester on an individual PCB [9].

The electrical connections between the energy harvester and the outside world are made with ball-wedge wirebonds and soldered wires. The wirebonds are used to create the intermediate electrical connection between the energy harvester and the PCB. The solder paste layer on the PCB is removed in order to wirebond directly to the copper trace. When material compatibility, bond adhesion and bond direction are a challenge, ball-wedge bonding is a more suitable choice. For example, for energy harvesters using platinum for electrode layers, the ball-wedge bond is more suitable to overcome the difficult adhesion to platinum. Additionally, ball-wedge bonding provides additional flexibility by allowing bonding at low ultrasonic bonding power, overcoming material compatibility issues such as fragile/rough structures, including the platinum/PZT/platinum layers produced in the research. The use of a ball bond allows for the decrease of applied ultrasonic energy to the platinum/PZT/platinum stack, preventing the possible delamination of electrodes. Generally, achieving good bond quality and adhesion is the major difficulty in wire bonding of energy harvesters. This can be significantly improved by tuning the bonding parameters to suit the electrode materials. These parameters include the bonding ultrasonic power, ultrasonic time, diameter of ball bond, and applied force [34]. Tuning these parameters for specific pad materials and conditions allows for the reliable and consistent bonding. With ball-wedge wirebonding, typically, the ball, or start of the bond is reserved for the pad that is more difficult to bond to. In this case, it was advantageous to use the wedge-end of the bond on the platinum/PZT/platinum stack. Heating applied to the bonding capillary was increased, in order to make the gold wire more malleable during bonding allowing for a significant decrease in applied ultrasonic energy required for the bond. This allowed for the wedge bond on the top platinum electrode to be completed with minimum ultrasonic energy applied, preventing delamination of the upper electrode. The connections to the outside world were provided by wires soldered to the PCB. These electrical connections do not to add a parasitic capacitance that would reduce electrical output. Examples of packaged harvesters can be seen in Figure 12.

Figure 12. Several examples of fully packaged Class II Harvesters ready for testing [9].

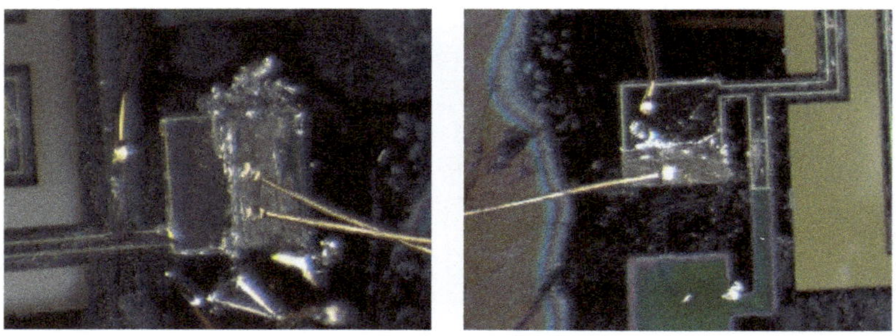

Figure 13. Examples of removing platinum and PZT from the remaining electrode of a damaged harvester to allow for packaging [9].

The wirebonding scheme developed in this work also allowed for the reclamation of the harvesters that were damaged through overetching encountered in the development of the microfabrication process. The overetch completely removed the exposed platinum-only electrodes of the harvesters leaving only the platinum/PZT/platinum stack electrode. To gain access to both electrodes, the upper platinum electrode and PZT material was carefully removed by hand. This allowed for access to both upper and lower electrodes for wirebonding, as shown in Figure 13. The remaining electrode pad area is large enough to allow for bonding to both upper and lower electrodes. The wirebonding procedures developed were not altered to fully package these salvaged harvesters. The ability to salvage damaged harvesters further demonstrates the robustness of this packaging scheme.

3.2. Overview of the Conditioning Circuitry for the Energy Harvesters

Conditioning circuitry is required to complete the integration of the energy harvester for characterization, testing, and later use in power management systems. In order to measure the output power of the harvester and determine the optimum load resistance, a simple circuit consisting of the harvester and a variable resistor, shown in Figure 14, was required.

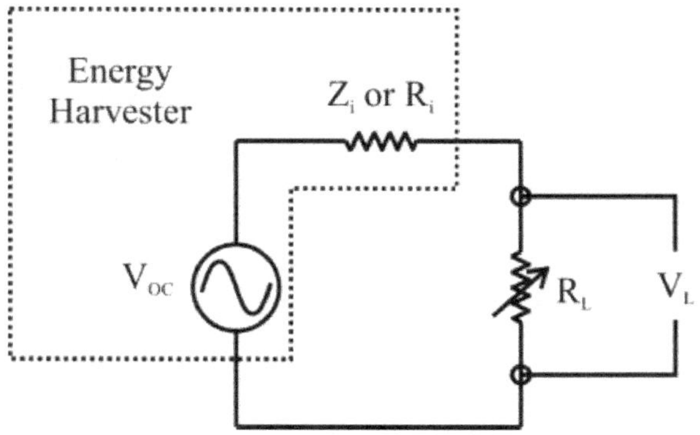

Figure 14. The simple circuit consisting of the energy harvester and a variable resistor used for initial characterization [9].

For maximum power transfer, the internal, or input resistance of the voltage source must match the load resistance [35]. As can be seen in Figure 14 the energy harvester can be idealized as an AC voltage source and a resistor symbolizing the real component of the internal impedance of the harvester (Z_i or R_i). The internal impedance of the harvester is a function of the operational frequency of the harvester and the capacitance formed across the deposited PZT material. When the load resistance matches the input resistance of the harvester, the harvester will transfer maximum output power to the circuit. The output voltage of the harvester can be measured across the load resistance (R_L) to calculate the total output power of the harvester. This circuit was used to confirm the operation of the harvesters and to determine the maximum output power available from the harvester.

For more complex characterization and integration into power management systems, a second conditioning circuit is required in order to convert the harvested AC signal into a storable DC voltage. Several conditioning circuits are reported in literature, divided into passive [36–38] and active [39–42] conditioning circuits. Passive conditioning circuitry uses a simple configuration of a small number of passive components for the AC/DC rectification. In contrast, active conditioning circuits require active components, such as Metal Oxide Semiconductor Field Effect Transistor (MOSFETs), to rectify the AC signal. Although the conversion efficiency of active conditioning circuitry is higher, additional components such as a feedback circuit, are required. In this feasibility study, a passive conditioning circuit was used. Figure 15 shows a complete energy harvesting system which consists of the energy harvester, a voltage doubler, a voltage regulator circuit and a Nickel Metal Hydride (NiMH) battery.

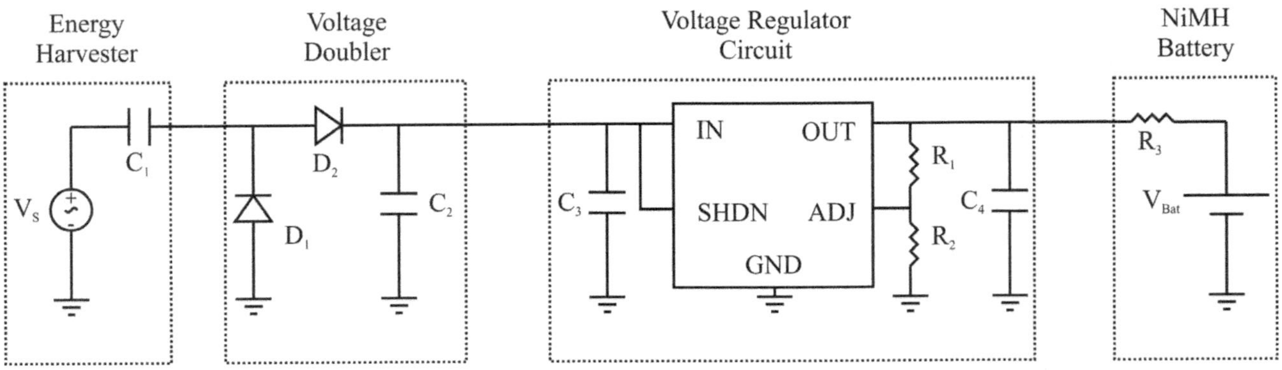

Figure 15. The full energy harvesting system developed in this study allowing for the storage of the power produced by the energy harvesters.

In the conditioning circuitry shown in Figure 15, the AC signal from the harvester is converted into a DC signal through a passive half-wave voltage doubler. The voltage doubler converts the two-phase AC signal produced by the harvester (positive and negative voltage signals) into a DC signal and doubles its magnitude. The diodes prevent signal cancellation from the production of equal and opposite voltages when an array of harvesters is used. A half-wave voltage doubler is chosen over a full wave rectifier due to the fewer diodes required, leading to lower diode voltage drop and higher rectification efficiency. The voltage doubler is shown consisting of two diodes D_1 and D_2, and an output capacitor C_2 to filter and smooth the output of the doubler.

In order to provide a suitably constant DC voltage to charge the NiMH battery, a voltage regulator is required. A LT3009 regulator (Linear Technology Inc., Milpitas, CA, USA) was chosen for this circuit due to its very low quiescent current (3 μA), the low minimum input voltage of 1.6 V, its adjustable output voltage, and the relatively low number of components required to make the integrated circuit (IC) operate. Input and output capacitors, C_3 and C_4, are required to filter the noise of the input and output signals to the regulator. The output voltage of the regulator can be adjusted by varying the ratio of the control resistors R_1 and R_2. For this study, R_1 set to 705 kΩ and R_2 is set to 604 kΩ to set the voltage regulator to 1.3 V.

The output of the voltage regulator must be current limited, with a limiting resistor R_3, to prevent damage to the NiMH battery as well as the regulator (maximum 20 mA only). A NiMH battery was chosen due to the low charging voltage (1.3 V–1.4 V per cell) than other types of batteries, such as Li-ion (3.7 V per cell), Li-polymer (4.2 V per cell), and Lead acid (2.4 V per cell) [43]. Although a NiMH battery reduces the required voltage threshold to charge the battery, limiting the energy generation requirements of the harvesters, it is the least evolved of the battery technologies available. NiMH comparatively has the lowest energy density, the largest degree of self-discharge, and the lowest life cycle in comparison to Li-ion and Li-polymer batteries.

3.3. Feasibility Study of the Conditioning Circuitry

A feasibility study of the conditioning circuitry was undertaken to overcome the challenges encountered in the development of the designed conditioning circuitry shown in Figure 15. These included maximizing the efficiency and charging current delivered to the battery requiring the use of low activation voltage and low current consumption components. Due to excessive losses, high voltage drops, and poor efficiency, components such as full wave rectifiers and battery management circuits with embedded logic were not suitable for this low power application. The feasibility study focused on the efficiency and battery charging current of conditioning circuitry in response to different input signals' characteristics such as, the input voltage(varied from 3 to 5 V peak-to-peak, the input frequency (varied from 100 to 300 Hz), and the NiMH battery voltage (varied from 1 to 1.25 V). The range of input frequency and voltage considered in this work complies with those reported in the literature. For example, common input vibrations from household appliances are between 60 and 200 Hz [3,44]. A comparison between different piezoelectric energy harvesters is shown in [4], and the operating voltage values are presented. Both simulation and experimental measurements were used to determine the battery charging current and overall efficiency for a variety of combinations of these variables.

The simulation was performed with LTSpice IV (Linear Technology Inc. Milpitas, CA, USA) to utilize standard models of the chosen voltage regulator from Linear Technology. It is a high performance SPICE simulator allowing user to view the regulator output waveforms in a few minutes. The experimental setup is shown in Figure 16, consisting of an 33220A AC supply (Agilent, Santa Clara, CA, USA), Agilent 34401A voltage meters, 189 RMS multimeters (Fluke, Mississauga, ON, Canada), and an HR-4U 1000 mAh NiMH battery (Sanyo, Newark, NJ, USA).

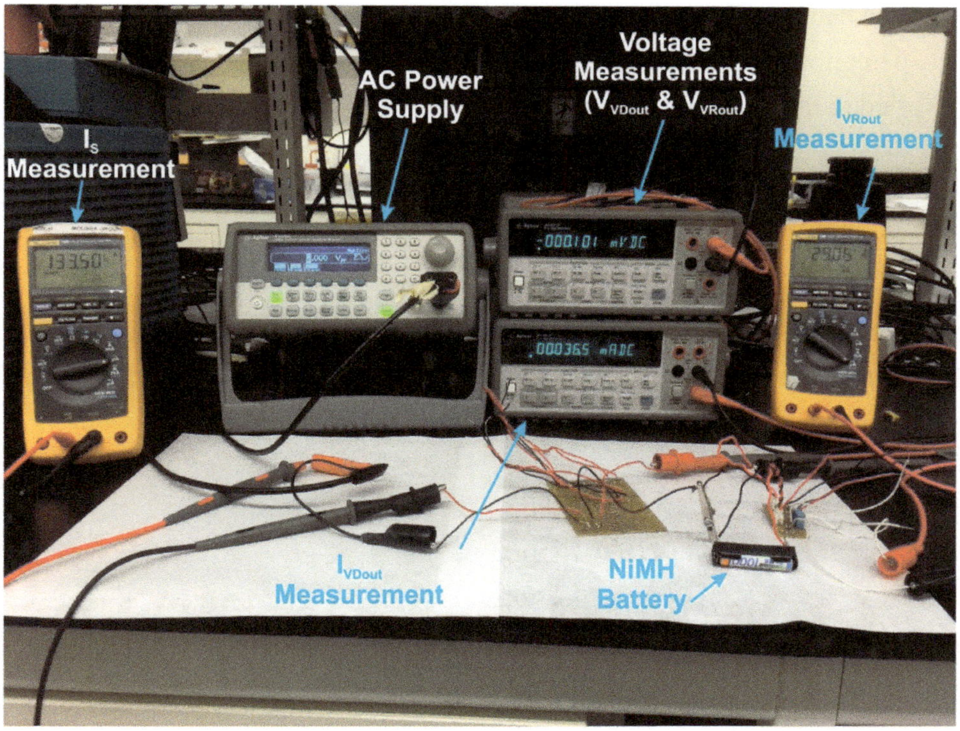

Figure 16. The experimental setup for the proposed electrical interface.

The initial stage of the feasibility study dealt with the effect of the battery voltage on the performance of the circuitry. In general, it was observed that the voltage of the NiMH battery was the only parameter in the study that affected the current delivered to the battery from the conditioning circuitry, as shown in Figure 17. The internal resistor of the NiMH is constant (approximate 0.17 Ω) though the entire discharge stage.

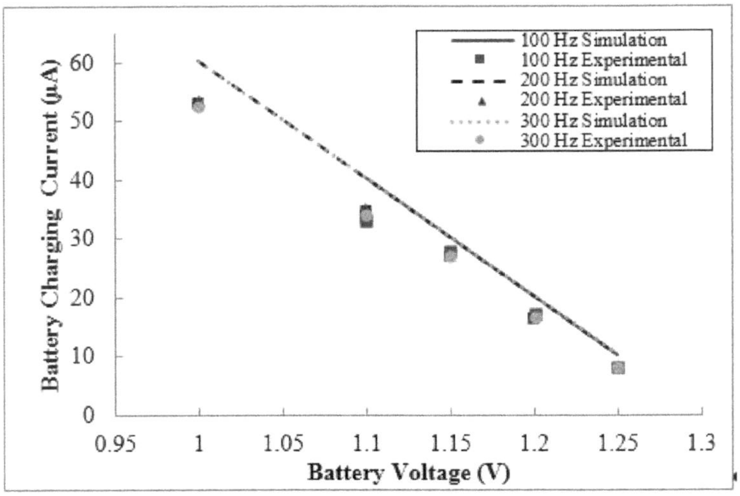

Figure 17. The simulated and experimental battery current behavior *versus* battery voltage for three input frequency cases (100 Hz, 200 Hz, and 300 Hz). The correlation of results for all input frequencies suggests that the battery current does not depend on input frequency.

As shown in Figure 17, any variation in the input frequency to the conditioning circuit does not affect the current delivered to the battery. The same conclusion was found when varying input frequency for set levels of input voltage. Therefore, it can be concluded that the current supplied to the battery is insensitive to the input signal supplied to the conditioning circuitry. This result is logical, since the voltage regulator and current limiting resistance would regulate the current and voltage supplied to the battery. As shown in Figures 17, the current delivered to the battery is maximized at low battery voltage due to Ohm's law. It was found that the maximum the maximum battery current was 51.7 µA, at an input frequency of 100 Hz, input voltage of 4 V p-p, and battery voltage of 1 V.

To examine the effect of battery voltage on the efficiency of the conditioning circuit, the circuit efficiency was evaluated for a variety of input voltages and frequencies. The efficiency behavior of the circuit *vs.* the battery voltage for varying input frequencies is shown in Figure 18.

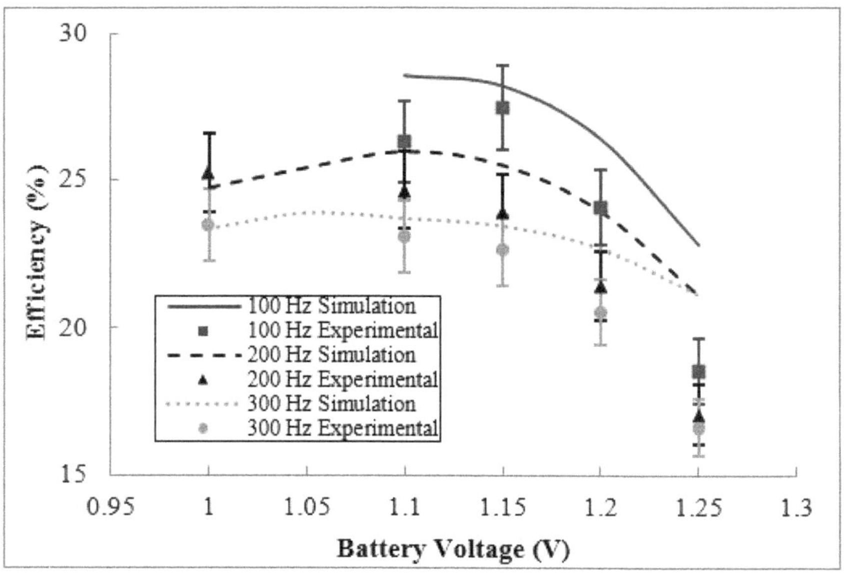

Figure 18. The simulated and experimental efficiency behavior *versus* battery voltage for three input frequency cases (100 Hz, 200 Hz, and 300 Hz).

As shown in Figure 18, the efficiency of the conditioning circuit was measured at 16.6% to 27.5% for the three input frequency cases. The overall circuit efficiency decreases as the input frequency increases. This occurs due to the higher DC voltage output from the voltage doubler leading to a higher voltage difference between the input and output voltages on the voltage regulator. In the ideal situation, the input power ($I_{in} \times V_{in}$) is equal to the output power ($I_{out} \times V_{out}$). Therefore, the ratio between I_{in} and the quiescent current of the voltage regulator affects the efficiency of the regulator. In the 100 Hz case, the circuit was not able to operate at battery voltages lower than 1.1 V due to the output DC voltage level from the voltage doubler dropping below the minimum 1.6 V required by the voltage regulator. To aid in the operation of the circuit at low input frequencies, the input voltage or the current limiting resistance should increase. This will not only allow for the operation of the circuit at low input voltage frequency, but will lead to a higher voltage doubler efficiency. The voltage regulator dominates the overall efficiency of the circuit, varying between 35% and 72% efficiency, while the voltage doubler varies between 41% and 52% efficiency. Additionally, the overall circuit efficiency can be improved by increasing the efficiency of voltage doubler and voltage regulator circuits. The discrepancy between the simulation and experimental results are expected, and it is mainly due to the tolerance of the different components used in the experiment, also due to un-modeled physical parameters as the NiMH battery internal resistance. In addition, experimental errors shown in Figure 18, and following figures, are calculated based on the measurement errors in each quantity required to capture the efficiency. Lastly, in general, the efficiency of the circuit decreases as the battery voltage increases for all frequency cases examined, due to the reduction of the battery charging current as seen in Figure 17. The efficiency behavior of the circuit *vs.* the battery voltage for varying input voltages is shown in Figure 19.

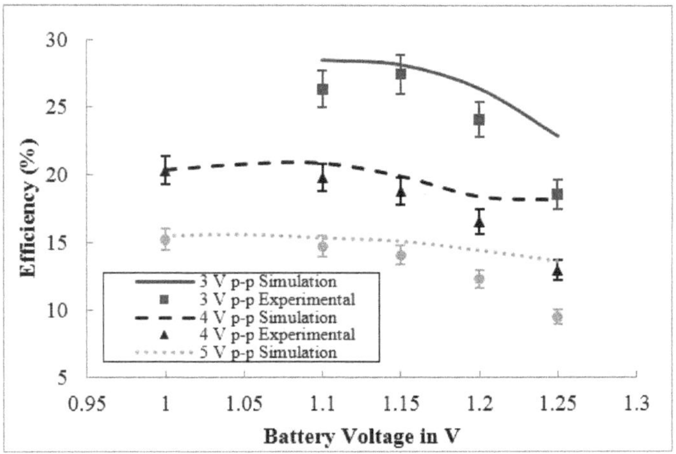

Figure 19. The simulated and experimental efficiency behavior *vs.* battery voltage for three input voltage cases (3 V p-p, 4 V p-p, and 5 V p-p).

Figure 19 shows the overall efficiency of the proposed circuit for varying battery voltages for three cases of input voltage to the circuit. The overall efficiency for the proposed circuit was measured between 9.4% and 27% for the examined input voltage cases. The efficiency of the circuit was consistently higher for lower applied input voltages. As with the case in efficiency study in Figure 18, this input voltage-dependent behavior occurs due to higher DC voltage output from the voltage doubler, leading to a higher quiescent current and lower regulator efficiency. Additionally, the voltage regulator dominates the overall efficiency of the circuit, varying between 20% and 73% efficiency, while the voltage doubler varies between 42%

and 50% efficiency. As with the frequency-based study, the low voltage case of 3 V peak-to-peak could not activate the conditioning circuit under a battery voltage of 1.1 V due an insufficient voltage supplied to the voltage regulator. An increase in input frequency would aid in the low voltage operation by increasing the voltage doubler output voltage. The discrepancy between the simulation and experimental results are due to the tolerance of the components used in the experiment and un-modeled NiMH battery internal resistance.

From the consistency of the efficiency behavior of the circuit for a variety of input frequency and input voltage cases in Figures 18 and 19 for varying battery voltage, it can be concluded that at a specific battery voltage, lower input frequency and voltage will lead to a higher circuit efficiency. To understand the interdependency of input frequency and voltage on the efficiency of the conditioning circuit, the efficiency of the circuit will be examined for varying input voltages and frequencies at a fixed battery voltage. The efficiency behavior for varying input voltage and frequency, at a fixed nominal battery voltage of 1.2 V can be seen in Figures 20 and 21, respectively.

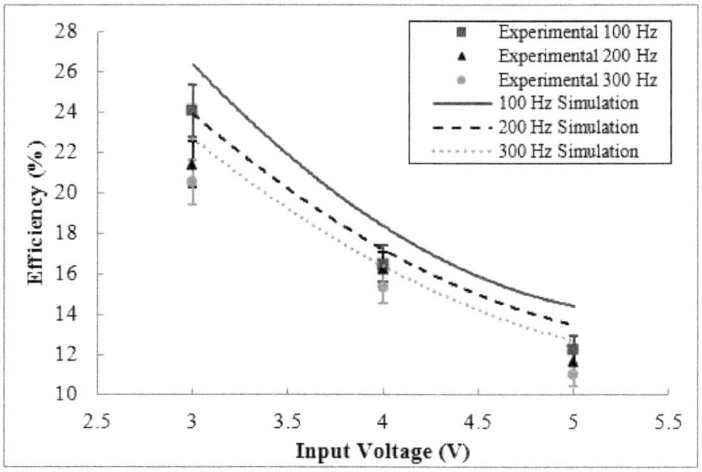

Figure 20. The simulated and experimental efficiency of the conditioning circuit for varying input voltage, for three input frequency cases (100 Hz, 200 Hz, and 300 Hz).

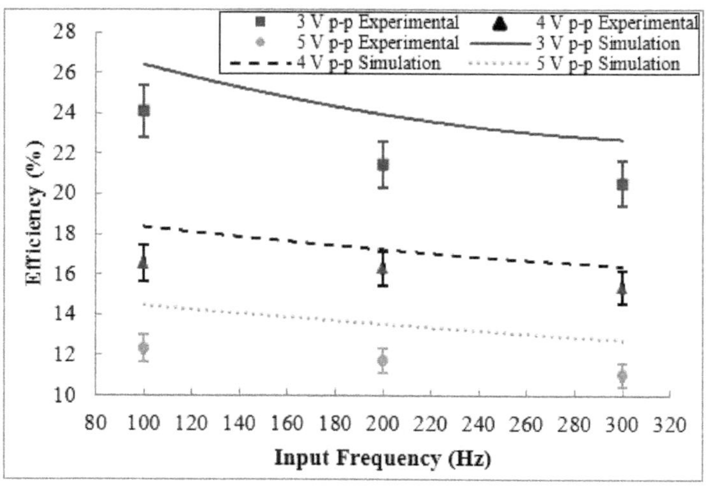

Figure 21. The simulated and experimental efficiency of the conditioning circuit for varying input frequency, for three input voltage cases (3 V p-p, 4 V p-p, and 5 V p-p).

As shown in Figure 20, at a constant input frequency, the circuit efficiency increases with decreasing input voltage due to the higher voltage regulator efficiency for lower DC voltages. Additionally, at a constant input voltage, the circuit efficiency increases with decreasing input frequency due to higher voltage ripples associated with lower frequency operating conditions lowering the DC voltage inputted into the voltage regulator, increasing regulator efficiency. As shown in Figure 21, the overall efficiency decreases as the input voltage and frequency increases due to the increased quiescent current absorbed by the voltage regulator. The maximum circuit efficiency for this study was 24.1% at an input voltage of 3 V p-p, an input frequency of 100 Hz, at the constant battery voltage of 1.2 V. Taking the two cases outlined in Figures 20 and 21, it is clear that the highest efficiency occurs at the lowest input voltage and frequency for constant battery voltages. Taking the dependence of circuit efficiency on the battery voltage into account, the highest possible efficiency would occur at the lowest operational battery voltage, input voltage, and input frequency.

4. Conclusions

The methodologies used and challenges encountered in the microfabrication and integration of fixed-fixed folded spring energy harvesters designed and characterized in [5] have been discussed in detail. A variety of challenges were overcome in the fabrication of the energy harvesters, such as the diagnosis and rectification of sol-gel PZT film quality and adhesion issues, including the use of more robust adhesion materials. Additionally, the use of a carrier wafer allowed for the use of a carrier wafer to promote non-invasive cleaving of the harvesters. A complete packaging and integration methodology was developed to facilitate testing the harvesters under a base vibration. This included developing a packaging methodology using printed circuit boards and wirebonding to allow for the electrical characterization of the harvester without relying on a traditional probe station.

Two conditioning circuits were developed to electrically characterize the harvester and to store the harvested energy in a NiMH battery. The initial simple conditioning circuit was useful to measure the total available power from the harvester. The second conditioning circuit developed allowed for the creation of a complete energy harvesting system, consisting of the harvester, a voltage doubler, a voltage regulator and a NiMH battery. The biggest challenge in the development of the conditioning circuitry is maximizing the efficiency and current stored in the battery. The maximum circuit efficiency is 27.5%, at an input frequency of 100 Hz, input voltage of 3 V p-p, battery voltage of 1.15 V, and current limiting resistance of 5 kΩ. The battery charging current for this case is 27.9 μA, with voltage doubler efficiency of 47.5%, and voltage regulator efficiency of 67.9%. However, the maximum battery charging current was found to be 51.7 μA at an input frequency of 100 Hz, input voltage of 4 V p-p, battery voltage of 1 V, and current limiting resistance of 5 kΩ. The overall efficiency for this case is 20.32%. The voltage doubler efficiency is 41.9% and the voltage regulator efficiency is 62.9%. It can be concluded that the maximum efficiency does not correlate to the maximum charging current supplied to the battery. Conversely, the maximum charging current can be achieved at the expense of the overall efficiency. The charging current supplied to the battery is only effected by the battery voltage, increasing with decreasing battery voltage. The efficiency of the designed conditioning circuit is increased with decreasing input voltage, input frequency, and battery voltage. The efficiency and charging current must be balanced to achieve a high output and a reasonable output current. Transitioning to active conditioning circuitry in future iterations of the energy harvestings system should increase efficiency and current supplied to the battery. The development of the complete energy harvesting system allows for the direct integration of the energy harvesting technology previously developed into existing power management schemes for wireless sensing.

Acknowledgments

The authors would like to acknowledge the support of the University of Alberta nanoFAB and Rick Conrad of the Mechanical Engineering Department of the University of Alberta for their technical support in the fabrication and testing of the energy harvesters. The financial support of the Natural Sciences and Engineering Research Council (NSERC) of Canada, the Micro Systems Technology Research Initiative (MSTRI), Alberta Innovates—Technology Futures (AITF), and nanoBridge through operating grants to W. Moussa is gratefully acknowledged.

Author Contributions

All authors contributed to the conception of the energy harvester system and methodology. The energy harvester design, microfabrication, and packaging work was completed by Jonathan Lueke. The initial simple conditioning circuitry work was initiated by Jonathan Lueke. The second conditioning circuit was designed, developed, and tested by Ahmed Badr and Edmond Lou. All authors contributed to the writing and revision of the manuscript.

Conflicts of Interest

The authors declare no conflict of interest.

References

1. Lueke, J.; Moussa, W.A. Mems-based power generation techniques for implantable biosensing applications. *Sensors* **2011**, *11*, 1433–1460.
2. Roundy, S.; Leland, E.S.; Baker, J.; Carleton, E.; Reilly, E.; Lai, E.; Otis, B.; Rabaey, J.M.; Wright, P.K.; Sundararajan, V. Improving power output for vibration-based energy scavengers. *IEEE Pervasive Comput.* **2005**, *4*, 28–36.
3. Roundy, S.; Wright, P.K. A piezoelectric vibration based generator for wireless electronics. *Smart Mater. Struct.* **2004**, *13*, 1131–1142.
4. Cook-Chennault, K.A.; Thambi, N.; Sastry, A.M. Powering mems portable devices—A review of non-regenerative and regenerative power supply systems with special emphasis on piezoelectric energy harvesting systems. *Smart Mater. Struct.* **2008**, *17*, doi:10.1088/0964-1726/17/4/043001.
5. Lueke, J.; Rezaei, M.; Moussa, W.A. Investigation of folded spring structures for vibration-based piezoelectric energy harvesting. *J. Micromech. Microeng.* **2014**, doi:10.1088/0960-1317/24/12/125011.
6. Lee, K.B. Statics. In *Principles of Microelectromechanical Systems*; John Wiley & Sons, Inc.: Hoboken, NJ, USA, 2011; pp. 64–143.
7. Lee, K.B. Static behavior of microstructures. In *Principles of Microelectromechanical Systems*; John Wiley & Sons, Inc.: Hoboken, NJ, USA, 2011; pp. 144–234.

8. Lee, K.B. Dynamics. In *Principles of Microelectromechanical Systems*; John Wiley & Sons, Inc.: Hoboken, NJ, USA, 2011; pp. 235–324.
9. Lueke, J. A Mems-Based Fixed-Fixed Folded Spring Piezoelectric Energy Harvester. Ph.D. Theis, University of Alberta, Edmonton, AB, Canada, 2014.
10. Boisseau, S.; Despesse, G.; Ricart, T.; Defay, E.; Sylvestre, A. Cantilever-based electret energy harvesters. *Smart Mater. Struct.* **2011**, *20*, doi:10.1088/0964-1726/20/10/105013.
11. Casset, F.; Cueff, M.; Suhm, A.; Le Rhun, G.; Abergel, J.; Allain, M.; Dieppedale, C.; Ricart, T.; Fanget, S.; Renaux, P.; *et al.* Pzt piezoelectric coefficient extraction by pzt-actuated micro-beam characterization and modeling. In Proceedings of the 2012 13th International Conference on Thermal, Mechanical and Multi-Physics Simulation and Experiments in Microelectronics and Microsystems (EuroSimE), Cascais, Portugal, 16–18 April 2012; pp. 1/4–4/4.
12. Ettelt, D.; Dodane, G.; Audoin, M.; Walther, A.; Jourdan, G.; Rey, P.; Robert, P.; Delamare, J. A novel microfabricated high precision vector magnetometer. In Proceedings of the 2011 IEEE Sensors, Limerick, Ireland, 28–31 October 2011; pp. 2010–2013.
13. Ettelt, D.; Rey, P.; Savoye, M.; Coutier, C.; Cartier, M.; Redon, O.; Audoin, M.; Walther, A.; Robert, P.; Zhang, Y.; *et al.* A new low consumption 3D compass using integrated magnets and piezoresistive nano-gauges. In Proceedings of the 2011 16th International Solid-State Sensors, Actuators and Microsystems Conference (TRANSDUCERS), Beijing, China, 5–9 June 2011; pp. 40–43.
14. Jeon, Y.B.; Sood, R.; Jeong, J.H.; Kim, S.G. Mems power generator with transverse mode thin film pzt. *Sens. Actuators A Phys.* **2005**, *122*, 16–22.
15. Ricart, T.; Lassagne, P.; Boisseau, S.; Despesse, G.; Lefevre, A.; Billard, C.; Fanget, S.; Defay, E. Macro energy harvester based on aluminium nitride thin films. In Proceedings of the 2011 IEEE International Ultrasonics Symposium (IUS), Orlando, FL, USA, 18–21 October 2011; pp. 1928–1931.
16. Yin, S.; Abergel, J.; Bontempi, A.; Ricart, T.; Le Rhun, G.; Defay, E.; Niu, G.; Vilquin, B.; Gautier, B.; Robach, Y. Single crystal pzt thin film membrane with highly conductive electrodes. In Proceedings of the 2012 IEEE International Ultrasonics Symposium (IUS), Dresden, Germany, 7–10 October 2012; pp. 792–794.
17. Elfrink, R.; Kamel, T.M.; Goedbloed, M.; Matova, S.; Hohlfeld, D.; van Andel, Y.; van Schaijk, R. Vibration energy harvesting with aluminum nitride-based piezoelectric devices. *J. Micromech. Microeng.* **2009**, *19*, doi:10.1088/0960-1317/19/9/094005.
18. Miller, L.M.; Halvorsen, E.; Dong, T.; Wright, P.K. Modeling and experimental verification of low-frequency mems energy harvesting from ambient vibrations. *J. Micromech. Microeng.* **2011**, *21*, 045029.
19. Rezaei, M.; Lueke, J.; Raboud, D.; Moussa, W. Challenges in fabrication and testing of piezoelectric mems with a particular focus on energy harvesters. *Microsyst. Technol.* **2013**, *19*, 1–25.
20. Williams, K.R.; Gupta, K.; Wasilik, M. Etch rates for micromachining processing—Part II. *J. Microelectromech. Syst.* **2003**, *12*, 761–778.
21. Zheng, K.; Lu, J.; Chu, J. A novel wet etching process of $Pb(Zr,Ti)O_3$ thin films for applications in microelectromechanical system. *Jpn. J. Appl. Phys.* **2004**, *43*, 3934–3937.
22. Ohring, M. Chapter 11—interdiffusion, reactions, and transformations in thin films. In *Materials Science of Thin Films*, 2nd ed.; Academic Press: San Diego, CA, USA, 2002; pp. 641–710.

23. Jensen, S.; Jensen, J.M.; Quaade, U.J.; Hansen, O. Uniformity-improving dummy structures for deep reactive ion etching (drie) processes. In *Progress in Biomedical Optics and Imaging—Proceedings of SPIE*; Maher, M.A., Stewart, H.D., Eds.; San Jose, CA, USA, 2005; pp. 39–46.
24. Laermer, F.; Schilp, A. Method of anisotropically etching silicon. U.S. Patent 5,501,893 A, 26 Mar 1996.
25. Sun, H.; Hill, T.; Schmidt, M.; Boning, D. Characterization and modeling of wafer and die level uniformity in deep reactive ion etching (drie). *Mater. Res. Soc. Symp. Proc.* **2004**, *782*, A1021–A1026.
26. Madou, M.J. *Fundamentals of Microfabrication: The Science of Miniturization*, 2nd ed.; CRC Press: Boca Raton, FL, USA, 2002; p. 723.
27. Senturia, S. *Microsystem Design*, 1st ed.; Kluwer Academic Publishers: Boston, MA, USA, 2001.
28. Zhu, H.; Miao, J.; Chen, B.; Wang, Z.; Zhu, W. Membrane microcantilever arrays fabrication with pzt thin films for nanorange movement. *Microsyst. Technol.* **2005**, *11*, 1121–1126.
29. Sakmen, U.; Stranz, A.; Fandling, S.; Wehmann, H.H.; Bandalo, V.; Bora, A.; Tornow, M.; Waag, A.; Peiner, E. Capabilities of icp-rie cryogenic dry etching of silicon: Review of exemplary microstructures. *J. Micromech. Microeng.* **2009**, *19*, doi:10.1088/0960-1317/19/10/105005.
30. Oliver, A.D.; Wallner, T.A.; Tandon, R.; Nieman, K.; Bergstrom, P.L. Diamond scribing and breaking of silicon for mems die separation. *J. Micromech. Microeng.* **2008**, *18*, doi:10.1088/0960-1317/18/7/075026.
31. Overstolz, T.; Clerc, P.A.; Noell, W.; Zickar, M.; de Rooij, N.F. A clean wafer-scale chip-release process without dicing based on vapor phase etching. In Proceedings of the IEEE International Conference on Micro Electro Mechanical Systems (MEMS), Maastricht, The Netherlands, 25–29 January 2004; pp. 717–720.
32. Sari, I.; Zeimpekis, I.; Kraft, M. A dicing free soi process for mems devices. *Microelectron. Eng.* **2012**, *95*, 121–129.
33. Rezaei, M., Lueke, J., Sameoto, D., Raboud, D., Moussa, A. A new approach to cleave mems devices from silicon substrates. *J. Mech. Eng. Autom.* **2013**, *3*, 731–738.
34. Benfield, D.; Lueke, J.; Moussa, W.A. Mems packaging using flexible printed circuit boards. *CMC Packag. Appl. Note* **2011**.
35. Irwin, J.D.; Wu, C.H. *Basic Engineering Circuit Analysis*, 6th ed.; Prentice-Hall, Inc.: Upper Sadle River, NJ, USA, 1999; p. 976.
36. Ramadass, Y.K.; Chandrakasan, A.P. An efficient piezoelectric energy harvesting interface circuit using a bias-flip rectifier and shared inductor. *IEEE J. Solid State Circuits* **2010**, *45*, 189–204.
37. Qiu, J.; Jiang, H.; Ji, H.; Zhu, K. Comparison between four piezoelectric energy harvesting circuits. *Front. Mech. Eng. China* **2009**, *4*, 153–159.
38. Szarka, G.D.; Stark, B.H.; Burrow, S.G. Review of power conditioning for kinetic energy harvesting systems. *IEEE Trans. Power Electron.* **2012**, *27*, 803–815.
39. Wang, X.; Lin, L. Dimensionless optimization of piezoelectric vibration energy harvesters with different interface circuits. *Smart Mater. Struct.* **2013**, *22*, 085011.

40. Chao, P.C.P. Energy harvesting electronics for vibratory devices in self-powered sensors. *IEEE Sens. J.* **2011**, *11*, 3106–3121.
41. Tang, L.; Yang, Y. Analysis of synchronized charge extraction for piezoelectric energy harvesting. *Smart Mater. Struct.* **2011**, *20*, 1–15.
42. Ottman, G.K.; Bhatt, A.C.; Hoffman, H.; Lesieutre, G.A. Adaptive piezoelectric energy harvesting circuit for wireless remote power supply. *IEEE Trans. Power Electron.* **2002**, *17*, 669–676.
43. Linden, D.; Reddy, T.B. *Handbook of Batteries*; McGraw-Hill: Blacklick, OH, USA, 2001.
44. Roundy, S.; Wright, P.K.; Rabaey, J. A study of low level vibrations as a power source for wireless sensor nodes. *Comput. Commun.* **2003**, *26*, 1131–1144.

9

Optical Detection of Paraoxon Using Single-Walled Carbon Nanotube Films with Attached Organophosphorus Hydrolase-Expressed *Escherichia coli*

Intae Kim [1], Geon Hwee Kim [1], Chang Sup Kim [2], Hyung Joon Cha [3] and Geunbae Lim [1,*]

[1] Department of Mechanical Engineering, Pohang University of Science and Technology, 77 Cheongam-Ro, Nam-Gu, Pohang, Gyeongbuk 790-784, Korea; E-Mails: one@postech.ac.kr (I.K.); q16164545@postech.ac.kr (G.H.K.)

[2] School of Biotechnology and Graduate School of Biochemistry, Yeungnam Univerisity, 280 Daehak-Ro, Gyeongsan, Gyeongbuk 712-749, Korea; E-Mail: cskim1409@ynu.ac.kr

[3] Department of Chemical Engineering, Pohang University of Science and Technology, 77 Cheongam-Ro, Nam-Gu, Pohang, Gyeongbuk 790-784, Korea; E-Mail: hjcha@postech.ac.kr

* Author to whom correspondence should be addressed; limmems@postech.ac.kr

Academic Editors: Gary R. Pickrell and Daniel Homa

Abstract: In whole-cell based biosensors, spectrophotometry is one of the most commonly used methods for detecting organophosphates due to its simplicity and reliability. The sensor performance is directly affected by the cell immobilization method because it determines the amount of cells, the mass transfer rate, and the stability. In this study, we demonstrated that our previously-reported microbe immobilization method, a microbe-attached single-walled carbon nanotube film, can be applied to whole-cell-based organophosphate sensors. This method has many advantages over other whole-cell organophosphate sensors, including high specific activity, quick cell immobilization, and excellent stability. A device with circular electrodes was fabricated for an enlarged cell-immobilization area. *Escherichia coli* expressing organophosphorus hydrolase in the periplasmic space and single-walled carbon nanotubes were attached to the device by our method. Paraoxon was hydrolyzed using this device, and detected by measuring the concentration of the enzymatic reaction product, *p*-nitrophenol. The specific activity of our device was calculated, and was shown to be over 2.5 times that reported previously for other whole-cell organophosphate sensors. Thus, this method for generation of whole-cell-based OP biosensors might be optimal, as it overcomes many of the caveats that prevent the widespread use of other such devices.

Keywords: organophosphates; biosensor; carbon nanotube; microbial immobilization

1. Introduction

Organophosphates (OPs) are widely employed in the agricultural industry as insecticides, however, their long-term toxicity toward both humans and the ecosystem is an important issue worldwide. OP pesticides deactivate the enzyme AChE, which is essential for nerve function in insects and animals, including humans [1,2]. Additionally, OPs are absorbable by all routes, including inhalation, ingestion, and through the dermis. Therefore, the detection and sensing of OPs is important for safety. Traditional methods, such as gas chromatography, high-performance liquid chromatography, and mass spectroscopy, and biological methods, such as immunoassays, have been used for the analysis of OPs [3,4]. However, such analyses are generally performed in the lab rather than in the field, require expensive analytical instruments and expertise, involve several sample preparation steps, and are time-consuming [5]. To overcome these shortcomings, a sensor that is fast and inexpensive with the ability of on-site detection is needed.

Organophosphorus hydrolase (OPH) hydrolyzes a wide range of OP compounds that are used as pesticides and chemical warfare agents. OPH can hydrolyze P-O, P-S, P-F, and P-CN bonds, resulting in a change in pH [6–12]. Hydrolysis products can be analyzed by spectrophotometry or electrochemical detection methods. Because the rate of product formation is proportional to the concentration of OP compound, the OP concentration can be determined by measuring the hydrolysis product [13]. This method has the following advantages over inhibition-based methods: (1) it does not require a continuous supply of the enzyme substrate, because the analyte itself is the substrate; and (2) enzymes such as AChE are inhibited by both OP and carbamate pesticides, rendering OPH more selective, as it only detects OPs [5,14].

Microbes are a common host for enzyme production. However, to obtain usable enzymes, the expressed enzymes need to be purified and, oftentimes, enzymes are expressed in the intracellular space as inclusion bodies. Recently, the secretion of enzymes into the periplasmic space or cell surface has been demonstrated [15,16]. With this technique, enzymes in the periplasmic space or cell surface can come into contact with reactants without the need for a purification process and, thus, microbes can be used as whole-cell catalysts to replace the use of purified enzymes [17–21]. Microbes that express OPH in the periplasmic space or cell surface have also been developed [15,19,22–24], and electrochemical sensors using *Escherichia coli (E. coli)* with surface-expressed OPH have been reported [17,18]. An optical microbial biosensor for the detection of OP compounds using *E. coli* expressing OPH in the periplasmic space has also been reported [21].

A whole-cell biosensor meets the requirements of a convenient OP-sensing tool, and is both more stable and less expensive than enzyme-based biosensors. Amperometric, potentiometric, and spectrophotometric detection methods have been used with OPH-expressing microbial biosensors. Spectrophotometric OP sensors measure the absorption wavelength of *p*-nitrophenol (PNP, 400 nm), which is produced by the hydrolysis of OP compounds [25–28]. Optical methods are advantageous in that they are not affected by electric and magnetic fields, have excellent selectivity, and can be manufactured to be portable by miniaturization of spectrophotometers. The detection limit of these methods using a standard sample container is approximately 5 μM, which, albeit higher than that of amperometric OP sensors, is sufficient for field usage (the rat oral LD_{50} is 1800 μg/kg, or 6.54 μM·L/kg) [29].

To use whole-cells as a biocatalyst, cells must be immobilized to ensure reusability by preventing their detachment. Various microbial immobilization methods, such as flocculation, adsorption, covalent bonding to a carrier, crosslinking of cells, encapsulation in a polymer gel, and entrapment in matrices or biofilms, have been devised and used in macro-scale systems [30–32]. We previously reported a microbe immobilization method, microbe-attached single-walled carbon nanotube (SWNT) film, for microelectromechanical system (MEMS)-based chips to apply microbes in a microfluidic system [33]. This process of microbe immobilization is simple and fast. The film is fabricated site-specifically and has a floating porous structure that enhances the efficiency of enzyme activity. This method is optimized to immobilize microbes over a small space at high density; however, it is not suited for large-scale (larger than several square centimeters) immobilization of microbes due to cost. Other whole-cell OP sensors use a standardized platform, either a 96-well microplate or a disposable cuvette, which have a 50- to 200-μL sample volume [21,25,26]. We expect that our method also has advantages over this sample volume range.

In the present paper, we demonstrate that microbe-attached SWNT film can be applied to whole-cell optical OP sensors and exhibits better performance compared to other whole-cell OP sensors. The device, which has a circular comb drive-shaped electrode, was fabricated for a large area of cell immobilization, and microbe-attached SWNT film was formed between electrodes using *E. coli* that express OPH in the periplasmic space. The concentration of paraoxon was measured using this device by spectrophotometry at 400 nm, and the advantages of this method were demonstrated by comparing it to other whole-cell OP sensors.

2. Experimental Section

2.1. Preparation of Carbon Nanotube (CNT) Solution

Acid treatment of CNTs by liquid-phase oxidation is a commonly used surface modification method for the introduction of carboxylic groups [34–36]. Both the metallic catalyst particles and the amorphous carbon particles are removed by acid treatment. Moreover, CNTs are dispersed by this process, because the introduced carboxylic groups repel each other and exhibit hydrophilic properties.

In this study, a SWNT solution was prepared by oxidation of SWNTs in a strong acid solution, with sonication to generate well-dispersed SWNTs [37,38]. SWNTs, produced by an arc discharge process, were purchased from Hanwha Chemical (Seoul, Korea). Five milligrams of SWNTs were suspended in concentrated H_2SO_4/HNO_3 (50 mL, 3:1, v/v) and sonicated in an ultrasonic bath for 3 h. The SWNT suspension was diluted with deionized (DI) water (2 L) and collected on a membrane filter (Millipore, Darmstadt, Germany) by vacuum filtration. The collected SWNTs were resuspended with 50 mL DI water so that the pH of the suspension was neutral. After centrifuging at 9000 *g* for 5 min, sediments were removed.

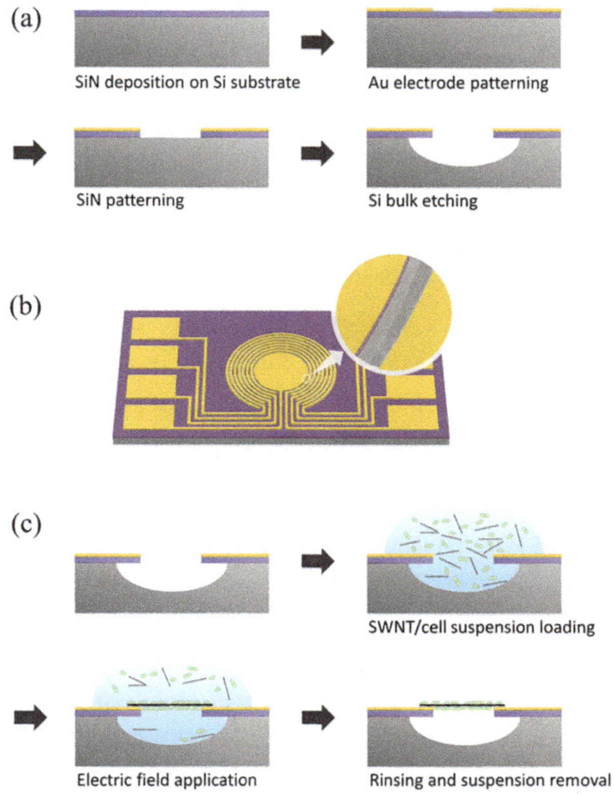

Figure 1. Device Preparation. Schematic representation for (**a**) the fabrication process of the device containing floating electrodes, (**b**) the complete device and an enlarged image of its bulk-etched gap, and (**c**) the fabrication process of *E. coli*-attached SWNT film.

2.2. Preparation of OPH-Expressing E. coli

Recombinant *E. coli* cells expressing both OPH in the periplasm and the chaperone GroEL/ES in the cytoplasm [21,23] were used as the target microorganism. Cells were cultured in Luria-Bertani (LB) medium (0.5% (w/v) yeast extract, 1% (w/v) tryptophan, and 1% (w/v) NaCl) containing 50 μg/mL ampicillin (Sigma-Aldrich, St. Louis, MO, USA) and 25 μg/mL chloramphenicol (Sigma-Aldrich). For efficient expression of recombinant OPH and GroEL/ES, 1 mM (final concentration) isopropyl-β-D-thiogalactopyranoside (IPTG; Sigma-Aldrich), 10 ng/mL tetracycline, and 0.5 mM $CoCl_2$ were added when the culture reached a cell density (OD_{600}) of 1.2. The recombinant cells were grown for an additional 12 h at 37 °C. The cells were harvested by centrifugation at 4000 *g* for 10 min. The harvested cells were resuspended in Tris-HCl buffer (pH 8.0) at various densities and diluted tenfold with SWNT suspension.

2.3. Fabrication Process for the Cantilever Electrodes

Figure 1a shows a schematic diagram of the fabrication process for the cantilever electrodes, which were fabricated using a MEMS. The cantilever structure was formed by depositing a low-stress nitride on a silicon (Si) substrate using low-pressure chemical vapor deposition (LPCVD). Gold (Au) electrodes were patterned on a silicon nitride (SiN) surface by a lift-off process (Cr/Au 20/200 nm). To construct the cantilever structure, the SiN was patterned by reactive-ion etching (RIE), and the Si was etched by isotropic wet etching. Lastly, an insulating layer was patterned on the Au electrodes and SiN surface using SU8 photoresist (MicroChem, Newton, MA, USA).

2.4. Fabrication of E. coli-Attached SWNT Film

Figure 1c shows the fabrication process of *E. coli*-attached SWNT film. Recombinant *E. coli* cells expressing OPH in the periplasm were used as a whole cell catalyst. The cell density of the suspension was adjusted to OD_{600} = 16. To prepare the *E. coli*-mixed SWNT suspension, the cell suspension and SWNT solution were mixed at a ratio of 1:9 (v/v). To fabricate the *E. coli*-attached SWNT film, an *E. coli*-mixed SWNT suspension was placed between the suspended electrodes, and an AC voltage with an amplitude of 7 V_{PP} and a frequency of 1 MHz was applied between the electrodes for 10 s. Then, SWNTs and *E. coli* were aligned between electrodes by dielectrophoresis force. The device was then rinsed in Tris-HCl buffer solution (pH 8.0, 10 mM).

2.5. Fabrication of Paraoxon-Sensing Platform

The platform of a microbe-attached SWNT film biosensor consists of four parts (Figure 2): a sensor holder, a sensor device, a chamber, and a test clip. The sensor holder was made with polyether ether ketone (PEEK) by a mechanical machining process. PEEK exhibits excellent mechanical and chemical resistance properties that are retained at high temperature. Furthermore, PEEK is an excellent insulator. These properties are suitable for the sensor holder material. The sensor holder has a shallow trench to fix the sensor device, and is pitted on both sides to connect a test clip. A chamber is needed to place the sample to the sensor surface, in which *E. coli* is immobilized. The chamber consists of three parts: a body to connect the chamber to the sensor holder, a tube to store the sample, and a polydimethylsiloxane (PDMS) gasket to seal between the chamber and the sensor device. The chamber was designed for a 500 μL sample and is autoclavable. The test clip was adopted to allow for an easy connection between electrodes in the sensor device.

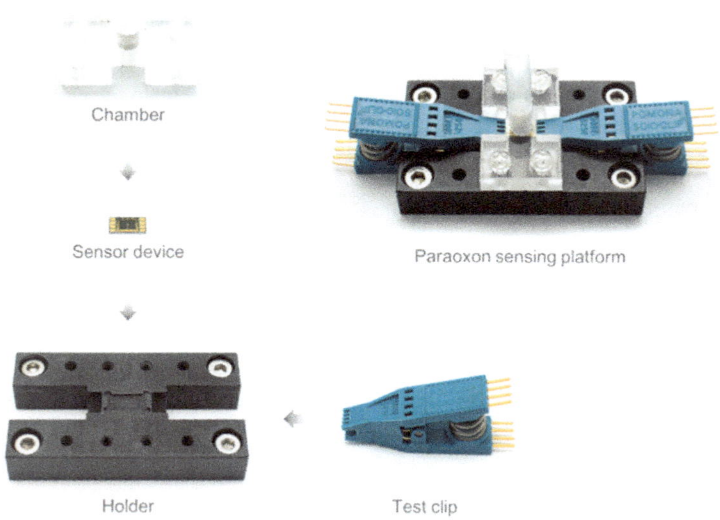

Figure 2. Fabrication process of a platform for a whole-cell OP sensor using *E. coli*-attached SWNT film.

2.6. Paraoxon Measurement Setup

Paraoxon solution was prepared at various concentration (20–500 μM) in 100 mM 2-(*N*-cyclohexylamino)ethanesulfonic acid (CHES) buffer (pH 9.0; Sigma-Aldrich). Paraoxon is hydrolyzed to *p*-nitrophenol (PNP) and diethylphosphoric acid by OPH. Therefore, the catalytic reaction of OPH can be measured by determining PNP density. Paraoxon (500 μL) was added to the chamber (Figure 2) and reacted for 50 min at 25 °C. The PNP density of the solution was then determined using a spectrophotometer to measure the absorbance of the solution at 400 nm (Mecasys, Daejeon, Korea), which is the absorption wavelength of PNP produced by the hydrolysis of OP compounds.

3. Results and Discussion

3.1. E. coli-Attached SWNT Film Fabrication

For the spectrophotometric whole-cell based OP sensor, immobilization of a sufficient amount of cells is required. Our cell immobilization method requires two cantilever electrodes that have an appropriate gap size (5–10 μm) for inducing an even dielectrophoresis (DEP) force. The area of cells immobilized is proportional to the area between the two cantilever electrodes. In our previous study, we fabricated a device containing two electrodes of 20-μm width and a 5-μm gap between the two electrodes [33]. In this study, we designed eight cantilever electrodes that have a circular comb drive shape for a larger area of cell immobilization than in the previous device (Figure 2). Eight Au electrodes, having widths of 20 μm, were placed parallel to each other with a 5-μm gap between adjacent electrodes. Consequentially, the area of the gaps between electrodes that was able to form the microbe-attached SWNT film was 0.043 mm^2, and the total area that contains the eight electrodes was 0.26 mm^2.

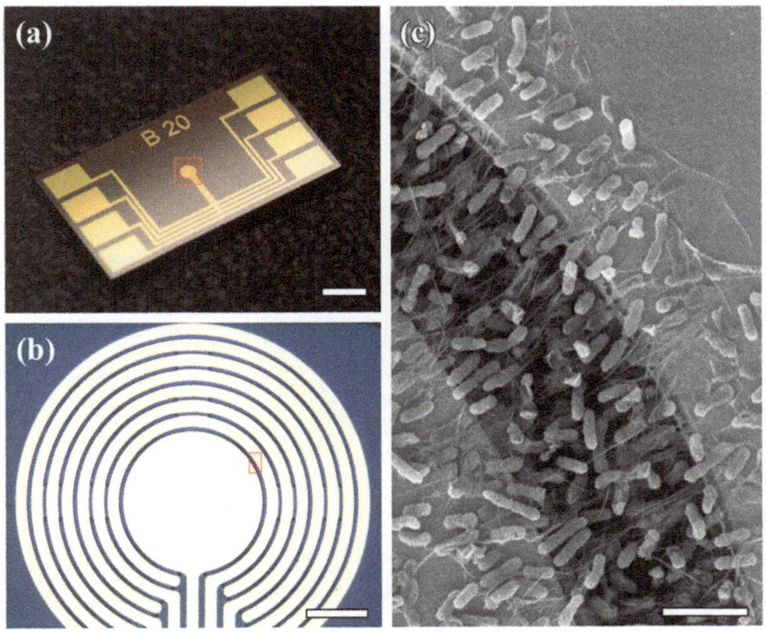

Figure 3. Device Imaging. Optical image of (**a**) the entire device (scale bar: 2 mm) and (**b**) a magnified view of the circular comb drive-shaped electrodes (scale bar: 50 μm). (**c**) SEM image of *E. coli*-attached SWNT film (scale bar: 5 μm). The red square in (a) indicates the image area of (b), and the red square in (b) indicates the image area of (c).

To fabricate an *E. coli*-attached SWNT film on the device, an AC voltage with an amplitude of 7 V_{PP} and a frequency of 1 MHz was applied between the two electrodes, across the mixed SWNT/*E. coli* suspension, for 10 s. This process was repeated in sequence from the innermost electrodes to the outermost electrodes to fabricate *E. coli*-attached SWNT film in every gap between the electrodes. Figure 3c shows a scanning electron microscopy (SEM) image of *E. coli*-attached SWNT film. Two cantilever electrodes are connected with SWNTs, and cells are attached to the SWNTs. Because the SWNTs are a one-dimensional structure, the attachment area of cells is minimized, while the reaction area of cells is maximized. In addition, *E. coli*-attached SWNT film is floated from the bottom of the substrate and contains spaces between cells that can pass substances. Therefore, mass transfer of the floated film is enhanced compared to the film attached to the floor of the substrate.

3.2. Structural Advantage of E. coli-Attached SWNT Film

For the optical density-based whole-cell OP sensor, it is advantageous to maximize the enzyme activity per unit area in which cells are immobilized. Our immobilization method has advantages in this aspect. First, cells can be immobilized at a high density. When the electric field conditions are 6 V_{pp} and 1 MHz over 5 s, and the *E. coli* density (OD_{600}) is 0.2, the cell density of the film is 0.619 cells/ μm^2 [33]. Although this value is not directly comparable with other studies, it is clear that our method can immobilize cells at a high density, considering the fact that the two-dimensional area of *E. coli* is about 1 μm^2. Second, the efficiency of enzymatic activity is increased. This is because mass transfer is enhanced by the flotation from the substrate and porous structure of the film and the reaction area of the cell is maximized by attaching cells to SWNTs of a one-dimensional structure.

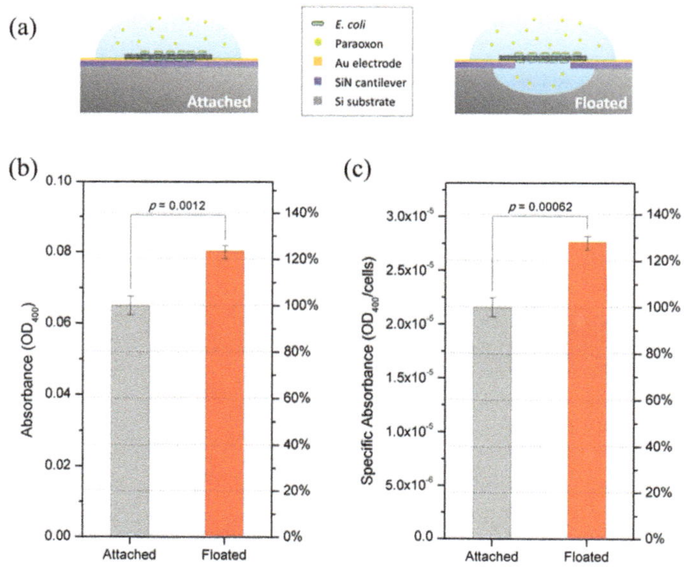

Figure 4. Device Activity. (**a**) Schematic representation of attached and floating *E. coli*-attached SWNT films. The floating film device has space for floating the *E. coli*-attached SWNT film from the surface of the device. On the other hand, the device for attached film did not undergo the etching process of silicon and silicon nitride to produce the film on the surface of the device; (**b**,**c**) show the comparison of enzymatic activity between the attached and floating films (b) per device and (c) per cell. Error bars indicate standard deviations.

To demonstrate the structural advantage of *E. coli*-attached SWNT film, two kinds of devices were prepared. They have the same electrode pattern, but the device for attached film did not undergo SiN etching and Si bulk etching in the fabrication process as did the device for floating film (Figure 4a). Each device was designed to have seven pairs of facing electrodes, with a width of 100 μm and a gap size of 5 um. To fabricate *E. coli*-attached SWNT film on the two devices, an AC voltage with amplitude of 7 V_{PP} and frequency of 1 MHz was applied between the two electrodes, across the mixed SWNT/*E. coli* suspension (OD_{600} = 0.4), for 5 s. Figure 4b,c show the activity of *E. coli*-attached SWNT films, comparing the one floated from the surface to the one attached on the surface of the substrate. *E. coli*-attached SWNT film that is floated showed activity that was improved by 23% over the attached one (Figure 4b). Because a greater number of cells are incorporated when the film is attached on the surface of the device compared to the floated one, the activity per cell was calculated. *E. coli*-attached SWNT film that is floated showed a 27% improvement in activity over the attached one (Figure 4c).

3.3. Paraoxon Measurement and Comparison with Other Whole-Cell OP Sensors

We measured the paraoxon concentration using *E. coli*-attached SWNT film by a spectrophotometric method, and compared our results with other whole-cell based OP sensors. Paraoxon was detected by measuring absorbance at 400 nm. Each density of paraoxon solution (500 μL, 100 mM CHES buffer, pH 9.0) was reacted and measured. Figure 5 shows quantitative detection using various concentrations of paraoxon: 20, 50, 100, 300, and 500 μM. Error bars indicate standard deviations. The absorbance increased linearly as the paraoxon concentration was increased. The coefficient of determination has a high value (R^2 = 0.9936) in the tested range of paraoxon concentrations.

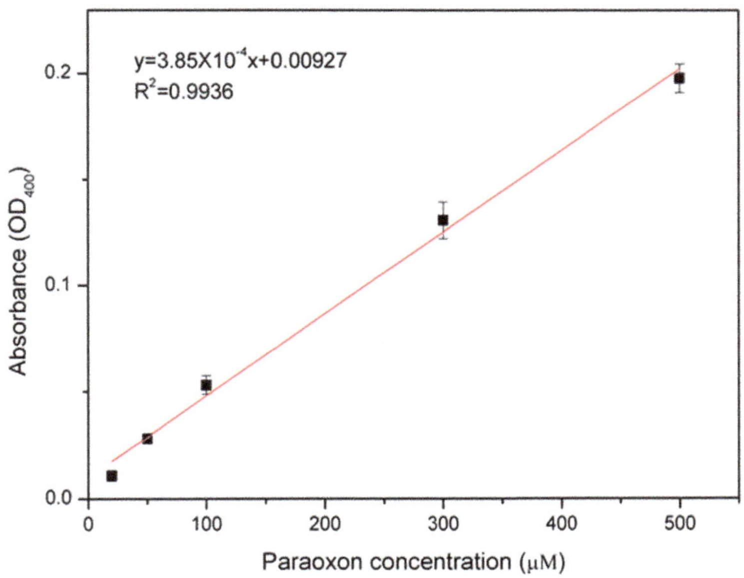

Figure 5. Detection of paraoxon using *E. coli*-attached SWNT film. Error bars indicate standard deviations. The absorbance increased linearly as the paraoxon concentration was increased.

In the absorbance-based whole-cell biosensor, enzyme activity per unit area is an important factor related to sensor performance. High enzyme activity per unit area can lead to enhanced sensitivity, faster measurement time, and a reduction in sample volume. For comparison with other studies on whole-cell based OP sensors, we defined enzyme activity per unit cell-immobilized area as specific activity (μmol·min^{-1}·mm^{-2}) and calculated its value when the paraoxon concentration is 100 μM. Specific activity is defined as:

$$Specific\ activity = c \times V \times \frac{1}{t} \times \frac{1}{a} \quad (1)$$

where c is the concentration of PNP, V is the sample volume, t is the enzymatic reaction time, and a is the area of immobilized cells. Because $A = \varepsilon cl$ (A is the absorbance, ε is the molar absorption coefficient, and l is the light path length) by the Beer-Lambert law, specific activity is conversed as follows:

$$Specific\ activity = \frac{AV}{\varepsilon lta} \quad (2)$$

The ε of PNP is 17,000 M^{-1}·cm^{-1}, and the other parameters are determined by the experimental methods of each study. Table 1 shows a comparison of whole-cell based OP sensors, including the one developed in this study, and our device shows the highest specific activity. The specific activity of our device shown in this study is more than 2.5 times the best previously reported result [25]. This is due to the ability of *E. coli*-attached SWNT film to immobilize cells at a high concentration and enhance the enzymatic activity due to structural advantages. Our immobilization method requires only 10 s for each pair of electrodes, and is able to be fabricated consecutively, so the required time for cell immobilization can be represented as: required time = (number of electrode pairs) ×10 s + (rinsing time). In this study, we used a device that has seven pairs of electrodes, so the total required time for cell immobilization is less than 100 s. The time for cell immobilization reported in other studies includes only the steps whose exact time duration is mentioned, and does not include the time required for drying, solution changing, and sample preparation and manipulation. In addition, our method exhibits the highest long-term storage stability. These results suggest that *E. coli*-attached SWNT film exhibits excellent performance, reliability, and a fast fabrication process compared to previously-reported cell immobilization methods for whole-cell based OP sensors.

Table 1. Characteristic comparison between this study and other whole-cell based OP sensors.

Immobilization Method	Cell Type	Substrate	Specific Activity (μmol·min^{-1}·mm^{-2})	Time for Cell Immobilization	Long-Term Storage Stability	Reference
MAP-based adhesion	*E. coli* BL21	Paraoxon	19.06	>2 h	28 d (80%)	[21]
Glutaraldehyde-based chemical crosslinking	*Sphingomonas* sp. JK1	4(methyl parathion)	43.18	>1 h	18 d (80%)	[25]
Glutaraldehyde-based chemical crosslinking	*Sphingomonas* sp. JK1	4(methyl parathion)	43.12	>1 h	32 d (90%)	[26]
Adsorption by Van der Waals force	*E. coli* BL21	Paraoxon	119.91	<100 s	37 d (90%), 67 d (80%)	This study

Table 2 shows the immobilized cell area required to achieve the performance of previously-reported whole-cell based OP sensors [21,25,26] using typical spectrophotometric sensing platforms. The detection limit and time were chosen as 5 μM and 5 min, respectively, because other whole-cell OP sensors are 4–5 μM and 5 min, respectively. The absorbance was chosen as OD_{400} = 0.01, because the noise level of conventional spectrophotometers is 0.0001–0.001 at OD_{400} = 0.5. We calculated the required area of immobilized cells from Equation (2), and the specific activity was adjusted to the concentration of paraoxon, because specific activity is proportional to the concentration of substrate. Most absorbance-based analyses use a spectrophotometer or microplate reader. Commercially-available equipment and common sample containers were selected for Table 2. Disposable cuvettes and 96-well microplates are the most commonly used, because they allow for multi-sample analysis and are disposable, standardized, and cheap. The required area for a disposable cuvette is 1.374 mm^2, and that for a 96-well microplate is 6.083 mm^2. If a larger area is applied to cell immobilization, higher performance can be expected compared to other whole-cell OP sensors.

Table 2. The cell immobilization area required to detect 5 μM of paraoxon in 5 min using typical spectrophotometric sensing platforms.

Equipment	Conventional Spectrophotometer	Conventional Spectrophotometer	Microplate Reader	Microvolume Spectrophotometer
Sample container	cuvette (disposable)	cuvette (quartz)	96-well microplate	Directly on the equipment
Sample volume (μL)	70	50	200	0.5
Light path length (mm)	10	10	6.5	0.5
Required area (mm^2)	1.374	0.981	6.038	0.196

4. Conclusions

In this study, we fabricated *E. coli*-attached SWNT film to immobilize cells on a MEMS-based chip, and demonstrated that *E. coli*-attached SWNT film can be applied as a spectrophotometric OP sensor. *E. coli*-attached SWNT film is a unique cell immobilization method for MEMS-based chips and has several advantages, including site-specificity, speed, and stability. As the sample volume is small, our method is the only way to immobilize microbes. However, the sample volume is greater than the approximately 50 μL that several methods, including chemical surface modification, crosslinking, and adhesive material coating, require. The important consideration is that the sample volume of 96-well microplates and cuvettes is 50–200 μL, and these standardized sample containers are used not only for optical OP sensors but also for most biological analyses. So, we designed a device that has circular electrodes for a larger immobilization area than our previously-reported device for larger sample volume. For comparison with other whole-cell OP sensors, the concentration of paraoxon was measured using our film, and the specific activity was calculated. The results showed that our method has 2.5 times the specific activity shown by other whole-cell OP sensors. This suggests that *E. coli*-attached SWNT film exhibits excellent performance compared to other methods. In addition, our method requires less time for cell immobilization and has better long-term stability.

We demonstrated that our microbe-attached SWNT film can be used not only for microfluidic systems, but also for 96-well microplates and cuvettes, by applying OP detection. We believe our method is optimal for enhancing performance and reducing the time required for cell immobilization in bio-application fields using microbes.

Acknowledgments

This work (2012R1A2A2A06047424) was supported by Mid-career Researcher Program through NRF grant funded by the MSIP

Author Contributions

I. Kim wrote the manuscript, and also designed and performed the experiments and data analysis. G.H. Kim and C.S. Kim performed the experiments and discussed the results. H.J. Cha contributed to discussion of results and gave a keen insight in this manuscript. G. Lim arranged all the work in this project.

Conflicts of Interest

The authors declare no conflict of interest.

References

1. Cremisini, C.; di Sario, S.; Mela, J.; Pilloton, R.; Palleschi, G. Evaluation of the use of free and immobilised acetylcholinesterase for paraoxon detection with an amperometric choline oxidase based biosensor. *Anal. Chim. Acta* **1995**, *311*, 273–280.
2. Zhang, S.; Zhao, H.; John, R. Development of a quantitative relationship between inhibition percentage and both incubation time and inhibitor concentration for inhibition biosensors—Theoretical and practical considerations. *Biosens. Bioelectron.* **2001**, *16*, 1119–1126.
3. Sherma, J. Pesticides. *Anal. Chem.* **1993**, *65*, 40R–54R.
4. Miller, J.K.; Lenz, D.E. Development of an immunoassay for diagnosis of exposure to toxic organophosphorus compounds. *J. Appl. Toxicol.* **2001**, *21*, S23–S26.
5. Musameh, M.M.; Gao, Y.; Hickey, M.; Kyratzis, I.L. Application of Carbon Nanotubes in the Extraction and Electrochemical Detection of Organophosphate Pesticides: A Review. *Anal. Lett.* **2012**, *45*, 783–803.
6. Ghanem, E.; Raushel, F.M. Detoxification of organophosphate nerve agents by bacterial phosphotriesterase. *Toxicol. Appl. Pharmacol.* **2005**, *207*, 459–470.
7. Cho, C.M.-H.; Mulchandani, A.; Chen, W. Functional analysis of organophosphorus hydrolase variants with high degradation activity towards organophosphate pesticides. *Protein Eng. Des. Sel.* **2006**, *19*, 99–105.
8. Mello, S.V.; Coutures, C.; Leblanc, R.M.; Cheng, T.-C.; Rastogi, V.K.; DeFrank, J.J. Interaction between organophosphorous hydrolase and paraoxon studied by surface chemistry in situ at air–water interface. *Talanta* **2001**, *55*, 881–887.
9. Aubert, S.D.; Li, Y.; Raushel, F.M. Mechanism for the Hydrolysis of Organophosphates by the Bacterial Phosphotriesterase. *Biochemistry (Mosc.)* **2004**, *43*, 5707–5715.
10. Dumas, D.P.; Caldwell, S.R.; Wild, J.R.; Raushel, F.M. Purification and properties of the phosphotriesterase from Pseudomonas diminuta. *J. Biol. Chem.* **1989**, *264*, 19659–19665.
11. Wong, K.-Y.; Gao, J. The Reaction Mechanism of Paraoxon Hydrolysis by Phosphotriesterase from Combined QM/MM Simulations†. *Biochemistry (Mosc.)* **2007**, *46*, 13352–13369.

12. Zheng, F.; Zhan, C.-G.; Ornstein, R.L. Theoretical studies of reaction pathways and energy barriers for alkaline hydrolysis of phosphotriesterase substrates paraoxon and related toxic phosphofluoridate nerve agents. *J. Chem. Soc. Perkin Trans 2* **2001**, 2355–2363.
13. Singh, A.K.; Flounders, A.W.; Volponi, J.V.; Ashley, C.S.; Wally, K.; Schoeniger, J.S. Development of sensors for direct detection of organophosphates. Part I: Immobilization, characterization and stabilization of acetylcholinesterase and organophosphate hydrolase on silica supports. *Biosens. Bioelectron.* **1999**, *14*, 703–713.
14. Van Dyk, J.S.; Pletschke, B. Review on the use of enzymes for the detection of organochlorine, organophosphate and carbamate pesticides in the environment. *Chemosphere* **2011**, *82*, 291–307.
15. Kang, D.G.; Lim, G.-B.; Cha, H.J. Functional periplasmic secretion of organophosphorous hydrolase using the twin-arginine translocation pathway in Escherichia coli. *J. Biotechnol.* **2005**, *118*, 379–385.
16. Choi, J.H.; Lee, S.Y. Secretory and extracellular production of recombinant proteins using Escherichia coli. *Appl. Microbiol. Biotechnol.* **2004**, *64*, 625–635.
17. Mulchandani, A.; Mulchandani, P.; Kaneva, I.; Chen, W. Biosensor for Direct Determination of Organophosphate Nerve Agents Using Recombinant Escherichia coli with Surface-Expressed Organophosphorus Hydrolase. 1. Potentiometric Microbial Electrode. *Anal. Chem.* **1998**, *70*, 4140–4145.
18. Mulchandani, A.; Kaneva, I.; Chen, W. Biosensor for Direct Determination of Organophosphate Nerve Agents Using Recombinant Escherichia coli with Surface-Expressed Organophosphorus Hydrolase. 2. Fiber-Optic Microbial Biosensor. *Anal. Chem.* **1998**, *70*, 5042–5046.
19. Richins, R.D.; Kaneva, I.; Mulchandani, A.; Chen, W. Biodegradation of organophosphorus pesticides by surface-expressed organophosphorus hydrolase. *Nat. Biotech.* **1997**, *15*, 984–987.
20. Shimazu, M.; Mulchandani, A.; Chen, W. Simultaneous degradation of organophosphorus pesticides and p-nitrophenol by a genetically engineered Moraxella sp. with surface-expressed organophosphorus hydrolase. *Biotechnol. Bioeng.* **2001**, *76*, 318–324.
21. Kim, C.S.; Choi, B.-H.; Seo, J.H.; Lim, G.; Cha, H.J. Mussel adhesive protein-based whole cell array biosensor for detection of organophosphorus compounds. *Biosens. Bioelectron.* **2013**, *41*, 199–204.
22. Kang, D.G.; Choi, S.S.; Cha, H.J. Enhanced Biodegradation of Toxic Organophosphate Compounds Using Recombinant Escherichia coli with Sec Pathway-Driven Periplasmic Secretion of Organophosphorus Hydrolase. *Biotechnol. Prog.* **2006**, *22*, 406–410.
23. Kang, D.G.; Kim, C.S.; Seo, J.H.; Kim, I.G.; Choi, S.S.; Ha, J.H.; Nam, S.W.; Lim, G.; Cha, H.J. Coexpression of molecular chaperone enhances activity and export of organophosphorus hydrolase in Escherichia coli. *Biotechnol. Prog.* **2012**, *28*, 925–930.
24. Kim, C.S.; Seo, J.H.; Kang, D.G.; Cha, H.J. Engineered whole-cell biocatalyst-based detoxification and detection of neurotoxic organophosphate compounds. *Biotechnol. Adv.* **2014**, *32*, 652–662.
25. Kumar, J.; D'Souza, S.F. An optical microbial biosensor for detection of methyl parathion using Sphingomonas sp. immobilized on microplate as a reusable biocomponent. *Biosens. Bioelectron.* **2010**, *26*, 1292–1296.
26. Kumar, J.; D'Souza, S.F. Immobilization of microbial cells on inner epidermis of onion bulb scale for biosensor application. *Biosens. Bioelectron.* **2011**, *26*, 4399–4404.

27. Kumar, J.; Jha, S.K.; D'Souza, S.F. Optical microbial biosensor for detection of methyl parathion pesticide using Flavobacterium sp. whole cells adsorbed on glass fiber filters as disposable biocomponent. *Biosens. Bioelectron.* **2006**, *21*, 2100–2105.
28. Michelini, E.; Roda, A. Staying alive: New perspectives on cell immobilization for biosensing purposes. *Anal. Bioanal. Chem.* **2012**, *402*, 1785–1797.
29. Richard, J.; Lewis, S. Sax's Dangerous Properties of Industrial Materials; Wiley: Hoboken, NJ, USA, 2012.
30. Cassidy, M.B.; Lee, H.; Trevors, J.T. Environmental applications of immobilized microbial cells: A review. *J. Ind. Microbiol.* **1996**, *16*, 79–101.
31. Park, J.K.; Chang, H.N. Microencapsulation of microbial cells. *Biotechnol. Adv.* **2000**, *18*, 303–319.
32. Costerton, J.W.; Lewandowski, Z.; Caldwell, D.E.; Korber, D.R.; Lappin-Scott, H.M. Microbial Biofilms. *Annu. Rev. Microbiol.* **1995**, *49*, 711–745.
33. Kim, I.; An, T.; Choi, W.; Kim, C.S.; Cha, H.J.; Lim, G. Site-specific immobilization of microbes using carbon nanotubes and dielectrophoretic force for microfluidic applications. *RSC Adv.* **2013**, *4*, 1347–1351.
34. Kanai, Y.; Khalap, V.R.; Collins, P.G.; Grossman, J.C. Atomistic Oxidation Mechanism of a Carbon Nanotube in Nitric Acid. *Phys. Rev. Lett.* **2010**, *104*, doi:10.1103/PhysRevLett.104.066401.
35. Lee, G.-W.; Kumar, S. Dispersion of Nitric Acid-Treated SWNTs in Organic Solvents and Solvent Mixtures. *J. Phys. Chem. B* **2005**, *109*, 17128–17133.
36. Marshall, M.W.; Popa-Nita, S.; Shapter, J.G. Measurement of functionalised carbon nanotube carboxylic acid groups using a simple chemical process. *Carbon* **2006**, *44*, 1137–1141.
37. An, T.; Kim, K.S.; Hahn, S.K.; Lim, G. Real-time, step-wise, electrical detection of protein molecules using dielectrophoretically aligned SWNT-film FET aptasensors. *Lab. Chip* **2010**, *10*, 2052–2056.
38. Cang-Rong, J.T.; Pastorin, G. The influence of carbon nanotubes on enzyme activity and structure: Investigation of different immobilization procedures through enzyme kinetics and circular dichroism studies. *Nanotechnology* **2009**, *20*, doi:10.1088/0957-4484/20/25/255102.

10

Strong Ferromagnetically-Coupled Spin-Valve Sensor Devices for Droplet Magnetofluidics

Gungun Lin [1,2,*], Denys Makarov [1] and Oliver G. Schmidt [1,2]

[1] Institute for Integrative Nanosciences, IFW Dresden, Helmholtzstr. 20, 01069 Dresden, Germany; E-Mails: d.makarov@ifw-dresden.de (D.M.); o.schmidt@ifw-dresden.de (O.G.S.)

[2] Material Systems for Nanoelectronics, Chemnitz University of Technology, Reichenhainerstr. 70, 09107 Chemnitz, Germany

* Author to whom correspondence should be addressed; E-Mail: g.lin@ifw-dresden.de

Academic Editor: Andreas Hütten

Abstract: We report a magnetofluidic device with integrated strong ferromagnetically-coupled and hysteresis-free spin valve sensors for dynamic monitoring of ferrofluid droplets in microfluidics. The strong ferromagnetic coupling between the free layer and the pinned layer of spin valve sensors is achieved by reducing the spacer thickness, while the hysteresis of the free layer is eliminated by the interplay between shape anisotropy and the strength of coupling. The increased ferromagnetic coupling field up to the remarkable 70 Oe, which is five-times larger than conventional solutions, brings key advantages for dynamic sensing, e.g., a larger biasing field giving rise to larger detection signals, facilitating the operation of devices without saturation of the sensors. Studies on the fundamental effects of an external magnetic field on the evolution of the shape of droplets, as enabled by the non-visual monitoring capability of the device, provides crucial information for future development of a magnetofluidic device for multiplexed assays.

Keywords: droplet microfluidics; spin valve; ferromagnetic coupling; high field sensing; ferrofluid

---

1. Introduction

Modern medical research calls for the development of new technologies or platforms. Droplet microfluidics is a prominent example of such new platforms and has picked up momentum in recent years [1]. Droplet microfluidics bears great potential for the practical realization and replacement of conventional medical instrumentations for, e.g., protein [2] and DNA assays [3], drug discovery [4] and cell studies [5]. Concurrently, the recent advances in droplet micro-magnetofluidics have witnessed growing interest in the utilization of magnetic materials [6,7]. For instance, droplets encapsulating ferrofluids allow on-chip manipulation of biochemical species [8] and remote actuation of liquid by coupling with an external magnetic field [9,10]. Liposome capsules or vesicles associated with magnetic particles have been broadly studied for targeted drug delivery [11,12]. Furthermore, magnetic particles have been utilized as droplet barcodes, either providing an additional functionality of manipulation [13,14] or enhancing the encoding capacity [15] for multiplexed assays. Hence, developing on-chip magnetic field sensing technologies in droplet microfluidic systems to actively sense and analyze droplets containing various amounts of magnetic materials is highly demanded and desirable.

Magnetoresistive sensing technologies have been adopted in biosensing applications since the end of last century [16]. Major advantages of the employment of magnetoresistive sensorics are associated with compact integration, direct signal readouts, the capability to detect magnetic labels, as well as providing a feedback control for on-chip actuation with magnetic materials. On-chip magnetoresistive sensing reveals great potential in microarray technologies [17,18], which relies on static detection of magnetic labels bound to the substrate's surface for molecular recognition [19]. Recently, there has been a trend towards in-flow detection of magnetically-functionalized objects in microfluidics [20], which is superior in terms of throughput compared with the conventional static detection schemes [21] when objects of interest are immobilized at the sensor's location for the detection event. While sensing of magnetic particles, beads or cells in the flow is rather advanced already [22], the development of magnetoresistive sensing technologies for droplet microfluidic systems is still in its infancy. To date, droplet-based micro-magnetofluidic systems have employed spin valve sensors [23] and giant magnetoresistive (GMR) multilayers [15,24,25] to determine the size and magnetic content of droplets.

Spin valve sensors are an advanced type of GMR sensorics due to its large sensitivity to small magnetic fields. However, spin valve sensors are commonly applied as switches, and they are characterized with a hysteresis that is not suitable for dynamic sensing in fluidics due to its vulnerability to random magnetic fields and magnetic noise [26]. Typical approaches to eliminate the hysteretic response of spin valves involve biasing a free layer's magnetization with shape anisotropy, which is orthogonal to a pinned layer's magnetization set by exchange bias [27] or designing a sensor with a yoke shape geometry to eliminate domain walls [26]. Alternatively, an external magnetic field is used to stabilize the free layer's magnetization during the measurements [28]. The free layer and the pinned layer are usually weakly coupled, so as to bring the sensing point to a near zero magnetic field [29]. In this respect, in order to detect superparamagnetic nanoparticles or beads, which are frequently used as the carrier materials for labeling or manipulation of species in fluidics, an external out-of-plane magnetic field is applied to induce net magnetic moments. Nonetheless, particularly for applications, such as to measure the size of ferrofluid droplets by a magnetofluidic device [23,24], an in-plane-dominant magnetic field is typically employed. In such cases, a stronger in-plane biasing field is crucial to achieve larger detection signals. The low field sensitive feature of spin valve sensors is vulnerable to saturation during the field biasing procedure and limits the further increase of detection signals. Therefore, a magnetic field sensor with a larger biasing point is advantageous and facilities the operation of the device in practical settings.

In this article, we report a magnetofluidic platform for in-flow detection of ferrofluid droplets in microfluidics. For this purpose, we optimize the performance of the integrated spin valve sensor device to be hysteresis free and to possess a biasing point shifted to a remarkable 70 Oe, which is at least five-times larger than applied previously [23]. A strong ferromagnetic coupling between the free layer and the pinned layer of spin valve sensors is achieved by reducing the spacer thickness between pinned and free layers of the spin valve stack, while a reduced hysteresis is achieved by tailoring the shape anisotropy and the coupling strength that is crucial for on-chip dynamic sensing. The optimized spin valve sensors integrated in a microfluidic device allow the detection of ferrofluid droplets with about six-times enhanced signal responses compared to the commonly-used GMR multilayers with the same measurement configuration [24]. In addition, the magnetofluidic device provides insights into the fundamental effects of an external magnetic field impact on the shape evolution of droplets in microfluidics. In turn, the fundamental understanding provides crucial information for the future design of micro-magnetofluidic devices for multiplexed magnetic barcoding assays.

2. Experimental Section

2.1. Fabrication of Spin Valve Sensors

The spin valve sensors are prepared by magnetron sputter deposition on thermally-oxidized silicon wafers. The fabrication process can be described as follows: Firstly, the substrate was cleaned with acetone and isopropanol, rinsed by DI water and placed on a hotplate at 120 °C to get rid of the absorbed vapor. Then, the surface of the substrate was patterned with photoresist by photolithography. For the lithography process, a photoresist (AZ5214E, Microchems) was coated on the substrate. The spin speed and time are 4500 rpm and 35 s. After the coating, the substrate was transferred on to a hotplate and baked at 90 °C for 4 min to remove residual solvents. Then, the substrate was transferred to a mask aligner (Karl Suss, MJB4) and exposed to UV light for 2 s. A photomask with designed chromium patterns of sensor geometries was used. Subsequently, the wafer was placed on a hotplate again, which was set at 120 °C for 2 min. Afterwards, it was exposed at the mask aligner without a photomask for 30 s and developed by a developer (MIF 726, Microchems) for 1 min to remove the uncrosslinked photoresist.

The substrate with patterned photoresist was introduced into the chamber with a base pressure 9.5×10^{-8} mbar, where the deposition of the GMR layer stack was carried out. The layer stack used in this study is: SiO_x/Ta (5 nm)/Py (4 nm)/CoFe (1 nm)/Cu (t nm)/CoFe (1 nm)/Py (4 nm)/IrMn (8 nm)/Ta (2 nm), with t the thickness of the Cu spacer and Py = $Ni_{81}Fe_{19}$. Argon was used as the sputter gas, with a sputter pressure of 9.5×10^{-4} mbar and a flow rate of 10 sccm kept constant during the deposition. To induce exchange bias in spin valve sensors, an external in-plane magnetic field of about 0.5 kOe was applied during the deposition by means of arrangements of permanent magnets behind the sample holder. The thickness of each deposited layer of spin valve sensors is controlled by adjusting the deposition time with pre-calibrated deposition rates. After the deposition of the sensor stack, a lift-off process was used to remove the photoresist. Then, another photolithography is performed to align the electrical contacts with the sensors. In this case, Ta (5 nm)/Cu (200 nm)/Ta (5 nm) were deposited as the electrical interconnects by magnetron sputtering.

2.2. Assembly of Microfluidic Devices

The microfluidic channel was fabricated by casting PDMS (SYLGARD 184, 1:10 wt% for curing agent and base polymers) with a SU-8 50 (Microchem) mold. The mold was prepared by coating the photoresist SU-8 50 on a silicon wafer at a spin speed of 1000 rpm for 35 s leading to the height of the channel of 100 μm. The photoresist SU-8 50 was cured at 60 °C for 5 min and at 90 °C for 15 min. Then, the wafer with cured SU-8 50 was illuminated by UV light with a mask aligner (MJB4, Karl Suss) and baked at 90 ° C for 15 min. The channel geometry was revealed by immersing the wafer in a solvent (mr-Dev 600, micro resist technology GmbH) to remove unexposed areas. The wafer was cleaned by isopropanol and dried by compressed air.

Before the assembly of the channel and the sensor chip, a layer of polyetherimide (PEI) of a thickness of about 300 nm was spin coated on top of the sensors as an insulation layers. Afterwards, a layer of SiO_2 of 200 nm was deposited on top of the PEI by e-beam evaporation in order to facilitate bonding with the PDMS channel. To achieve permanent bonding, the PDMS microfluidic channel and the sensor chip were activated in an oxygen plasma (40 mW, O_2 flow rate: 20 sccm) for 30 s. They were brought in touch and aligned with each other under a microscope.

2.3. On-Chip Formation of Ferrofluid Droplets

The microfluidic channel is composed of a T-junction to produce droplets. Ferrofluid magnetic nanoparticles (EMG 700 series, purchased from Ferrotec) are used as a disperse phase, and mineral oil (Sigma-Aldrich, M8410) with 5% surfactant (SPAN 80) are used as a continuous phase. The ferrofluids are dispersions of superparamagnetic nanoparticles with a nominal size of 10 nm. Different concentrations (5.0 mg/mL and 7.5 mg/mL) of ferrofluids were prepared by diluting the original stock solutions with DI water. The oil and ferrofluid particles are separately loaded in two syringes and pumped into the microfluidic channel by a syringe pump (Cetoni GmbH, NEMESYS). Water-in-oil emulsion droplets encapsulating ferrofluid particles are thus produced at a T-junction directly on-chip. For the present study, the flow rates of oil and ferrofluids were kept in total at 15 nL/s with a flow speed of droplets about 1 mm/s. To produce droplets of different sizes (volume: from 200 pL to a few nanoliters), the ratio of the flow rates of the two different fluids was adjusted accordingly.

2.4. Magnetoelectrical Characterizations of Spin Valve Sensors

For magnetoelectrical characterizations, a sample holder with four pins was used to measure the resistance of sensors by the standard four-point method. The sensor was clamped in the holder, which was placed in between the pole-shoes of an electromagnet. An external uniform magnetic field was applied on the sample, and the sample resistance was recorded for each magnetic field by a multimeter (Keithley Model 2000). For the measurement, the magnetic field was cycled between ±300 Oe to measure the full range GMR curves.

2.5. Real-Time Electrical Measurements of Droplets in Microfluidics

An electrical measurement setup was built up to detect droplets in the microfluidic channel. A lock-in amplifier (SRS 830) was used to produce AC measuring current of about 1 mA to a Wheatstone bridge circuit. The bridge circuit was formed by the spin valve sensor for the detection of droplets together with other resistors. The Wheatstone bridge allows us to reduce the background signal level, so that the sensitivity can be enhanced. For signal amplification, the differential voltage of the bridge was picked up by the lock-in amplifier. In this case, a reference sinusoidal signal was used to modulate the AC voltage signal at a high frequency up to 1 kHz to increase the signal-to-noise ratio. After signal processing by the lock-in amplifier, the analog output voltage was fed into a data acquisition box (NI-USB 6009) with a sampling rate of 5 kHz. As the ferrofluid is made of superparamagnetic particles, an external magnetic field was applied to magnetize the particles by using a permanent magnet placed below the sensor. The position of the magnet was carefully adjusted to bias the sensor to the most sensitive region via monitoring the real-time detection signals of droplets by a LabVIEW program.

3. Results and Discussion

3.1. Optimization of Spin Valve Sensors

To achieve a hysteresis-free sensor response, the magnetization of the free layer should reverse via a coherent rotation process [27]. The strength of interlayer exchange coupling, as well as the anisotropy configuration play a crucial role in the magnetization reversal mode of spin valve sensors [30]. The former is determined by the thickness of the spacer layer of spin valve sensors [31,32]. To optimize the coupling strength aiming at the increase of the biasing point of the sensor as needed for dynamic detection, we prepared a series of spin valve sensors with different thicknesses of the Cu spacer in the range from 1.8 nm to 3.0 nm (Figure 1a). For the experiments on the sensor optimization, the layer stack was patterned into rectangular stripes with a size of 1×16 mm^2. The exchange bias direction of the sensors was set during the deposition, which is either along the long axis or the short axis of the sensor stripes (Figure 1b). The interlayer exchange coupling between the free layer and the pinned layer thus creates an induced uniaxial anisotropy, which is parallel to the pinning direction (the exchange bias direction). On the other hand, the large aspect ratio of the sensor stripe induces a shape anisotropy along the long axis of the sensor stripe. Thus, according to the arrangements of the exchange bias direction with respect to the shape anisotropy, two configurations of spin valve sensors with crossed or parallel anisotropies can be realized.

The GMR curves of spin valve sensors with the two different anisotropy configurations are shown in Figure 1c. With the reduction of the thickness of Cu from 3.0 nm to 1.8 nm, the exchange coupling strength, as characterized by the shift of the free layer's response curves, of both series of spin valve sensors increases (Figure 2a). As the shift of the free layer's response curves is opposite the pinning direction (in the negative direction of the magnetic field), the exchange coupling is ferromagnetic. It was observed that the ferromagnetic coupling fields for the sensors pinned along the short axis of the stripes (crossed anisotropy configuration) are smaller than those pinned along the long axis of the stripes (parallel configuration), which can be attributed to magnetic stringing fields emanating from the pinned layer.

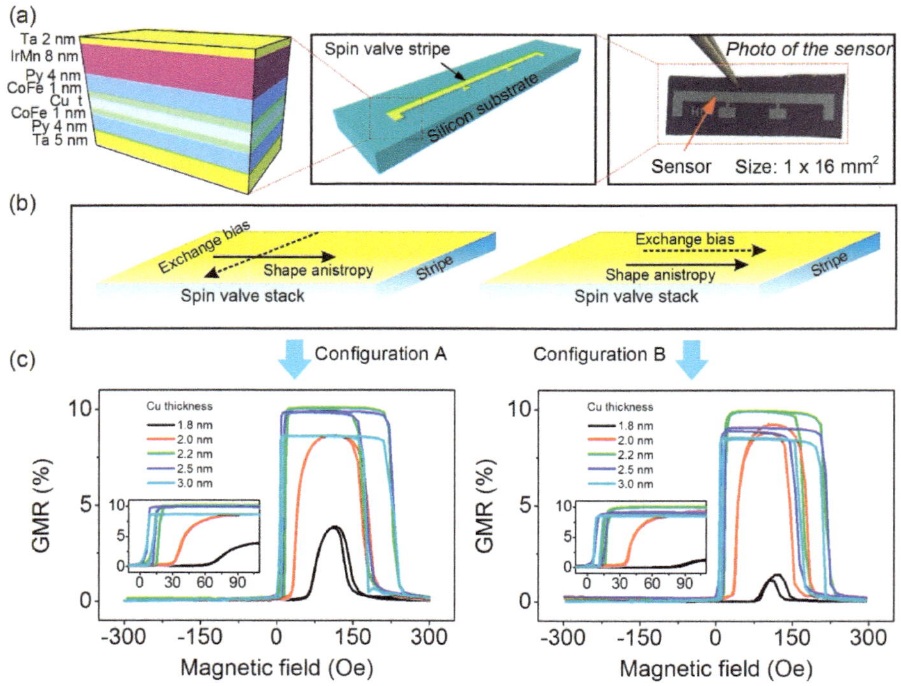

Figure 1. (**a**) Schematic representation of the layer stack of spin valve sensors (left) and a sensor stripe on a silicon substrate (middle), as well as the photograph of a spin valve sensor patterned with a size of 1 × 16 mm² (right); (**b**) Schematics of two anisotropy configurations of spin valve sensors, which are achieved by setting the direction of exchange bias with respect to shape anisotropy. For Configuration A (crossed anisotropy), the exchange bias is along the short axis of the sensor stripe, and hence, it is perpendicular to the direction of the shape anisotropy. For Configuration B (parallel anisotropy), it is along the long axis of the sensor stripe parallel to the direction of the shape anisotropy; (**c**) Giant magnetoresistive (GMR) curves for the spin valve sensors prepared with crossed and parallel anisotropy configurations. The spin valve sensors are of the same layer stack structure as shown in (a) and the thickness of Cu ranges from 1.8 nm to 3.0 nm. Inset: The GMR response of the free layer. The x- and y-axis correspond to those in the main figure.

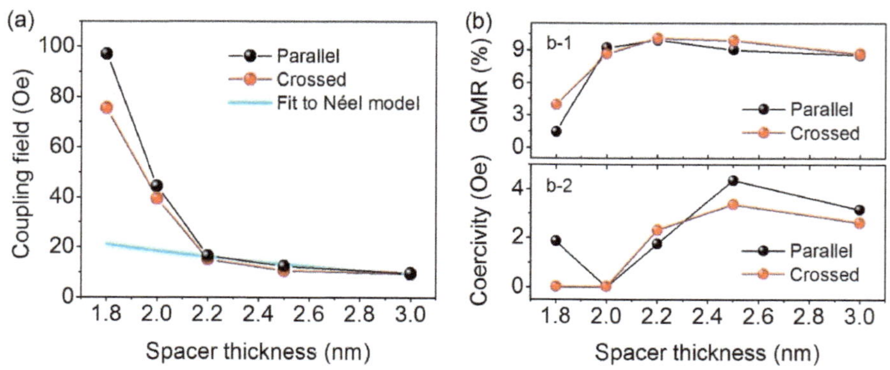

Figure 2. (**a**) Ferromagnetic coupling field as a function of the spacer thickness for spin valve sensors with the parallel anisotropy configuration, as well as the fit of the data to the Néel model. Lines connecting the symbols are a guide to the eyes; (**b**) GMR ratio (b-1), as well as the coercivity of the free layer (b-2) as a function of the spacer thickness for spin valve sensors of parallel and crossed anisotropy configurations.

The ferromagnetic coupling field typically arises from two contributions: orange peel coupling and direct coupling through pinholes. The orange peel coupling can be described in the frame of the Néel model [33]:

$$H_{coupling}=M\frac{\pi^2}{\sqrt{2}}\frac{\gamma^2}{\lambda}\frac{1}{t_F}\exp\left(\frac{-2\pi\sqrt{2}t_s}{\lambda}\right) \quad (1)$$

with M being the magnetization of the pinned layer, t_F and t_s being the thickness of the free layer and the spacer layer, λ and γ being the wavelength and the magnitude of correlated roughness, respectively. The Néel model can be successfully applied to fit to the experimental data with spacer thickness ≥2.2 nm (Figure 2a) for sensors designed with a parallel anisotropy configuration (without stringing field coupling). A wavelength $\lambda \sim 13$ nm and a roughness $\gamma \sim 0.23$ nm are determined, which are close to the same parameters extracted from similar spin valve sensors [34]. For Cu thicknesses smaller than 2.2 nm, a large deviation from the Néel fit is observed. Hence, the strong coupling field could be ascribed to the formation of pinholes in the Cu spacer [34].

With the increased ferromagnetic coupling strength, a hysteresis-free sensor response of the free layer is achieved (Figure 2b). Although characterized by a smaller GMR ratio (Figure 2b-1) compared with other hysteresis-free sensors of larger spacer thickness (2 nm), the spin valve sensor with the Cu thickness 1.8 nm and crossed anisotropy shows the largest coupling field (about 70 Oe) and a rather broad sensing range. These features are advantageous for magnetic in-flow detection, *i.e.*, a large external magnetic field can be applied to magnetize the particles to induce a larger magnetic moment and, thus, to enhance the detected signals without saturating the sensor.

3.2. Dynamic Monitoring of Droplets in Microfluidics by Spin Valve Sensors

For magnetic detection, the optimized spin valve sensor of a crossed anisotropy configuration and Cu thickness of 1.8 nm was integrated in a microfluidic channel (Figure 3a). It has been demonstrated previously [24,35] that the real-time detection of the peak patterns of droplets loaded with ferrofluids using a GMR sensor carries the signature of the emulsion droplets. The signature can be resolved in terms of the signal amplitude and peak width to analyze encapsulated magnetic content, as well as droplet size, respectively. A magnetoresistive emulsion analyzer [35] with the size of the GMR sensor along the fluid flow direction being 1 mm is capable of analyzing bivariant droplets, but with the size limited to larger than 1 mm (volume: about 110 nL). This limitation is due to the fact that the peak width is no longer sensitive to the change of droplet size when the droplet is smaller than the sensor size. To address the need to analyze bivariant nanoliter or even picoliter-droplets in microfluidics, we fabricated spin valve sensors with a size of 6×100 μm^2 (6 μm along the fluid flow direction), which is smaller than the length of droplets typically produced in microfluidics. The aspect ratio of the integrated spin valve sensors is close to the above sensors used for optimization; hence, similar performance is expected. The transfer curve of the integrated spin valve sensor (Figure 3b) shows that the sensor retains a GMR ratio of about 3.5% close to the optimized spin valve stack of the same spacer thickness and possesses a maximum sensitivity of about 0.08%/Oe at a large field of about 70 Oe. The sensitivity of the sensor is comparable to previously reported spin valve sensors (with a sensitivity of 0.077%/Oe) employed for the detection of ferrofluid droplets [23]. However, the ferromagnetic coupling field is about five-times larger than the reported value of about 15 Oe [23].

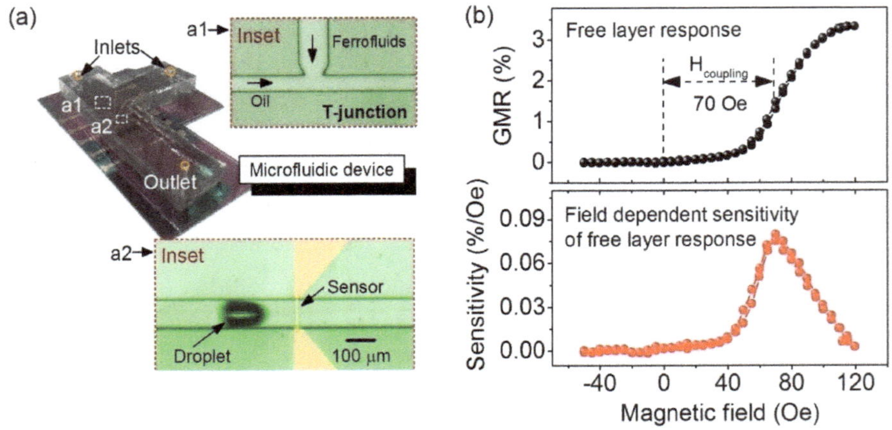

Figure 3. (**a**) Photograph of an assembled microfluidic device with an integrated on-chip T-junction (a1) and spin valve sensors (a2). (a1) and (a2) are micrographs corresponding to the dashed squares indicated on the photograph of the device; (**b**) GMR response of the free layer of an integrated spin valve sensor with a spacer thickness of 1.8 nm with a crossed anisotropy configuration (top). The sensitivity is plotted as a function of the magnetic field (bottom).

Figure 4a shows the real-time detection results of ferrofluid emulsion droplets produced with different lengths. Each peak represents the detection event of an emulsion droplet. Only local stray fields are detected by the sensor. The local maximum and minimum of the detection peaks are attributed to the local maximum and minimum of the stray field distribution around the sensor. The amplitude of the detection signals, e.g., while measuring droplets of a volume ~1 nL containing ferrofluids of 7.5 mg/mL is about six-times that of a previously reported microfluidic device (about 5 μV) with integrated GMR multilayers in detecting the same amount of ferrofluids based on the same measurement configuration, which possesses a maximum sensitivity of 0.4%/Oe at a low biasing field of 12 Oe [24].

The temporal interval between the rising and falling edges of detection signals is referred to as the detection peak width. The values of the peak width can be robustly extracted by a peak search algorithm without artifacts [36]. Figure 4b shows that the length of droplets (with a volume ranging from 200 pL to a few nanoliters) is proportional to the detection peak width, which is consistent with previously reported results [24]. In contrast to the expectation, the signal amplitude changes with the droplet length, namely we observe that it increases with the increase of the droplet length with a clear trend to saturate at the higher range. The droplet length of >100 μm along the travel direction is larger than the sensor size (6 μm). Hence, the change of the signal amplitude with the droplet length is ascribed to the stretching of droplets by an external magnetic force, leading to the shrink of droplets, as determined by optical measurements of the droplet width transverse to the droplet travel direction. The external magnetic force results from an external magnetic field used to magnetize the superparamagnetic nanoparticles (Figure 4a,b). The effect of droplet deformation under an external magnetic field has been studied previously on sessile microliter-droplets [37]. It was shown that the deformation is determined by the balance of magnetic surface force, gravitational force and surface tension. Surface tension is an intrinsic property of a liquid's interface with another contacting medium, which is referred to as the interfacial force per length or interfacial energy per area (density). In microfluidic measurements with nanoliter-droplets, magnetic force and interfacial energy are the main contributions to the final shape of droplets. With a constant external magnetic field of about 70 Oe, relatively large magnetic force is exerted on the droplets carrying larger concentrations of ferrofluids thus, giving rise to substantial shrinkage of the droplets (in-plane) perpendicular to the droplet travel direction. This is evidenced by the decrease of the droplet width when the concentration of ferrofluids is increased from 5.0 mg/mL to 7.5 mg/mL (Figure 4c). It is to be noted that the initial surface area of longer droplets is larger than the smaller ones. The interfacial energy to be overcome in order to stretch larger droplets to an equal extent is larger than smaller droplets. This mechanism has led to the effect of lateral shrinkage being almost extinguished for longer droplets (Figure 4c).

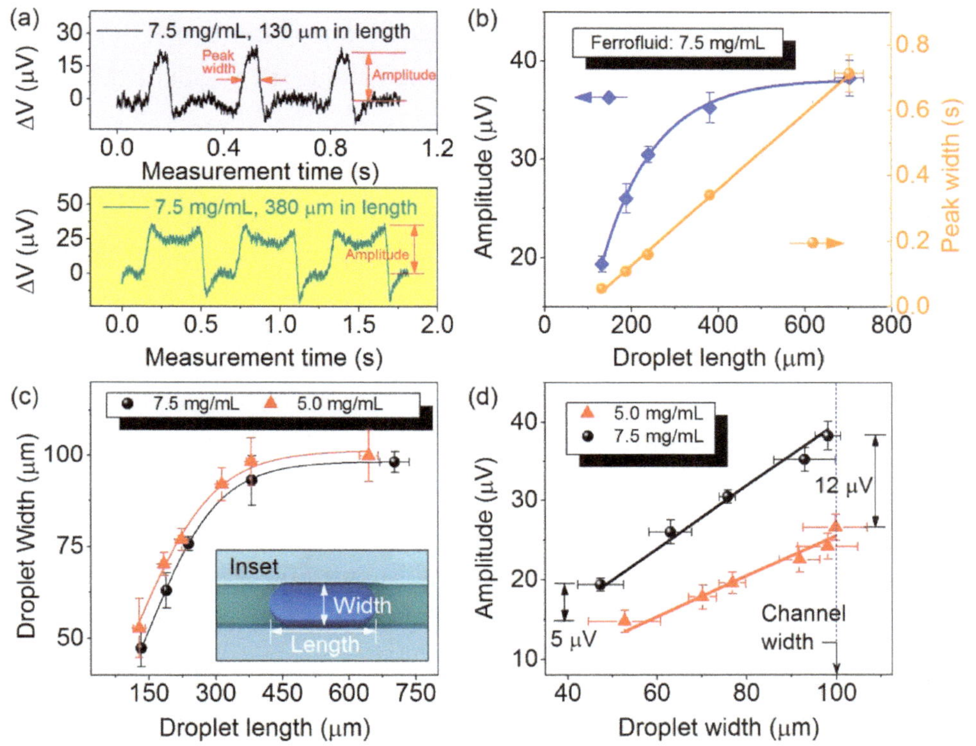

Figure 4. (**a**) Real-time detection of emulsion droplets with different lengths (130 μm and 380 μm), but the same concentration (7.5 mg/mL) of loaded ferrofluids; (**b**) The detection signal amplitude (square) and peak width (sphere) as a function of droplet length (with volume ranging from 200 nL to a few nanoliters). Lines are a guide to the eye; (**c**) The droplet width as a function of the droplet length for different concentrations of encapsulated ferrofluids. Lines are a guide to the eye. The inset depicts the schematics to measure the width and length of droplets; (**d**) The signal amplitude as a function of the droplet width for different concentrations of encapsulated ferrofluids. Lines are linear fittings to the data. Error bars are standard deviations of measured data.

Figure 4d shows a linear relationship of the signal amplitude with the droplet width (Figure 4d). The ratio of the slope of the two curves (1.56 ± 0.05) obtained from droplets encapsulating ferrofluids of 5.0 mg/mL and 7.5 mg/mL can be determined by linear fits to the data. The value agrees well with the ratio of the different concentrations of ferrofluids encapsulated in the droplets (equal to 1.5). Considering the fact that magnetic stray fields decay quickly over distance, as well as the distance of the sensor (about 500 nm) is far smaller than the height of the channel (100 μm), only the magnetic stray fields generated from the lower part of the droplets (close to the location of the sensor) contribute mainly to the final detected signals. Thus, the detected signals are less dependent on the whole volume of droplets, but rather the surface coverage of magnetic nanoparticles, which is analogous to the linear dependence of detection signals on the surface coverage of magnetic particles by a GMR sensor in a static detection scheme [28]. In other words, the measured signal is determined by the area of the sensor exposed to the magnetic field stemming from the magnetized droplet. As droplets are close to the sensor, the area exposed to the field can be estimated by the area of the sensor covered by the droplet. Because the sensor along the flow direction is only 6 μm, it is completely covered by the droplets. Therefore, the surface coverage of particles is determined by the width of droplets, which is perpendicular to the flow direction. In agreement with this discussion and based on our experimental data (Figure 4d), the signal amplitude can be given by $\Delta V \sim c \cdot w$, with c the concentration of ferrofluids and w the width of the droplet, namely the size perpendicular to the travel direction.

The fundamental effect of the lateral shrinkage of droplets under an external applied magnetic field is crucial for magnetofluidic applications based on multiparametric analytics relying on magnetic droplet barcodes [36]. As the extent of shrinkage is a function of the droplet length and encapsulated concentration of ferrofluids, analysis based on two parameters, such as the amplitude and the peak width, to analyze the size and concentration of encapsulated magnetic content is still feasible, but needs a prior calibration with optical measurements of the size. The effect determines the minimum volume of droplets that can be still distinguished between droplet groups of two different concentrations. For instance, the difference in the signal amplitude between two groups of droplets carrying ferrofluids of 5.0 mg/mL and 7.5 mg/mL drops from 12 μV to 5 μV (Figure 4d), which may eventually cause the overlap of signal amplitudes when the size of droplets becomes even smaller (about 100 μm in length). Reducing the external magnetic field may alleviate this effect, but this will result in the reduction of the signal amplitude. However, our results suggest that the lateral shrinkage of droplets under an external magnetic field can be extinguished when droplets become longer (>500 μm) (Figure 4c) and only the lower part of droplets is contributing to the detected signals. Thus, the fundamental understanding of these influencing factors provides a solution for future development of a magnetofluidic platform to analyze and distinguish droplets of even smaller volumes, that is by integrating GMR sensors into a microfluidic channel with an even smaller cross-section to geometrically elongate droplets into longer plugs in order to avoid this issue.

4. Conclusions

We designed a microfluidic droplet analyzer based on a hysteresis-free spin valve sensor with a strong ferromagnetic coupling field. The coupling field of the spin valve sensor is 70 Oe, which is about five-times larger than that of conventional hysteresis-free spin valve sensors and GMR multilayers used for magnetic in-flow detection. The strongly ferromagnetically-coupled spin valve sensors with reduced hysteresis bring key advantages for magnetic in-flow detection: (i) a large in-plane/out-of-plane magnetic field can be applied to magnetize ferrofluid particles to induce large magnetic moments for sensing; (ii) it facilitates the operation of the device without easily saturating the sensor; and (iii) it allows improving the signal-to-noise ratio. We showed that with this device, the signal-to-noise ratio can be enhanced by six-times that of state-of-the-art realizations based on GMR multilayers. The device has been able to analyze bivariant droplets produced in microfluidics and to resolve the evolution of droplet size (volume: from ~200 pL to a few nanoliters) under an external applied magnetic field of up to 70 Oe. The studies are crucial for the development of a new generation of magnetoresistance sensors, such as tunneling magnetoresistance sensors for high-field magnetic sensing, and the future design of a magnetofluidic platform for highly-sensitive detection and analysis of ferrofluid emulsion droplets.

Acknowledgments

We thank Irina Fiering (IFW Dresden) for support in sample deposition and Dmitriy Karnaushenko (IFW Dresden) for development of the PEI encapsulation layer. The work is financially supported in part by the DFG Research Group for 1713 and the European Research Council under the European Union's Seventh Framework Programme (FP7/2007-2013)/ERC Grant Agreement No. 306277.

Author Contributions

G. Lin performed the experiments and analyzed the data. G. Lin and D. Makarov wrote the paper with input from O.G. Schmidt.

Conflicts of Interest

The authors declare no conflict of interest.

References

1. Mazutis, L.; Gilbert, J.; Ung, W.L.; Weitz, D.A.; Griffiths, A.D.; Heyman, J.A. Single-cell analysis and sorting using droplet-based microfluidics. *Nat. Protoc.* **2013**, *8*, 870–891.
2. Teste, B.; Ali-Cherif, A.; Viovy, J.L.; Malaquin, L. A low cost and high throughput magnetic bead-based immuno-agglutination assay in confined droplets. *Lab Chip* **2013**, *13*, 2344–2349.
3. Abate, A.R.; Hung, T.; Sperling, R.A.; Mary, P.; Rotem, A.; Agresti, J.J.; Weiner, M.A.; Weitz, D.A. DNA sequence analysis with droplet-based microfluidics. *Lab Chip* **2013**, *13*, 4864–4869.
4. Miller, O.J.; El Harrak, A.; Mangeat, T.; Baret, J.-C.; Frenz, L.; El Debs, B.; Mayot, E.; Samuels, M.L.; Rooney, E.K.; Dieu, P.; *et al.* High-resolution dose-response screening using droplet-based microfluidics. *Proc. Natl. Acad. Sci. USA* **2012**, *109*, 378–383.
5. Brouzes, E.; Medkova, M.; Savenelli, N.; Marran, D.; Twardowski, M.; Hutchison, J.B.; Rothberg, J.M.; Link, D.R.; Perrimon, N.; Samuels, M.L. Droplet microfluidic technology for single-cell high-throughput screening. *Proc. Natl. Acad. Sci. USA* **2009**, *106*, 14195–14200.
6. Nguyen, N.-T. Micro-magnetofluidics: Interactions between magnetism and fluid flow on the microscale. *Microfluid. Nanofluidics* **2011**, *12*, 1–16.
7. Misuk, V.; Mai, A.; Giannopoulos, K.; Alobaid, F.; Epple, B.; Loewe, H. Micro magnetofluidics: Droplet manipulation of double emulsions based on paramagnetic ionic liquids. *Lab Chip* **2013**, *13*, 4542–4548.
8. Sander, J.S.; Erb, R.M.; Denier, C.; Studart, A.R. Magnetic transport, mixing and release of cargo with tailored nanoliter droplets. *Adv. Mater.* **2012**, *24*, 2582–2587.

9. Nguyen, N.-T.; Ng, K.M.; Huang, X. Manipulation of ferrofluid droplets using planar coils. *Appl. Phys. Lett.* **2006**, *89*, doi:10.1063/1.2335403.
10. Kurtoğlu, E.; Bilgin, A.; Şeşen, M.; Mısırlıoğlu, B.; Yıldız, M.; Acar, H.F.Y.; Koşar, A. Ferrofluid actuation with varying magnetic fields for micropumping applications. *Microfluid. Nanofluidics* **2012**, *13*, 683–694.
11. Sabaté, R.; Barnadas-Rodríguez, R.; Callejas-Fernández, J.; Hidalgo-Alvarez, R.; Estelrich, J. Preparation and characterization of extruded magnetoliposomes. *Int. J. Pharm.* **2008**, *347*, 156–162.
12. Martina, M.-S.; Fortin, J.-P.; Ménager, C.; Clément, O.; Barratt, G.; Grabielle-Madelmont, C.; Gazeau, F.; Cabuil, V.; Lesieur, S. Generation of superparamagnetic liposomes revealed as highly efficient MRI contrast agents for in vivo imaging. *J. Am. Chem. Soc.* **2005**, *127*, 10676–10685.
13. Zhao, Y.; Xie, Z.; Gu, H.; Jin, L.; Zhao, X.; Wang, B.; Gu, Z. Multifunctional photonic crystal barcodes from microfluidics. *NPG Asia Mater.* **2012**, *4*, doi:10.1038/am.2012.46.
14. Zhao, Y.; Shum, H.C.; Chen, H.; Adams, L.L.A.; Gu, Z.; Weitz, D.A. Microfluidic generation of multifunctional quantum dot barcode particles. *J. Am. Chem. Soc.* **2011**, *133*, 8790–8793.
15. Lin, G.; Makarov, D.; Medina-Sánchez, M.; Guix, M.; Baraban, L.; Cuniberti, G.; Schmidt, O.G. Magnetofluidic platform for multidimensional magnetic and optical barcoding of droplets. *Lab Chip* **2015**, *15*, 216–224.
16. Baselt, D.R.; Lee, G.U.; Natesan, M.; Metzger, S.W.; Sheehan, P.E.; Colton, R.J. A biosensor based on magnetoresistance technology. *Biosens. Bioelectron.* **1998**, *13*, 731–739.
17. Gaster, R.S.; Hall, D.A.; Wang, S.X. nanoLAB: An ultraportable, handheld diagnostic laboratory for global health. *Lab Chip* **2011**, *11*, 950–956.
18. Gaster, R.S.; Hall, D.A.; Nielsen, C.H.; Osterfeld, S.J.; Yu, H.; Mach, K.E.; Wilson, R.J.; Murmann, B.; Liao, J.C.; Gambhir, S.S.; *et al*. Matrix-insensitive protein assays push the limits of biosensors in medicine. *Nat. Med.* **2009**, *15*, 1327–1332.
19. Osterfeld, S.J.; Yu, H.; Gaster, R.S.; Caramuta, S.; Xu, L.; Han, S.-J.; Hall, D.A.; Wilson, R.J.; Sun, S.; White, R.L.; *et al*. Multiplex protein assays based on real-time magnetic nanotag sensing. *Proc. Natl. Acad. Sci. USA* **2008**, *105*, 20637–20640.
20. Loureiro, J.; Ferreira, R.; Cardoso, S.; Freitas, P.P.; Germano, J.; Fermon, C.; Arrias, G.; Pannetier-Lecoeur, M.; Rivadulla, F.; Rivas, J. Toward a magnetoresistive chip cytometer: Integrated detection of magnetic beads flowing at cm/s velocities in microfluidic channels. *Appl. Phys. Lett.* **2009**, *95*, doi:10.1063/1.3182791.
21. Fernandes, A.; Duarte, C.; Cardoso, F.; Bexiga, R.; Cardoso, S.; Freitas, P. Lab-on-Chip Cytometry Based on Magnetoresistive Sensors for Bacteria Detection in Milk. *Sensors* **2014**, *14*, 15496–15524.
22. Loureiro, J.; Andrade, P.Z.; Cardoso, S.; da Silva, C.L.; Cabral, J.M.; Freitas, P.P. Magnetoresistive chip cytometer. *Lab Chip* **2011**, *11*, 2255–2261.
23. Pekas, N.; Porter, M.D.; Tondra, M.; Popple, A.; Jander, A. Giant magnetoresistance monitoring of magnetic picodroplets in an integrated microfluidic system. *Appl. Phys. Lett.* **2004**, *85*, doi:10.1063/1.1825059
24. Lin, G.; Makarov, D.; Melzer, M.; Si, W.; Yan, C.; Schmidt, O.G. A highly flexible and compact magnetoresistive analytic device. *Lab Chip* **2014**, *14*, 4050–4058.

25. Mönch, I.; Makarov, D.; Koseva, R.; Baraban, L.; Karnaushenko, D.; Kaiser, C.; Arndt, K.-F.; Schmidt, O.G. Rolled-up magnetic sensor: Nanomembrane architecture for in-flow detection of magnetic objects. *ACS Nano* **2011**, *5*, 7436–7442.
26. Fermon, C.; Pannetier-Lecoeur, M.; Biziere, N.; Cousin, B. Optimised GMR sensors for low and high frequencies applications. *Sens. Actuators A Phys.* **2006**, *129*, 203–206.
27. Buschow, K.H.J. *Handbook of Magnetic Materials*; Elsevier: Amsterdam, the Netherlands, 2003; Volume 15.
28. Wang, S.X. Advances in Giant Magnetoresistance Biosensors With Magnetic Nanoparticle Tags: Review and Outlook. *IEEE Trans. Magn.* **2008**, *44*, 1687–1702.
29. Tamanaha, C.R.; Mulvaney, S.P.; Rife, J.C.; Whitman, L.J. Magnetic labeling, detection, and system integration. *Biosens. Bioelectron.* **2008**, *24*, 1–13.
30. Labrune, M.; Kools, J.C.S.; Thiaville, A. Magnetization rotation in spin-valve multilayers. *J. Magn. Magn. Mater.* **1997**, *171*, 1–15.
31. Parkin, S. Giant magnetoresistance in magnetic nanostructures. *Annu. Rev. Mater. Sci.* **1995**, *25*, 357–388.
32. Dieny, B. Giant magnetoresistance in spin-valve multilayers. *J. Magn. Magn. Mater.* **1994**, *8853*, 335–359.
33. Lee, C.-L.; Bain, J.A.; Chu, S.; McHenry, M.E. Separation of contributions to spin valve interlayer exchange coupling field by temperature dependent coupling field measurements. *J. Appl. Phys.* **2002**, *91*, doi:10.1063/1.1451598.
34. Teixeira, J.M.; Ventura, J.O.; Fermento, R.P.; Araujo, J.P.; Sousa, J.B.; Freitas, S.C.; Freitas, P.J. Interlayer Coupling and Magnetoresistance of MnIr-Based Spin Valves: Dependencies on Deposition Rate, Spacer Thickness, and Temperature. *IEEE Trans. Magn.* **2007**, *43*, 3143–3145.
35. Lin, G.; Baraban, L.; Han, L.; Karnaushenko, D.; Makarov, D.; Cuniberti, G.; Schmidt, O.G. Magnetoresistive emulsion analyzer. *Sci. Rep.* **2013**, *3*, doi:10.1038/srep02548.
36. Lin, G.; Fomin, V.M.; Denys, M.; Schmidt, O.G. Supervised discriminant analysis for droplet-based micro-magnetofluidics. *Microfluid. Nanofluidics* **2015**, Accepted.
37. Zhu, G.-P.; Nguyen, N.-T.; Ramanujan, R.V; Huang, X.-Y. Nonlinear deformation of a ferrofluid droplet in a uniform magnetic field. *Langmuir* **2011**, *27*, 14834–14841.

11

Novel Monitoring Techniques for Characterizing Frictional Interfaces in the Laboratory

Paul A. Selvadurai * and Steven D. Glaser

University of California, Berkeley, CA 94720, USA; E-Mail: glaser@berkeley.edu

* Author to whom correspondence should be addressed; E-Mail: pa.selvadurai@berkeley.edu

Academic Editor: Thomas Schumacher

Abstract: A pressure-sensitive film was used to characterize the asperity contacts along a polymethyl methacrylate (PMMA) interface in the laboratory. The film has structural health monitoring (SHM) applications for flanges and other precision fittings and train rail condition monitoring. To calibrate the film, simple spherical indentation tests were performed and validated against a finite element model (FEM) to compare normal stress profiles. Experimental measurements of the normal stress profiles were within -7.7% to 6.6% of the numerical calculations between 12 and 50 MPa asperity normal stress. The film also possessed the capability of quantifying surface roughness, an important parameter when examining wear and attrition in SHM applications. A high definition video camera supplied data for photometric analysis (*i.e.*, the measure of visible light) of asperities along the PMMA-PMMA interface in a direct shear configuration, taking advantage of the transparent nature of the sample material. Normal stress over individual asperities, calculated with the pressure-sensitive film, was compared to the light intensity transmitted through the interface. We found that the luminous intensity transmitted through individual asperities linearly increased 0.05643 ± 0.0012 candelas for an increase of 1 MPa in normal stress between normal stresses ranging from 23 to 33 MPa.

Keywords: multicontact interface; pressure-sensitive film; normal stress distributions; finite element modeling; photometry; acousto-optical method

---

1. Introduction

Structural health monitoring (SHM) using non-destructive techniques (NDT) is a continually developing field of study with immediate and important benefits to the general population. The field is driven by balanced studies using experimental and theoretical aspects. In most cases, the field is driven by new sensors and sensing techniques, which allow us to develop a better understanding of the external forcing facing structures in either a continuous or iterative manner. In this study, we develop a novel method to better understand frictional processes and their degradative influences on a structure's health. A better understanding of friction and the means to assess (1) wear (surface to surface interactions); (2) attrition (erosion by friction) and (3) the proper assemblage of structural components (e.g., pressure vessels, pipe flanges, structural joints) could have long-term impacts on the structural health and efficacy of operations.

Friction from the contacts occurring between solids involves both the interaction of mechanical and geometric properties [1]. Accurate understanding of this problem has eluded physicist, engineers and scientists since the first well-known experiments conducted by Leonardo da Vinci (*ca.* 1500) [2]. While the simplistic versions of Amontons–Coulomb laws of friction were sufficient until the late 1950s, it was not until the following developments of Bowden and Tabor [3] and Rabinowicz [4], who discussed the importance of plasticity in contact, Archard [5] and Greenwood and Williamson [6], who characterized the surface roughness and interactions using statistical mechanics, that contacts along the multicontact interface (MCI) were shown to form in three mechanical constitutive behaviors: (I) elastic; (II) elasto-plastic; and (III) fully plastic [7]. For dry, unlubricated friction, the only manner for stress to be transmitted across the interface is via the junction regions that are dispersedly located along the fault, so the mesoscopic motions of the interface must be described by the interaction and movements of multiple asperities at the asperity-level.

In the late 1970s, Dieterich [8] and Ruina [9] developed the so-called rate and state constitutive laws (RS) to describe the manner in which mesoscopic motions are dictated by the numerous contacts formed along the MCI. Their field of study was rock mechanics, and the RS law was developed in the laboratory with the goal of understanding the earthquake cycle in a mechanical manner. Their model has been shown to capture the most prominent features of the seismic cycle, but its applicability is still widely debated [10,11]. The promising features of the RS model are that it is a unifying law, largely material-independent (applies for paper-paper, Lucite-Lucite, wood-wood and rock-rock interfaces among others [12]) and applicable down to the microscopic level. However, one of the goals and reasons for the continued study of dry frictional sliding at low-velocities in the laboratory is the search for mechanistic constitutive laws. Feedback and improvement are allowed by improvement in sensors and sensing techniques in the laboratory.

In this study, we present methods to improve our ability to understand how asperities are formed along the multicontact interface (MCI). Here, we develop unique techniques to characterize contact interactions, specifically their stress states. Our experiment consists of two PMMA samples, a slider and a base block (Figure 1a), which when pressed together, form an MCI due to the interaction of the rough surfaces (Detail A in Figure 1d). Two methodologies have been developed to understand the intricacies of asperity formation under simple normal loads. We first use a pressure-sensitive film to characterize the normal stresses on the contacting asperities. The pressure-sensitive film provides the detail of both the heterogeneity along singular asperities and the relative distances between larger asperities, which dominate the transfer of stress across the interface. Secondly, we use the transparent nature of the PMMA to our advantage. The interface could be visualized (Figure 1b) through a video camera focused on the interface through the side of the PMMA slider block. Dieterich and Kilgore [12–14] used a similar technique to image individual contacts on transparent PMMA samples. They observed that light was transmitted through the asperities, which appeared brighter in images (Figure 1d, top) than the non-contacting background. We use a similar technique, but focus on the subtle variations in transmitted light through the asperities, which is indicative of variations in the normal stress field at the contact junction (*i.e.*, the millimetric length scales).

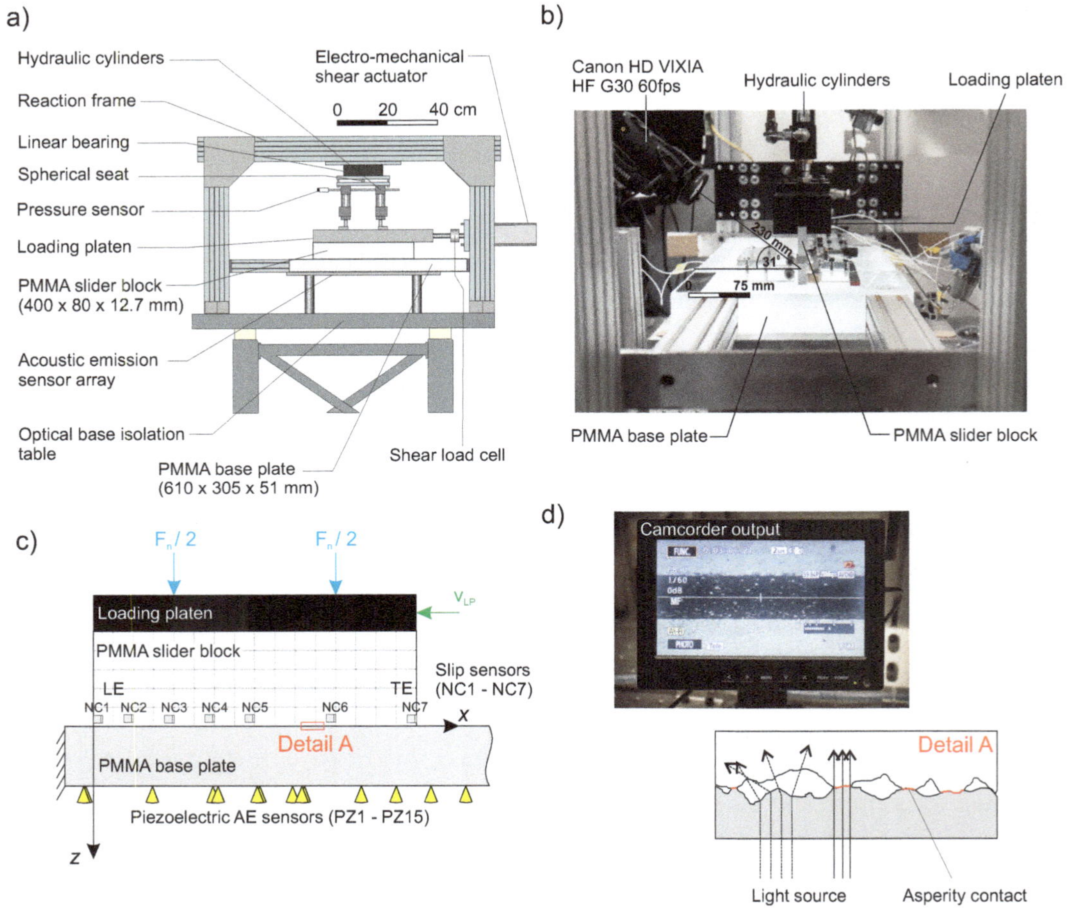

Figure 1. (**a**) An overall schematic representation of the direct shear friction apparatus and its major components; (**b**) a photograph of the apparatus showing the location of the video camera in relation to the frictional multicontact interface (MCI); (**c**) general configuration of the direct shear apparatus from the side view with general locations for the non-contact eddy current (NC1–NC7) and acoustic emission (AE) piezoelectric sensors (PZ1–PZ15). These sensors were used to detect dynamic changes as the fault began to fail; (**d**) (top) A photograph of the monitor used to display the video camera images in real time. Asperities appear as the bright spots within the darker, interface region. (bottom) A schematic representation of Detail A from (**c**) depicting the theory, which describes the transmission of light through asperities (solid lines) and diffracted light along the void space (dashed lines).

We found that as higher normal stress is observed over singular asperities, a "better" bond is achieved, allowing for better transmission of light, which can be detected using the video camera. We could then compare the light transmitted through individual asperities whose normal stresses were also calculated beforehand using the pressure-sensitive film. This allows quantification of changes in transmitted light to changes in normal stress, an indicator of the asperities' shear strength in terms of contact mechanics [7]. This is important when trying to quantitatively understand the local variations in stress states causing individual asperities to fail dynamically when the interface is slowly sheared [15–17]. Results for the slow shearing of the interface are presented in the latter sections of this study.

We performed a full direct shear experiment with along-fault strain changes monitored using slip and the foreshocks released by asperity slip recorded by an array of Glaser-type acoustic emission sensors. A camera was focused on the region where sudden changes in asperity luminosity were corollary to the failure (or partial failure) of singular asperities. The failure of these asperities were located in space and time with respect to the video camera images, and the methodology used to capture this transition is discussed.

2. Experimental Facilities

2.1. General

In the experiments presented here, dry and unlubricated conditions along the fault were carefully controlled during each experiment. The tests consisted of loading two samples of PMMA in the direct shear apparatus detailed in Figure 1a. To create the frictional interface, a far-field normal force F_n was applied via two hydraulic balanced cylinders (Parker H3LLT28A) to the PMMA slider block (400 mm $\times$ 80 mm $\times$ 12.7 mm) through a rigid loading platen. An in-line pressure transducer (OMEGADYNE PX329-2KG5V) was connected to the cylinders and was used to measure the pressure, then converted to a force. In the first part of the experiments conducted here, the interface was not sheared, and the rigid loading platen was held stationary (V_{LP} = 0 mm/s) using the shear actuator (Exlar Tritex II T2X115). During the shearing of the interface, discussed in the final part of this study, the shear actuator drove the rigid loading platen at V_{LP} = 0.007 mm/s.

2.2. Instrumentation

The general arrangement of sensors used in this investigation is shown in Figure 1c. We used non-contact eddy current induction sensors (Shinkawa VS-020L-1) at seven locations along the fault (NC1–NC7) to measure the macroscopic longitudinal strains along the fault. Two eddy current sensors were placed at the leading (LE) and trailing edge (TE; the edge in contact with the thruster) of the slider block in the y-direction. These allowed us to index the slider block with respect to the base plate and measure for any twisting during failure of the interface. Aluminum targets were used to change the inductance of the magnetic field, which was linearized using a displacement converter (Shinkawa VC-202N), over frequency ranges from 0 to 20 kHz (-3 dB), and showed a linearity of $\sim$444.52 $\pm$ 0.11 μm/V between 0 and 2000 μm with a resolution of 0.1 μm for each sensor. Mounting of the targets to the sample was done with a minimal amount of cyanoacrylate in a similar manner to Ohnaka and Shen [18].

An array of 15 Glaser-type conical acoustic emission sensors (PZ1–PZ15) was placed on the underside of the base plate (Figure 1c). The detailed layout of the acoustic emission array and non-contact sensor array can be found in Selvadurai and Glaser [17]. These sensors were absolutely calibrated beforehand using the techniques of McLaskey and Glaser [19,20]. They detect surface normal displacements from a near-field vantage point with a frequency bandwidth of ~5 kHz–2.5 MHz (−3 dB) with a ±1 pm noise floor [19]. An ELSYS TraNET EPC 16-bit dynamic range, 40-MHz sample rate, high-speed digitizer measured the high-frequency stress waves emitted during the impulsive destruction of asperities.

2.3. Sample Preparation

The surfaces of the PMMA slider and base are machined flat (~0.001" over 16"), then professionally sandblasted using 40–60 grit Al_2O_3 (440–220 μm) to create roughness profiles [21] similar to those found in nature [22]. Before any reported testing, the interface was sheared for a cumulative slip of ~36.1 mm. This ensured the formation of a mature frictional interface; at this point, the changes from abrasive and adhesive wear [11] become minimal and do not control friction [23], which is associated with a mature faulting surface.

2.4. PMMA Properties

Polymethyl methacrylate is a glassy polymer [24] with physical characteristics at room temperature as follows: density $\rho = 1180$ kg/m^3, shear modulus $G = 1.8$ GPa and a Poisson's ratio of $\nu = 0.3$. The body and shear wave velocities are 2700 m/s and 1390 m/s, respectively. Using digital scanning calorimeter (DSC), we determined the glass transition to be $T_g = 114$ °C. We determined all of the above values independently, and they fell within the range of values of PMMA material properties found in the literature [24,25].

2.5. Photography of Interfacial Asperities

The unique setup and transparent properties of the "glassy" PMMA allowed us to directly observe the interface during the experiments. Figure 1b depicts the location of the video camera (VIXIA HF G30 CMOS) in relation to the faulting plane. The video camera was focused at a 31° angle to the horizontal and a distance of 230 mm from the frictional plane. Detail A in Figure 1d illustrates the theory of operation [13]: (I) two nominally flat interfaces only touch asperities; (II) the light was transmitted more effectively through these contacting asperities; and (III) light was diffracted through the void space in which no contact occurred. The high definition video camera operated at a video frame rate of 60 fps (Δt_{frame} = 1/60 s ~ 16.7 ms), and images were obtained at a focal length of 3.67 mm. To improve image resolution without the use of digital enhancement, a CANON 58 mm Close-up Lens 250D was attached on the camera directly. The field of view was 10 mm, and images are 2136 × 1362 pixels per frame, making the resolution 0.005 mm/pixel (~5 μm/pixel). No image correction was performed for the optical distortion; instead 200 μm-thick lines were drawn along the interface at 5-mm spacing, forming a stereographic grid to account for the orthoscopic distortions from the camera, close-up optical lens attachment and light passing through the PMMA sample simultaneously.

2.6. Pressure-Sensitive Film

We employed a FUJI Prescale pressure-sensitive film to detect and measure the contacting asperities formed between the interacting surfaces. The film is polyethylene based and has a thickness b of approximately 90 μm. The film has embedded microcapsules (5-μm resolution) that, when compressed, release ink with colors proportional to the applied pressure ($\pm$1.5 Pa). The film is rated for normal pressures between 12 to 50 MPa according to the manufacturer. At stresses higher than 50 MPa, the microcapsules do not further discolor, and the intensity is saturated ($I \sim 0.3$ candelas (cd)). After loading in the fault, the film was digitized using an image scanner and algorithms in MATLAB to detect, size and catalog all contacting asperities in the static state, under normal load and not subjected to shear. The film was calibrated using a finite element model (FEM) in the commercially available software ABAQUS, and the results are presented in Section 3.

3. Calibration of the Pressure-Sensitive Film

3.1. Indentation Test

To test the pressure-sensitive film, we used the spherical indenter apparatus shown in Figure 2a. The pressure-sensitive film was used to measure the normal stress profiles induced from a spherical steel indenter (radius R = 9.5 mm), which compressed the film against a 50 mm-thick PMMA base plate. The substrate was assigned a Young's modulus E = 1.8 GPa [25] and a Poisson's ratio ν = 0.3 [24]. The apparatus was capable of delivering a normal force F_n from 0–70 N (compression), which was measured using an in-line load cell (ELPF-T3M-500N). A new piece of pressure-sensitive film was used for each level on normal force, since once the film discolored, it no longer had the ability to measure lower normal stress levels. A small section of 5.5 mm $\times$ 6.1 mm surrounding the discolored contact patch resulting from contact was digitized using an HD scanner (MUSTEK SE A3 USB 2400 Pro) at 24-bit RGB color. Once digitized, the image was imported into MATLAB [26] and each pixel converted from an RGB color scheme to the luminous intensity color scheme [27]. Luminous intensity is an SI (International System) photometric measurement unit that has a value between 0 and 1 cd and refers to an average measurement of light intensity across the visible spectrum. This metric is known as photometry and was also employed for the video camera in the following sections. Figure 2b shows the luminous intensity for three indentation tests at $F_n \approx$13, 25 and 40 N. The brighter region (outside the contact patch) has a higher value of $I \sim 0.65$ cd, while the darker contact region varies from $I \sim 0.3$–0.6 cd.

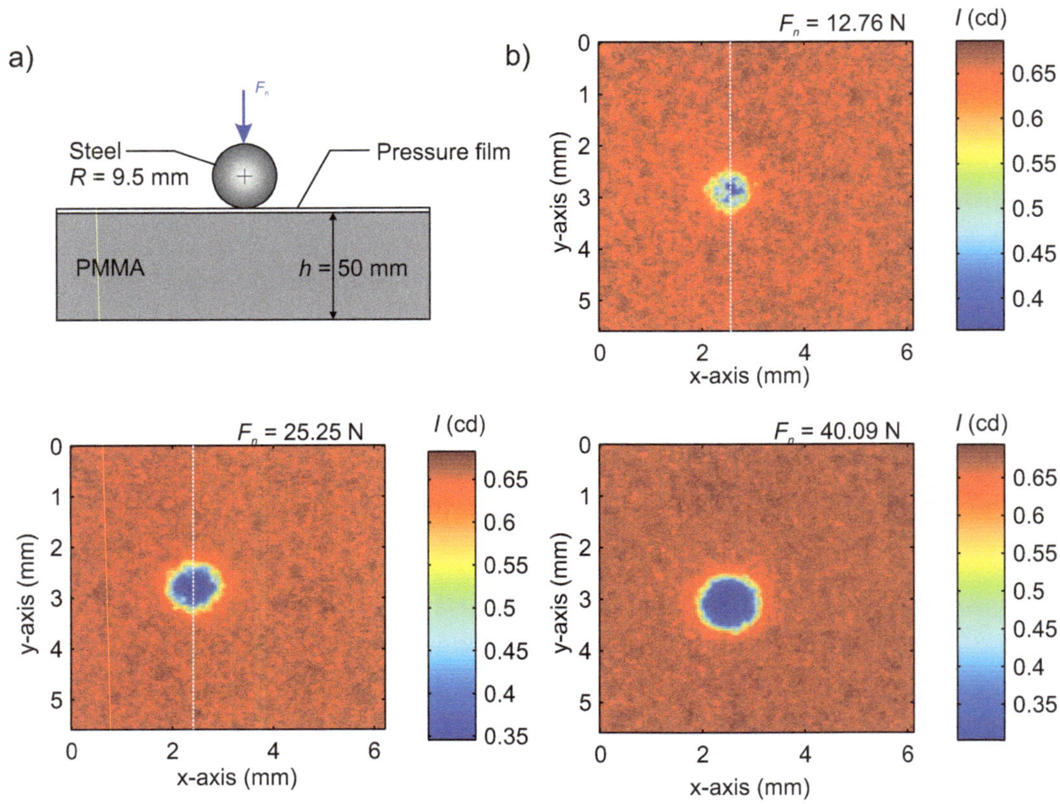

Figure 2. (**a**) Indentation test configuration used to calibrate the pressure-sensitive film; (**b**) digitized images of the pressure-sensitive film for various levels of normal force applied through the steel indenter. Digitized images were converted to luminous intensity and had an absolute range of 0–1 candelas (cd).

3.2. Numerical Modeling

To better understand the normal stress distributions seen in Figure 2b, an axisymmetric, 2D finite element model was created using the commercially available ABAQUS software. The model excluded the effects of the film and frictionless contact assumed between the rigid indenter and PMMA substrate. At room temperature (~25 °C), we do not expect the material to yield (or experience any softening) until ~78 MPa [24], and a linear elastic model was used. The deformations were modeled as classical Hookean isotropic elastic [28]. Incremental elastic strains are given by,

$$d\epsilon_{ij} = \frac{d\sigma_{ij}}{2G} - \frac{\lambda^* \cdot d\sigma_{kk}\delta_{ij}}{2G(3\lambda^* + 2G)} \tag{1}$$

where λ^* is the Lamé's first parameter and G is the shear modulus, and summation over the repeated indices is implied. The model is composed of 46,687 quad elements and was refined to a length scale of 100 μm in the region of contact; it is presented in Figure 3a, as are the boundary conditions. Detail C in Figure 3a is enhanced and shows the z-direction component of displacement ($u_z = u_2$) (Figure 3b) and the normal stress component (σ_{zz}) (Figure 3c) for an applied normal force F_n = 49.74 N.

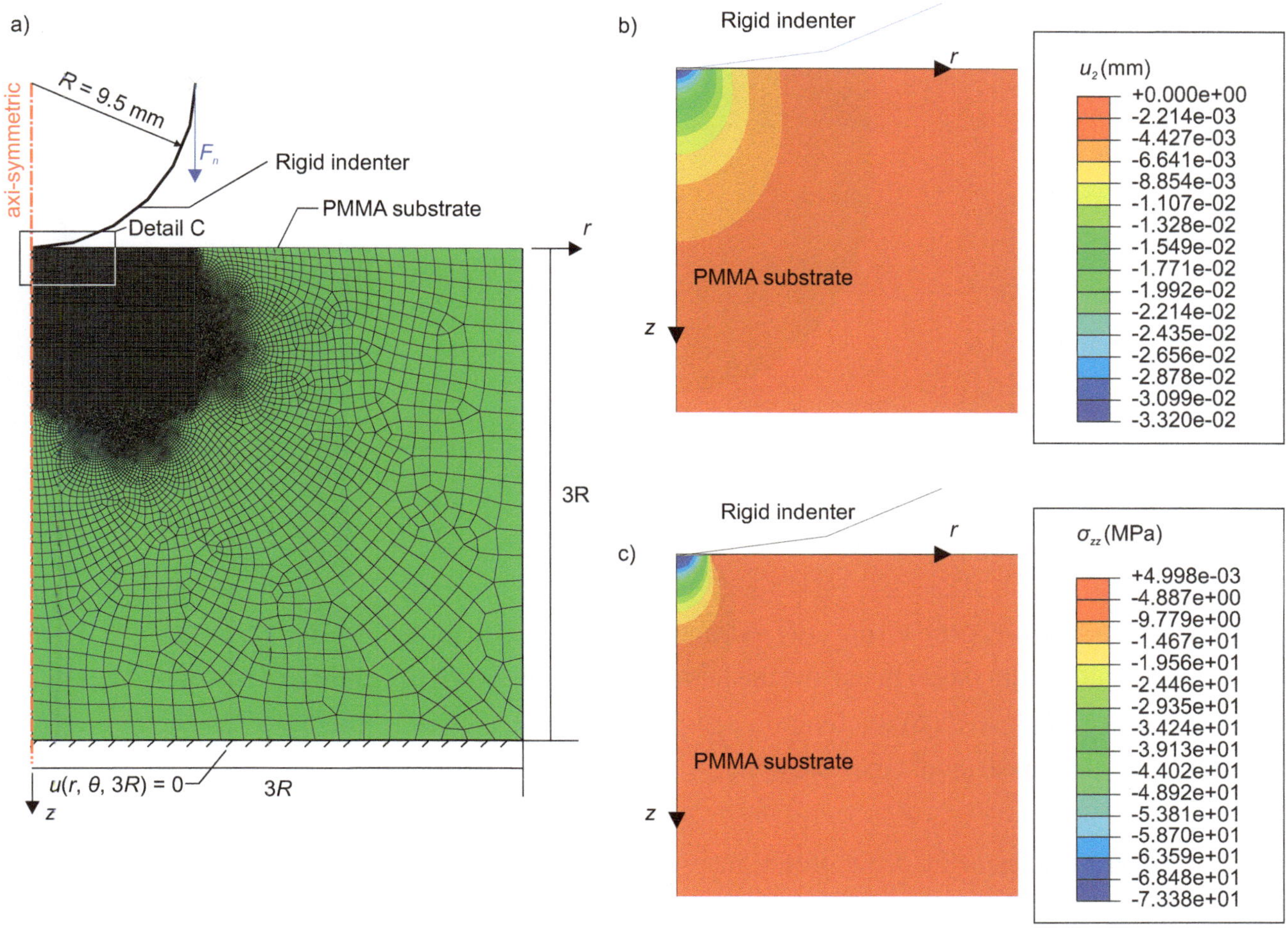

Figure 3. (**a**) Numerical model used to model the pressure distributions measured by the pressure-sensitive film; (**b**) displacements in the z-direction ($u_z = u_2$); and (**c**) normal stress (σ_{zz}) from Detail C in (a). The results in (b) and (c) are for an applied normal force of 49.74 N.

3.3. Numerical Modeling vs. Experimental Observations

The distribution of pressures calculated by the model is shown in Figure 4a (dashed lines) for various normal loads for comparison with the measured one (solid lines). Transects were drawn through the centroid of contacts from the experimental pressures and compared with image intensity (Figure 4b). The centroid of contact was determined using the image detection algorithm in MATLAB. We see that the film saturates at 50 MPa, as was expected from the manufacturer's specifications. We calculated the errors between the experimental measurements and numerical results of the normal stress profiles at 12 MPa and 50 MPa and took the average error across the centroid in both the x- and y-directions for all three applied normal forces seen in Figure 4a. The maximum percentage error in the normal stress profiles ranged from −5%–6.6% at 12 MPa and ranged from −7.7%–0% at 50 MPa.

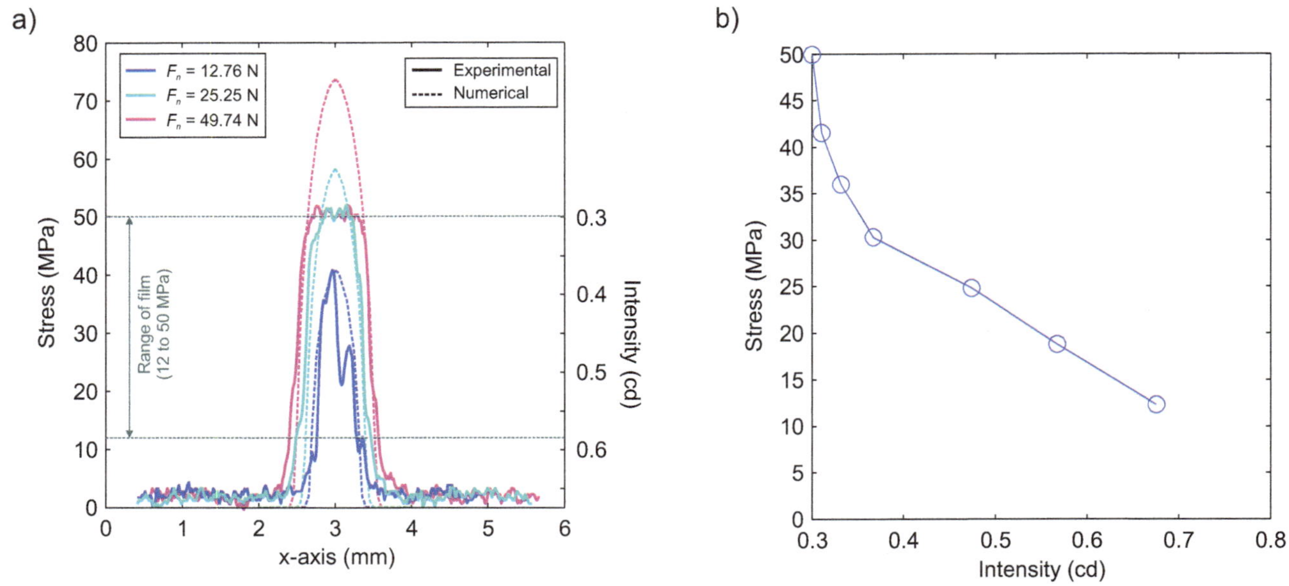

Figure 4. (**a**) Comparison of normal stress distributions from experimental pressure-sensitive film (solid lines) and the numerical model shown in Figure 3 (dashed lines) for various normal loads; (**b**) the relation between normal stress calculated numerically and the light intensity measured experimentally.

We suspect that some error in stress profile measurements may be induced by the fact that both the PMMA substrate and the steel indenter have roughness profiles at some length scale. Figure 5 shows images of normal stress distributions for indentation tests conducted at two normal load levels. Three different substrates were used in combination with the same steel indenter: (I) a polished steel substrate (S-S; Figure 5a); (II) a smooth PMMA substrate (S-P/S; Figure 5b); and (III) a rough PMMA substrate (S-P/R; Figure 5c). The rough PMMA substrate was sandblasted using 40–60 grit Al_2O_3 (440–220 μm), which had a similar preparation to the experimental interface used in the direct shear configuration shown in Figure 1a. Figure 5d shows the summation of pixels below light intensities of $I < 0.65$ cd at a wider range of applied normal force F_n. The contact area was calculated knowing that each pixel had an area of 25 μm^2 (5 μm × 5 μm). Archard [5] examined the effect of surface roughness on the elastic contact formed along an interface and found that the relation of the real contact area A_r varied proportional to the applied normal force F_n, and the relation is given as:

$$A \propto (F_n)^{\Omega} \tag{2}$$

where Ω ranges from 2/3 to 1 for smoother to rougher surfaces, respectively. Figure 5d shows that Archard's [5] theoretical relation was comparable to the experimental measurements taken from the pressure-sensitive film. The steel-steel interface has $\Omega = 0.78$ ($R^2 = 0.9962$); the steel-PMMA/smooth interface has $\Omega = 0.70$ ($R^2 = 0.9949$); and the steel-PMMA/rough interface has $\Omega = 0.9721$ ($R^2 = 0.9834$).

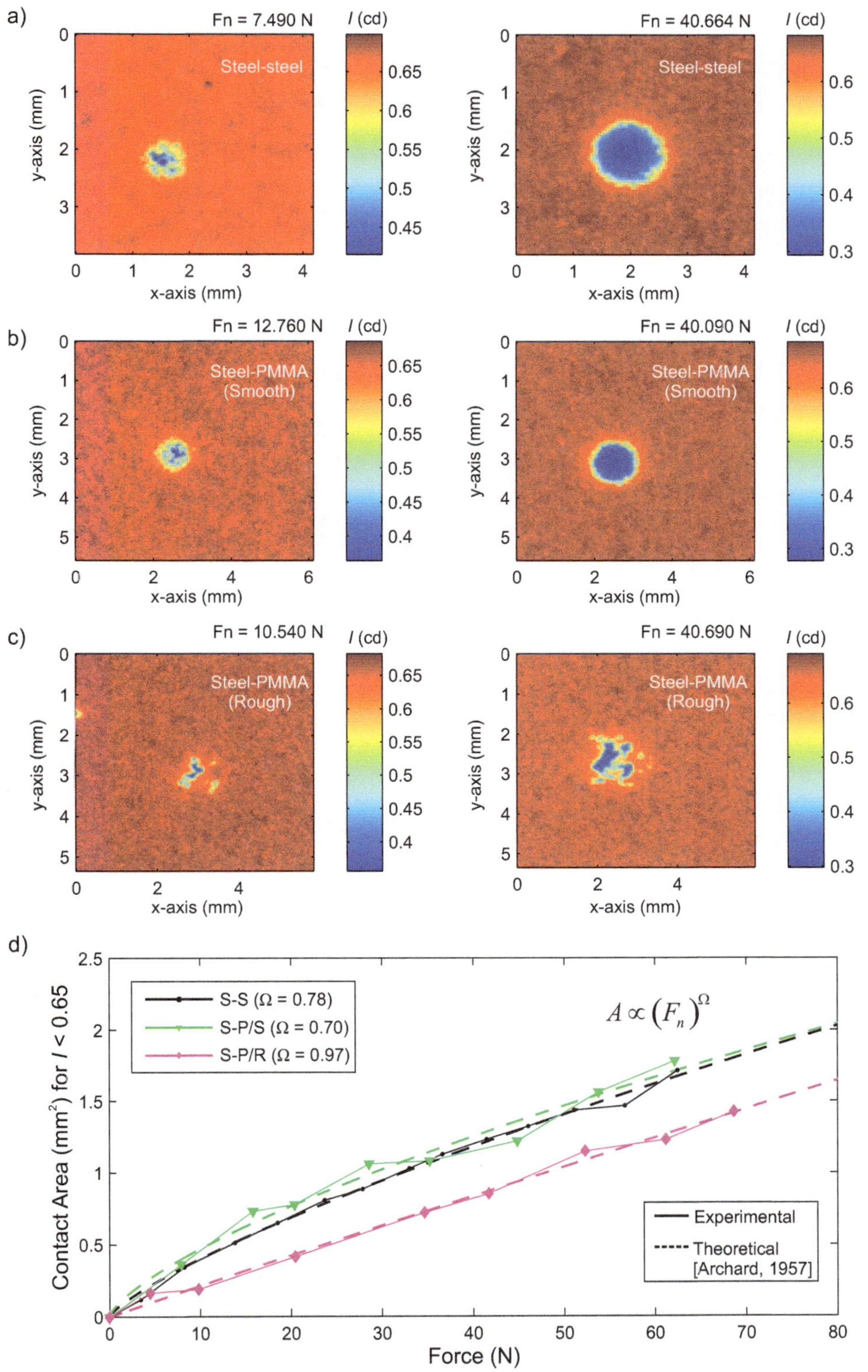

Figure 5. Light intensity (cd) obtained from the pressure-sensitive film for variable substrates pressed against with the same steel indenter. For a lower (~10 N) and higher (~40 N) applied normal force, we see the variations due to the: (**a**) steel substrate (S-S); (**b**) the smooth PMMA substrate (S-P/S); and (**c**) the rough PMMA substrate (S-P/R); (**d**) the comparison of the area, defined by pixels below the intensity threshold I = 0.65 cd, *versus* the applied normal force F_n. The results are compared to the theoretical estimates made by Archard [5].

4. Interfacial Measurements of Transmitted Light Using Photometric Methods

As mentioned in Section 2.5, we used a video camera to monitor the light transmitted through contacting asperities in the direct shear experimental configuration. In Figure 6a, we show an unprocessed image taken along the interface. The interface appears darker, and within the darkened section appear small populations of bright asperities. The edges of the image are distorted due to the lens' aperture effect. We focused our analysis on the portion of the image in the red box highlighted in Figure 6a.

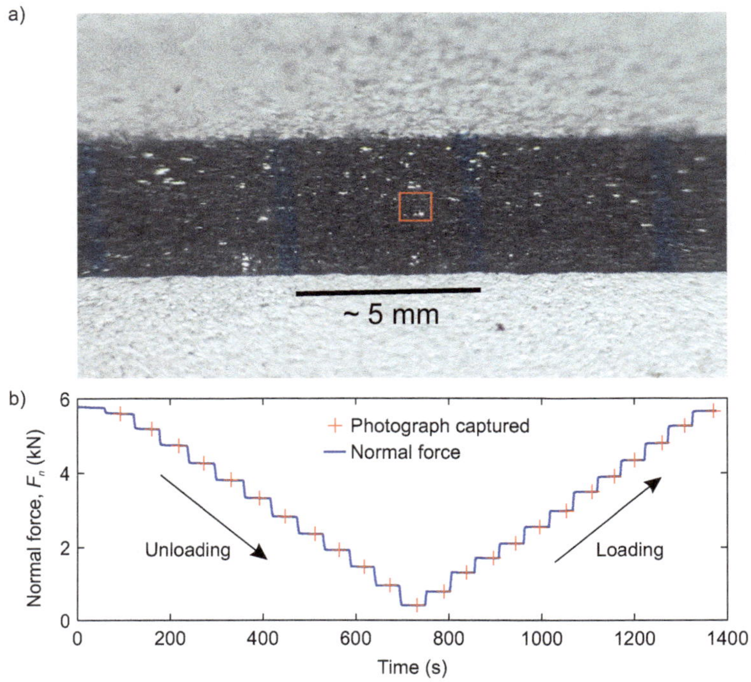

Figure 6. (**a**) An unprocessed image (2136 × 1362 pixels) from the video camera focused on the interface (darker region). Within the dark region are small bright regions that are asperities that transmit light more easily; (**b**) Test performed to measure the changes in light intensity and its relation to the applied normal force F_n.

4.1. Variations in Normal Load and Its Effect on Local Luminous Intensity

A test was performed to analyze the effect of light transmitted through asperities due to an increased far-field normal force F_n in the direct shear setup shown in Figure 1a. The normal force was varied incrementally in a step-like manner, as seen in Figure 5b. At each level of normal force, a 2-s video (*i.e.*, 120 picture frames) was captured (red crosses in Figure 5b). The hypothesis is that as the far-field normal force increases, the normal stress at the asperity junction level increases. The normal force was first unloaded from $F_n \sim 5.8$ kN to 0.4 kN and then reloaded back to $F_n \sim 5.8$ kN. A total of 12 loading steps were performed for both the loading and unloading portions of this experiment. For each video, the central frame was used and processed to examine the changes in light intensity *versus* normal force. The unprocessed images captured from the camera (RGB format) were converted to luminous intensity (I) in MATLAB. An identical algorithm was used as for the pressure-sensitive film discussed in Section 3.

To detect and catalog individual asperities, a lower threshold was set (I_{lower} = 0.5 cd) for the same image detection algorithms applied to the pressure-sensitive film.

4.2. Results for the Normal Force Unloading-Loading Test

We focused on the small section of the interface shown in the red box of Figure 6a. The threshold for light intensity was set to I_{thresh} = 0.5 cd. Examples of the images used are shown in the inset of the graph, and arrows indicate which data point they are associated with. We first examine the number of pixels above the threshold *versus* the applied normal force F_n. The absolute number of pixels is not an indication of area, since the images are taken at an angle of $\sim 310^0$. However, we expect that the sum of light intensity above I_{thresh} is related to the real contact area. If we assume that asperities form in a circular manner (e.g., Figure 5), the area will be proportional to the radius in the x-direction, and increasing the far-field normal force F_n should cause asperities to grow [13], in turn leading to an increase in pixels transmitting light above I_{thresh}. The added size of the asperity must balance the increase in F_n for elastic, elasto-plastic or fully plastic asperity growth [7].

In Figure 7a, we show how a small population of asperities form in relation to increasing the normal force shown in Figure 6b. Within the highlighted red box (see Figure 6a), the image processing algorithm (see MATLAB function *regionprops* [27]) allowed us to measure: (I) the number of regions (*i.e.*, the number of asperities); (II) the number of pixels in each individual region; and (III) the distribution of light intensity in each region. To eliminate any background noise, a region (asperity) was defined as having a minimum of 10 pixels, above the intensity threshold I_{thresh}, in direct contact with each other. The inset images in Figure 7a are snapshots of the highlighted region for respective levels of applied normal force F_n. The study of asperities using their transmitted light intensity and our ability to catalog them individually using post-processing software has many benefits and can provide a number of interesting metrics to better understand the random process models of rough surfaces [10,29,30].

Since the seminal paper by Greenwood and Williamson [6], the study of asperity formation between randomly rough surfaces (GW model) have been extensively employed in numerous engineering and scientific endeavors. However, Greenwood has recently come forward [31,32], criticizing some of the assumptions made in this model and recent works. Primarily, the field of tribology is attempting to better understand these key criticisms. One critical assumption of the GW model is that asperities form independent of each other. The spacing between larger asperities is great enough that they do not interact as the normal force F_n is increased. While interaction was believed to happen only when higher normal loads were applied, in truth, the process of asperity interaction is seen as prevalent. Studying the additional stresses introduced from the elastic interactions of asperities through the substrate are beginning to become important [10]. Newer studies on this subject are primarily computational interpretations [33–37], and experimental studies [38] are at the early stages of development.

Since the seminal paper by Greenwood and Williamson [6], the study of asperity formation between randomly rough surfaces (GW model) have been extensively employed in numerous engineering and scientific endeavors. However, Greenwood has recently come forward [31,32], criticizing some of the assumptions made in this model and recent works. Primarily, the field of tribology is attempting to better understand these key criticisms. One critical assumption of the GW model is that asperities form independent of each other. The spacing between larger asperities is great enough that they do not interact as the normal force F_n is increased. While interaction was believed to happen only when higher normal loads were applied, in truth, the process of asperity interaction is seen as prevalent. Studying the additional stresses introduced from the elastic interactions of asperities through the substrate are beginning to become important [10]. Newer studies on this subject are primarily computational interpretations [33–37], and experimental studies [38] are at the early stages of development.

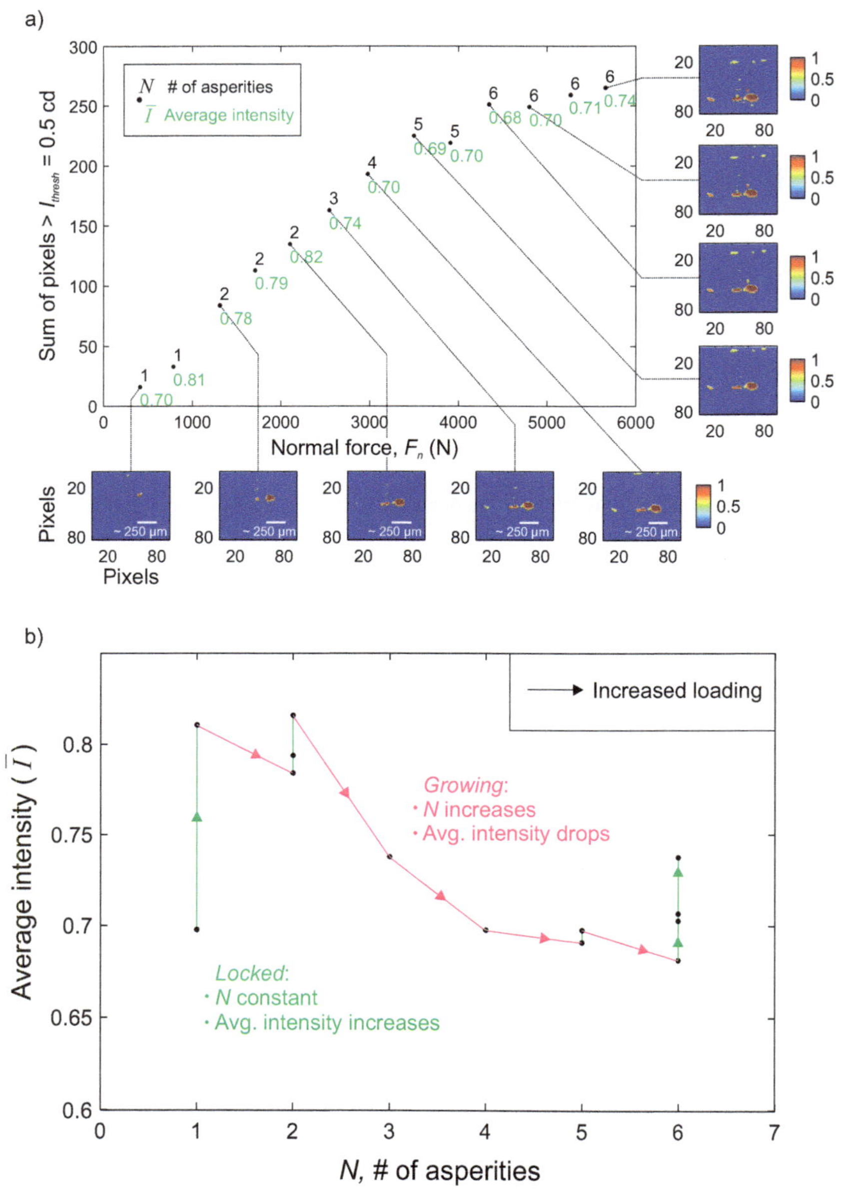

Figure 7. (**a**) Total pixels above the threshold light intensity I_{thresh} *versus* the applied far-field normal force F_n. The images along the axes are the processed images from the red box displayed in Figure 6a at various normal loads. For each loading step, the number of asperities N (black value) were calculated, using the image detection algorithm, and the average light intensity $\bar{I}$ over those asperities are given in green. (**b**) The average luminous intensity transmitted through the population of N asperities *versus* the number of total asperities in the population.

5. Combining Pressure Film and Photometric Measurements

We use the now calibrated pressure-sensitive film (Section 3) to examine the relation between luminous intensity across an asperity junction using the photometric methods (Section 4). Firstly, we employed the pressure-sensitive film in the direct shear configuration shown in Figure 1a. The fault is positioned at a datum location between the slider and base plate sample using the eddy current sensor array and fine-threaded screws [39]. Prior to applying a far-field normal force F_n, a new strip of pressure-sensitive film was placed in between the two interface surfaces. The fault was loaded using a known applied far-field force F_n, causing the pressure-sensitive film to develop. The pressure-sensitive film was developed for t_{hold} = 900 s, and then, the fault was unloaded and the film carefully extracted and digitized. The calibration used in Section 3 was applied to determine the normal stress field along the fault. The frictional fault was carefully re-indexed to the same location using the eddy current sensor array and fine-threaded screws [39] and loaded at the same normal load for an identical amount of time. At this point, photographs of the interface were taken at the central region of the frictional fault. We avoided the end regions of the interface (*i.e.*, x = 0–100 mm and x = 300–400 mm) to avoid the combination of shear and normal stress conditions that arise from the direct shear configuration [40].

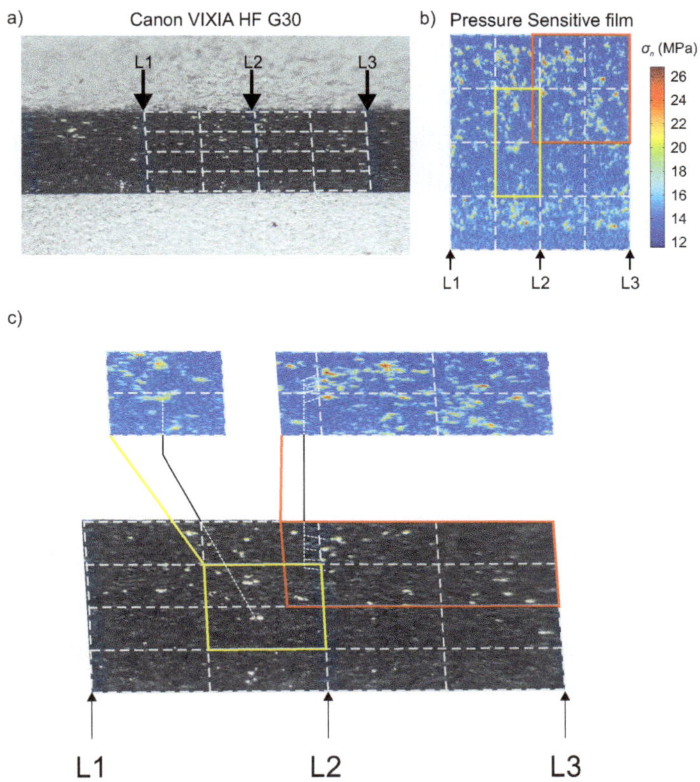

Figure 8. (**a**) A still frame obtained from the camcorder at one loading step. A grid was superimposed and followed the lines physically drawn on the interface. The lines L1, L2 and L3 were used for reference; (**b**) Results from the pressure-sensitive film (converted to stress) in the same location as in (a). Reference lines L1, L2 and L3 are shown. (**c**) The pressure-sensitive film was distorted using COREL DRAW to match the transformed coordinate system in the photograph (top). In the highlighted boxes, we see that the pressure measurements and the light transmitted occur at similar locations along the grid (connecting lines).

In Figure 8, we present a pre-processed photograph of the interface (Figure 8a) and the post-processed image of the same location using the pressure-sensitive film (Figure 8b) for the same applied normal load F_n. The three lines, physically drawn along the interface (*i.e.*, L1, L2 and L3), were used to construct a grid (white dashed lines) along the interface that accounted for the orthoscopic distortions imposed from the camera, additional lens and PMMA sample. This grid was superimposed over the pressure film measurements in Figure 8b and referenced using lines L1, L2 and L3. Within the grid, two locations are highlighted in the pressure measurements (Figure 8b) using a yellow and red box. The pressure-sensitive film measurements were distorted using an image processing software (COREL DRAW X4), so that the distorted coordinate system matched those from the photographic images. In Figure 8c, locations where light is highly transmitted (in the photographic frame) are compared to locations where the normal stress appears to be higher (from the pressure film measurements). While not all regions exhibiting high levels of normal stress were concomitant, this was most likely due to experimental error and our inability to perfectly recreate the interface between when the film is extracted and when the photometric measurements are taken.

5.1. Calibration of Luminous Intensity to Normal Stress along Larger Asperities

A total of 40 asperities that both transmitted light and had simultaneous film measurements of normal stress were analyzed for the same far-field normal load of $F_n \sim 5.6$ kN. These asperities were hand-picked, since the process of locating identical asperity distributions was not trivial. Figure 9a shows the normal stress distribution on a single asperity and the accompanying stress histogram. For the same asperity, we show the photometric measurement taken in Figure 9b with its accompanying histogram. For each asperity histogram, the maximum and mean normal stress and light intensity were recorded. Figure 8c shows the relation between normal stress and light intensity. Both the mean (black) and maximum (red) data points were plotted. Linear estimates of the maximum (R^2 = 0.82) and mean (R^2 = 0.75) relation between the normal stress and light intensity were plotted. We used the average of the two maximum and mean slopes and estimated the relation between light intensity and normal stress as follows: an increase of one MPa in normal stress caused the light intensity to increase 0.05943 ± 0.0012 cd between the normal stress range of 23–32 MPa with an error of ± 1.45 MPa within a 95% confidence interval (dashed lines). In similar studies [13], the light transmitted was assumed only to occur on asperities that formed due to plastic contact, *i.e.*, the stress levels were uniform over the entire asperity in accordance with the plastic contact limit p_m [41]. We clearly see in each histogram (Figure 9a,b) that the range varies, a phenomena that can only be explained by elastic or elasto-plastic contact. The experimental error observed is most likely introduced when the pressure-sensitive film is extracted, and reconstituting identical asperity populations (in order to measure the light transmitted) was not fully achieved.

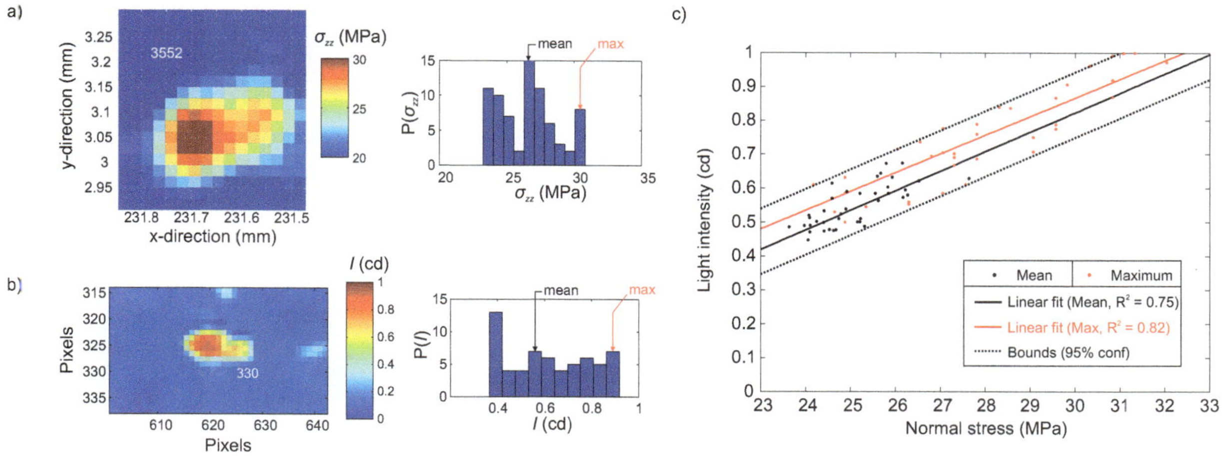

Figure 9. (**a**) Normal stress along an asperity measured using the pressure-sensitive film (left). On the right is the histogram of the stress for the same asperity; (**b**) Light intensity transmitted by the asperity in (a) measured using the video camera. The histogram for the light intensity is shown on the right. (**c**) The relation between normal stress and light intensity.

6. Application: Laboratory Earthquakes

Figure 10 shows a typical result for the direct shear experiment and the sensor measurements taken using the non-contact (NC) and acoustic emission (PZ) sensor array. Phase I of the direct shear experiment (see Figure 10a) consisted of loading the slider block against the base plate under a constant normal load F_n = 4400 N for t_{hold} = 900 s with no shear force F_s applied (V_{LP} = 0 mm/s). During Phase II, the normal load was maintained constant, and the rigid loading platen was driven at a constant velocity of V_{LP} = 0.007 mm/s using the electro-mechanical shear actuator. Steady motion of the rigid loading platen was used to simulate the motion of tectonic plates, causing an accumulation of shear force F_s along the interface. In Figure 10a, the normal force F_n (blue) remained constant, while an observable increase in shear force F_s (red) occurs due to the motion of the rigid loading plate. As time continues, the bulk shear force F_s increases until a stick-slip event (SS). Figure 10b shows slip displacements from one non-contact sensor (NC4) during three SSs (SS1, SS2 and SS3) that occurred during Phase II. Stick-slip events consisted of a sudden drop in shear force drops accompanied by rapid relative displacement across the interface. Figure 10c shows, in detail, the period leading up to a stick-slip, where slow slip built up prior to rapid sliding (instability). "Slow slip" was defined to be when local slip velocities are lower than v_{local} > 10 mm/s (star symbol in Figure 10c), which was evaluated by differentiating the raw slip sensor data *a posteriori*. Superimposed on Figure 10c are the acoustic emission (magenta) measurements that occurred prior to SS1. These events are referred to as foreshocks (FS) and were present prior to the rapid sliding phase. Figure 10d shows the high-resolution AE signals (PZ1–PZ13) just prior to the SS3. Here, a total of 24 FS were detected prior to rapid sliding. Figure 10e shows a detailed view of FS8, emphasizing the pulse-like P and S wave arrivals intrinsic to all of these FS signals and indicative of the sudden release of stress waves from local dynamic failing of a single asperity [16,42].

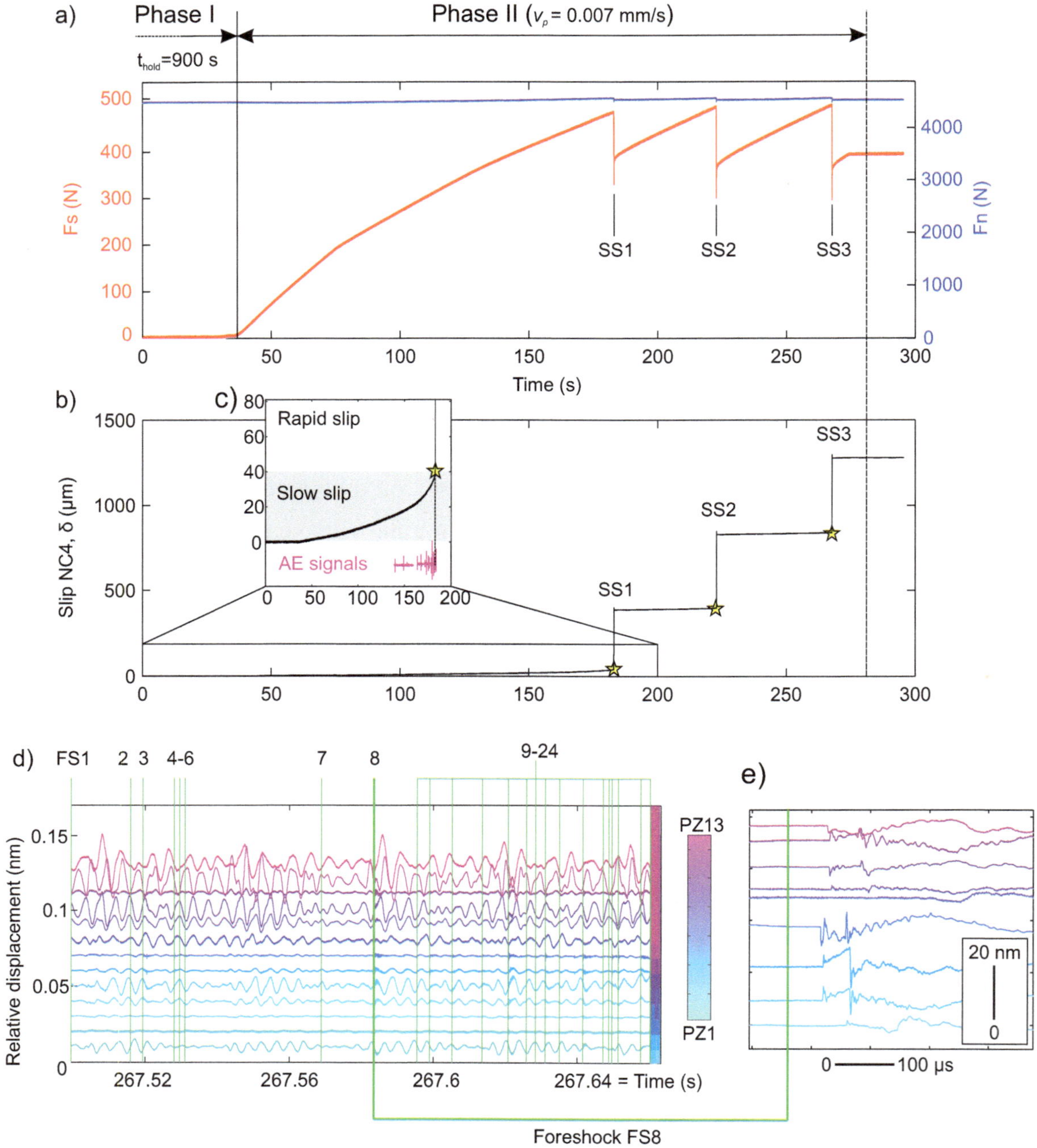

Figure 10. (**a**) Normal (blue) and shear (red) loads applied to the slider block during Phase II of the experiment. The rigid loading platen moved at V_{LP} = 0.007 mm/s until a stick-slip event (SS). During an SS, the shear force dropped, while slip increased rapidly, as seen in (**b**). (**c**) Prior to an SS, we observed an accumulation of slow slip and localized, dynamic acoustic emission signals (AE) measured just prior to rapid slip. (**d**) The AE signals from piezoelectric sensors (PZ1–PZ13) are shown 209.7 ms before SS3. A total of 24 foreshocks were detected and located using the P wave components, such as that seen in (**e**).

6.1. Slow Slip Distribution and Foreshock Locations

It has been found that the real distribution of asperities controls fault behavior for a dry friction fault [16]. Figure 11a shows estimated asperity locations measured using the pressure-sensitive film.

While, in reality, asperities seem to form convoluted shapes [39,43], they are represented here as a circular patch with an equivalent area. Graphically, we represented a region of subdued slip by the hatched region shown in Figure 11a. Foreshocks occurred only in this hatched region, as seen in Figure 11c. Figure 11c shows the foreshock catalog for SS2 (red triangles) and SS3 (blue squares). The field of view (FOV) of the video camera is shown here in gray. The size of the symbols are proportional to the peak ground displacements (PGD) averaged over the closest three AE sensors used to locate the foreshock [16]. Figure 11b presents the accumulation of slow slip using the seven slip sensors prior to the first stick-slip event (SS1). Measurements on all seven slip sensors (NC1–NC7) were interpolated at 50-ms time intervals. The time associated with the onset of rapid slip according to our test, *i.e.*, $t_{fail} = 0$ s, was the time that any sensors breached the slip velocity threshold of $v_{local} > 10$ mm/s. The grid composed of slip values on all seven sensors (NC1–NC7) was then interpolated in time to provide visual insight into the slow-slip distributions. From Figure 11b, we see that slip began accumulating from the trailing edge (TE) region of the sample and propagated (relatively) slowly into the fault over the 10 s before tfail. This was likely caused by larger and denser distributions of asperities and the preferential loading conditions.

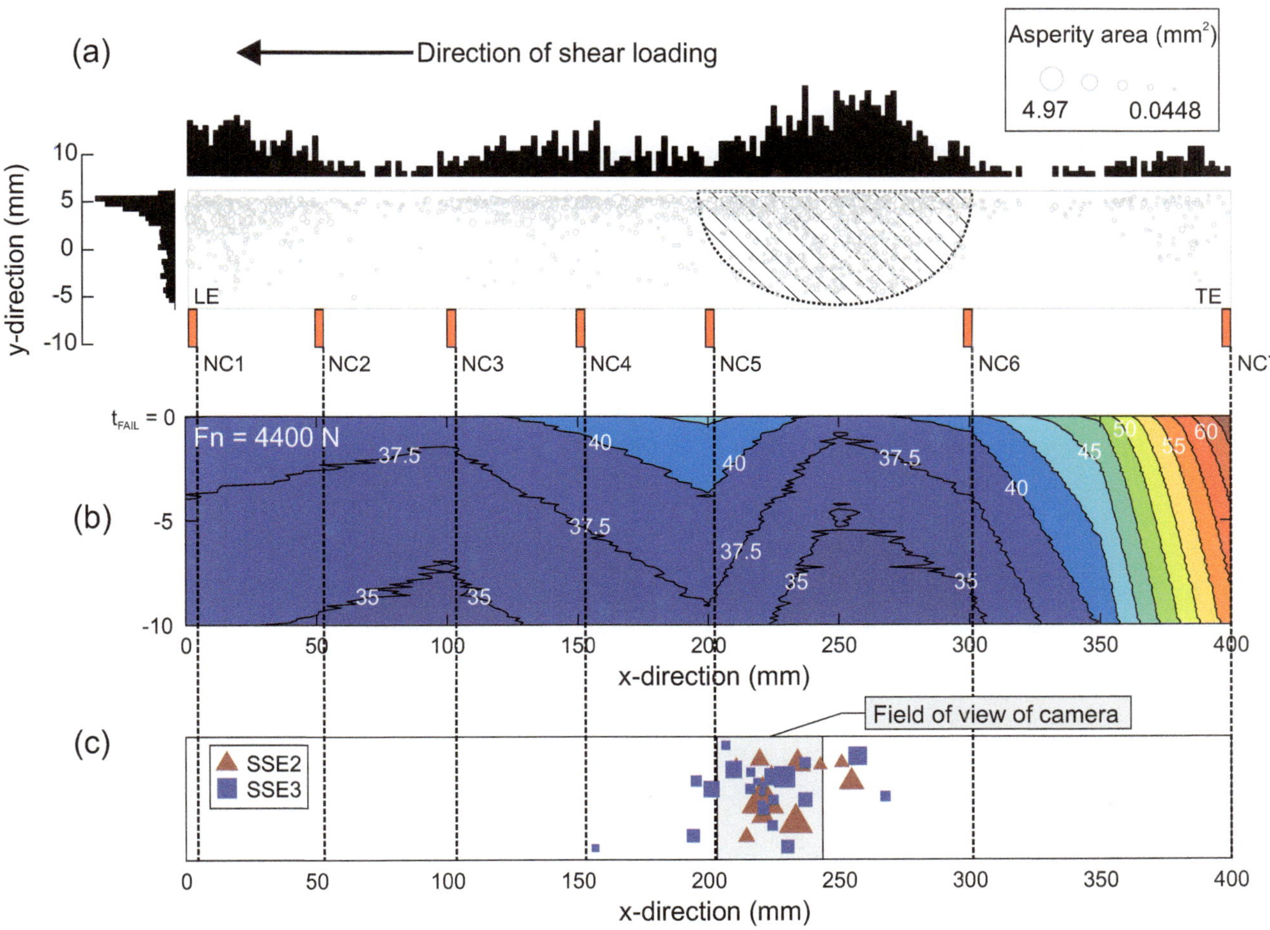

Figure 11. (**a**) Asperity locations are shown as circular patches on the interface with areas equivalent to that measured using the pressure-sensitive film; (**b**) slow slip measured along the fault using the slip non-contact sensors (NC1–NC7) over the last 10 s before rapid slip; (**c**) locations of foreshocks for SS2 (red triangles) and SS3 (blue squares). The sizes of the symbols are proportional to the magnitude of shaking from the event.

6.2. Optical Results in Relation to Localized Acoustic Events

The timing of the camera and the AE data acquisition was not synchronized *a priori* in this experiment. We rely solely on the frame-to-frame images and the changes in asperity transmitted luminous intensity between them. During the slow slip portion of the experiment, changes in the FOV occur slowly enough that images remain focused. During the transition and acceleration of shear rupture, the image suddenly becomes blurred. The first frame that was blurred was called the failure frame F_{tfail}. The frames prior to these are referred to as $F_{tfail} - 1$, $F_{tfail} - 2$, $F_{tfail} - 3$, *etc.*, that occurred at approximate times of $t_{fail} - 33.3$ ms, $t_{fail} - 66.6$ ms and $t_{fail} - 99.9$ ms. We then examine the changes in light intensity in frames leading up to failure (t_{fail}) during which foreshock sequences were observed (see Figure 11d). We consider only the SS2 foreshock (Figure 11c). The AE data showed that foreshock sequence SS2 occurred over a time span of 161.222 ms. Dividing by the frame rate of the camera (Δt_{frame}), we get 4.8 and 6.1, indicating that there was a minimum of four frames for the foreshock period.

Figure 12 shows the failure frame, F_{tfail} (for SS2), converted from a true-color image RGB to luminous intensity I in the same manner as described in Section 4 and inverted for clarity. The distorted grid accounted for orthoscopic distortions and was constructed along the interface using the same software as in Figure 8. In these images, the direction of shear loading was from left to right, and the location of the interface on the x-axis, referring to the coordinate system in Figure 11, was displayed above the grid. Only the foreshocks present in the field of view (FOV) are shown in Figure 11d. They are represented as black triangles, and their size is proportional to the peak ground displacement of each event [16]. The locations of foreshocks (FS), which aligned with asperities displaying higher luminous intensity, were analyzed in more detail.

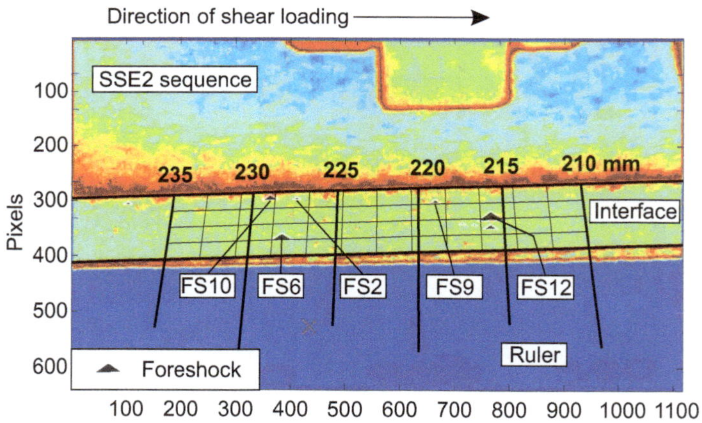

Figure 12. Images from the video camera at frame F_{tfail} for SS2. A non-uniform grid was constructed using image-processing software, and foreshock locations were superimposed on each image. Shear loading was from left to right.

Now, we must examine the specific regions on the interface where the foreshocks were located, which is shown in Figure 13. For example, foreshocks FS6 and FS9 occur between frame F_{tfail} and $F_{tfail} - 1$ (according to Figure 13b), and there is a definitive change in light intensity near the location of the event. Now that an FS event from frame-to-frame (concomitant with the acoustic location in space) is located, we can move backwards and forwards in time, shifting the camera timing a maximum of

$\Delta t_{frame} - \Delta$(FS6−FS9), *i.e.*, 33 − (9) = 24 ms. This methodology is referred to as clapboard timing, used to appropriately align the timing of the video camera frames with the dynamic foreshocks measured acoustically. The clapboard timing approach gives a possible upper estimate on the true length scale of the region causing this dynamic signal (Figure 6a). Estimates of the x-direction length scale are shown using the white boxes in Figure 14 between the frame-to-frame changes. Note that more foreshocks spaced over a longer time interval would provide more accurate constraints on the clapboard timing of the video camera.

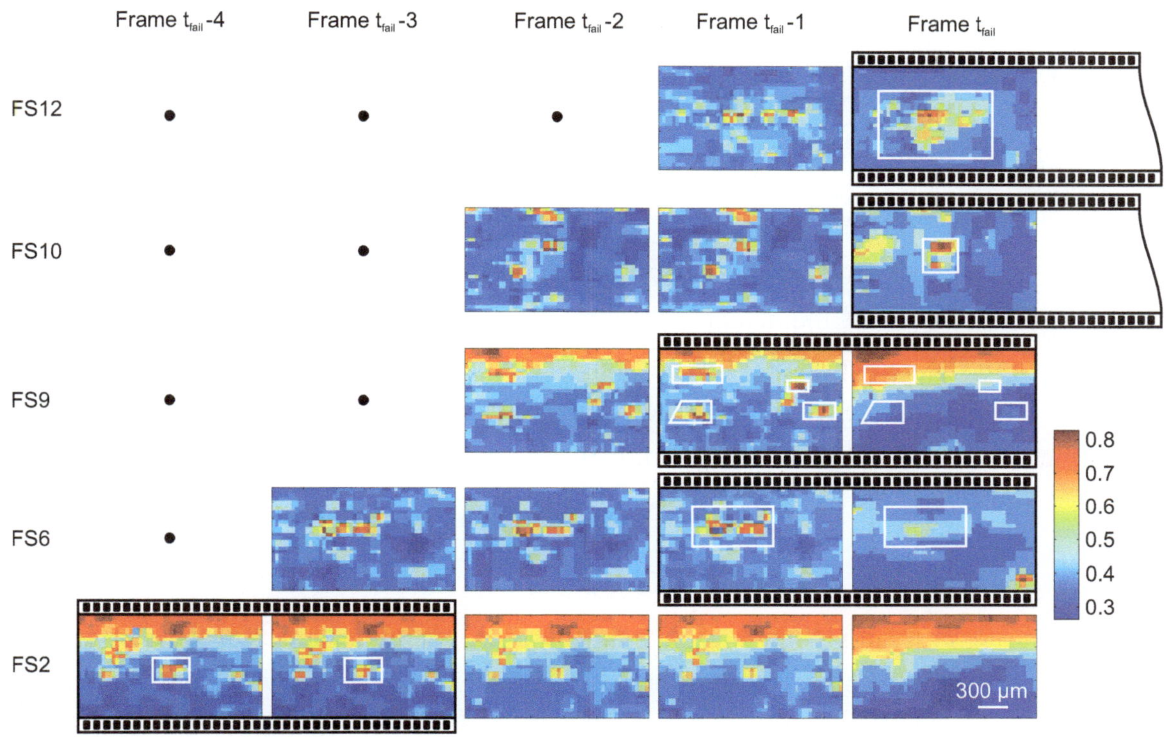

Figure 13. Final five frames prior to SS2 detailing the images taken in the location of each foreshock shown in Figure 12a. Changes in light intensity occurring between frames are highlighted by the bordered black boxes. Frame F_{tfail} +1 was not shown, since it was blurred. White boxes give an estimate of the regions in which light intensity changed.

We focus on the capabilities (and limitations) of associating the sudden frame-to-frame changes in individual asperity luminous intensity with the location (and timing) of events detected using the acoustic array. In Figure 14, we look at the acoustic signals and attempt to reconcile the timing of foreshocks with respect to the images from the video camera. Figure 14a shows the AE signals from sensor PZ5 for five foreshocks (FS2, FS6, FS9, FS10 and FS12) within the cameras field of view during SS2. These foreshocks have also been marked spatially in Figure 12 (top). In Figure 14b, the right-hand side shows the timing of foreshocks determined using the acoustic data, and the time between each foreshock was shown (Δt). To the left are the images taken from the camera from frame F_{tfail} to F_{tfail} 4, and the spacing between each frame was Δt_{frame} (33 ms). Foreshock signals have been aligned about the peak P-wave arrival and in ascending order of occurrence along the increasing y-axis. We assumed that these signals are created from the sudden failure of asperities along the interface [17,42]. If this is the case, we expect to see sudden frame-to-frame changes in the luminous intensity passing through the asperity after the foreshock due to either the destruction (or partial destruction) of the asperity.

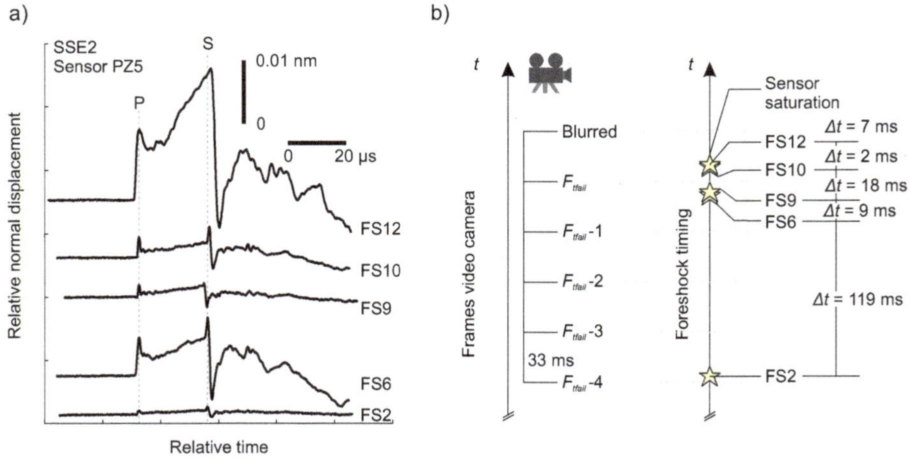

Figure 14. (**a**) AE signals from PZ5 for five foreshocks in the SS2 sequence, which were located in the field of view of the camera; (**b**) the left-hand side shows a schematic of the frame separation of the camera along time in the *y*-direction. Foreshocks shown in (a) are placed on the time axis on the right-hand side. Changes in light intensity due to foreshocks between frames allowed us to initialize the timing of the camera.

7. Conclusions

We presented three novel methodologies to better understand random contact processes occurring when an interface is created between two rough surfaces. The ability to develop a better understanding depends entirely on the sensors employed and the user's ability to properly interpret (*i.e.*, calibrate) the signals produced. Friction caused between two contacting surfaces is a problem that is misunderstood and widely studied in both engineering and scientific disciplines. In this study, we used a simple contact problem and determined that the pressure-sensitive film is a valid sensor to measure the normal stress distribution over a relatively small spatial resolution (micron length scale). Once validated, the film was used to estimate the relation between photometric light transmitted through contacting asperities and its dependence on the local normal stress. We found that increased normal stress caused increased transmission of luminous intensity for a known normal stress range. Photometry of the asperities along the interface becomes a valid tool for investigating dynamic stress changes in more intricate tests. The calibration of the pressure-sensitive film and photometric light measurements was done under simple normal loading.

We utilized both the non-contact sensor and the acoustic arrays during a direct shear experiment. Before the test, the pressure-sensitive film was used to characterize the interface, and during the test, the video camera took asperity photometric measurements from a specific "region of interest". Shear stress accumulated along the fault until failure, *i.e.*, a stick-slip event, where slow slip accrued non-uniformly along the fault. Within the "region of interest" (focused on by the camera), we observed decreased amounts of slow slip and foreshock events measured acoustically. The camera, focused on this region beforehand, capturing sudden changes in asperity luminosity occurring from frame-to-frame in the video prior to failure. These changes were concomitant, in both space and time, with the localized foreshocks measured acoustically. Using the acousto-optical method presented here, we display our ability to measure the sudden changes in light intensity caused from the sudden failure (or partial failure) of asperities. This is the first time that an acoustic emission event has been absolutely identified that was visually concomitant with respect to a physical event having known dynamics on a frictional fault.

The fact that the prescale pressure-sensitive film has been calibrated independently may move experimenters to apply it to their projects. While possibly of less utility, the optical method can be an important tool to calibrate the contact surface being worked with. Our results show the film's ability to estimate roughness characteristics along surfaces. We believe that the film can be employed to assess the health of a structural component that may be susceptible to wear or attrition. It is especially useful for the fitting and testing of flanges on pipelines and other precision jointings. For SHM, this allows quantification of corrosion. The film is ideal for quantifying track conditions, important for high-speed rail. Equipment that requires very smooth surfaces, such as turbines and ship props, experiences pitting caused by cavitation of the fluid near its surface that can be captured by the film. We believe that the methods presented in this paper have great possibilities for many researchers.

Acknowledgments

All datasets required to reproduce the results presented here are freely available to the reader upon request. Please contact the corresponding author for access. Funding for this research was provided by the National Science Foundation Grant (CMMI-1131582) to the University of California and the Jane Lewis Fellowship (University of California, Berkeley) and the National Science and Engineering Research Council of Canada (PGSD3-391943-2010) provided to Paul Selvadurai.

Author Contributions

Preparation of all experimental facilities, laboratory testing, data processing, numerical modeling and manuscript preparation was performed by Paul Selvadurai. Steven Glaser provided vital feedback and a number of full revisions of the overall manuscript. Funding was acquired by Steven Glaser via a National Science Foundation Grant (CMMI-1131582).

Conflicts of Interest

The authors declare no conflict of interest.

References

1. Baumberger, T. Contact dynamics and friction at a solid-solid interface: Material versus statistical aspects. *Solid State Commun.* **1997**, *102*, 175–185.
2. Pitenis, A.A.; Dowson, D.; Gregory Sawyer, W. Leonardo da Vinci's Friction Experiments: An Old Story Acknowledged and Repeated. *Tribol. Lett.* **2014**, *56*, 509–515.
3. Bowden, F.P.; Tabor, D. *The Friction and Lubrication of Solids*; Oxford University Press: New York, NY, USA, 2001.

4. Rabinowicz, E. Autocorrelation Analysis of the Sliding Process. *J. Appl. Phys.* **1956**, *27*, 131–135.
5. Archard, J.F. Elastic Deformation and the Laws of Friction. *Proc. R. Soc. Lond. A* **1957**, *243*, 190–205.
6. Greenwood, J.A.; Williamson, J.B.P. Contact of Nominally Flat Surfaces. *Proc. R. Soc. Lond. A* **1966**, *295*, 300–319.
7. Johnson, K.L. *Contact Mechanics*; Cambridge University Press: Cambridge, UK, 1985.
8. Dieterich, J.H. Modeling of rock friction: 1. Experimental results and constitutive equations. *J. Geophys. Res.* **1979**, *84*, 2161–2168.
9. Ruina, A. Slip instability and state variable friction laws. *J. Geophys. Res.* **1983**, *88*, 10359–10370.
10. Persson, B.N.J. Contact mechanics for randomly rough surfaces. *Surf. Sci. Rep.* **2006**, *61*, 201–227.
11. Scholz, C.H. *The Mechanics of Earthquakes and Faulting*, 2nd ed.; Cambridge University Press: Cambridge, UK, 2002.
12. Dieterich, J.H.; Kilgore, B.D. Direct observation of frictional contacts: New insights for state-dependent properties. *Pure Appl. Geophys.* **1994**, *143*, 283–302.
13. Dieterich, J.H.; Kilgore, B.D. Implications of fault constitutive properties for earthquake prediction. *Proc. Natl. Acad. Sci. USA* **1996**, *93*, 3787–3794.
14. Dieterich, J.H.; Kilgore, B.D. Imaging surface contacts: Power law contact distributions and contact stresses in quartz, calcite, glass and acrylic plastic. *Tectonophysics* **1996**, *256*, 219–239.
15. McLaskey, G.C.; Kilgore, B.D. Foreshocks during the nucleation of stick-slip instability. *J. Geophys. Res.* **2013**, *118*, 2982–2997.
16. Selvadurai, P.A.; Glaser, S.D. Laboratory-developed Contact Models Controlling Instability on Frictional. *J. Geophys. Res.* **2014**, in press.
17. Selvadurai, P.A.; Glaser, S.D. Insights into dynamic asperity failure in the laboratory. In Proceedings of the 48th US Rock Mechanics/Geomechanics Symposium, Minneapolis, MN, USA, 1–4 June 2014.
18. Ohnaka, M.; Shen, L.F. Scaling of the shear rupture process from nucleation to dynamic propagation: Implications of geometric irregularity of the rupturing surfaces. *J. Geophys. Res.* **1999**, *104*, 817–844.
19. McLaskey, G.C.; Glaser, S.D.; Grosse, C.U. Beamforming array techniques for microseismic monitoring of large structures. *J. Sound Vib.* **2010**, *329*, 2384–2394.
20. McLaskey, G.C.; Thomas, A.M.; Glaser, S.D.; Nadeau, R.M. Fault healing promotes high-frequency earthquakes in laboratory experiments and on natural faults. *Nature* **2012**, *491*, 101–105.
21. Schmittbuhl, J.; Chambon, G.; Hansen, A.; Bouchon, M. Are stress distributions along faults the signature of asperity squeeze? *Geophys. Res. Lett.* **2006**, *33*, doi:10.1029/2006GL025952.
22. Candela, T.; Renard, F.; Bouchon, M.; Schmittbuhl, J.; Brodsky, E.E. Stress Drop during Earthquakes: Effect of Fault Roughness Scaling. *Bull. Seismol. Soc. Am.* **2011**, *101*, 2369–2387.
23. Baumberger, T.; Caroli, C. Solid friction from stick-slip down to pinning and aging. *Adv. Phys.* **2006**, *55*, 279–348.

24. Arruda, E.M.; Boyce, M.C.; Jayachandran, R. Effects of strain rate, temperature and thermomechanical coupling on the finite strain deformation of glassy polymers. *Mech. Mater.* **1995**, *19*, 193–212.
25. Boyce, M.C.; Parks, D.M.; Argon, A.S. Large inelastic deformation of glassy polymers. Part I: Rate dependent constitutive model. *Mech. Mater.* **1988**, *7*, 15–33.
26. MATLAB. *Version 8.3 (R2014a)*; The MathWorks Inc.: Natick, MA, USA, 2014.
27. Gonzalez, R.C.; Woods, R.; Eddins, S.L. *Digital Image Processing Using MATLAB*; Gatesmark Publishing: Knoxville, TN, USA, 2009.
28. Timoskenko, S.P.; Goodier, J.N. *Theory of Elasticity*; McGraw-Hill: New York, NY, USA, 1970.
29. Nayak, P.R. Random process model of rough surfaces in plastic contact. *Wear* **1973**, *26*, 305–333.
30. Persson, B.N.J. Sliding Friction. *Surf. Sci. Rep.* **1999**, *33*, 89–119.
31. Greenwood, J.A. A simplified elliptic model of rough surface contact. *Wear* **2006**, *261*, 191–200.
32. Greenwood, J.A. A note on Nayak's third paper. *Wear* **2007**, *262*, 225–227.
33. Carbone, G.; Bottiglione, F. Asperity contact theories: Do they predict linearity between contact area and load? *J. Mech. Phys. Solids* **2008**, *56*, 2555–2572.
34. Ciavarella, M.; Greenwood, J.; Paggi, M. Inclusion of interaction in the Greenwood and Williamson contact theory. *Wear* **2008**, *265*, 729–734.
35. Afferrante, L.; Carbone, G.; Demelio, G. Interacting and coalescing Hertzian asperities: A new multiasperity contact model. *Wear* **2012**, *278–279*, 28–33.
36. Putignano, C.; Afferrante, L.; Carbone, G.; Demelio, G. The influence of the statistical properties of self-affine surfaces in elastic contacts: A numerical investigation. *J. Mech. Phys. Solids* **2012**, *60*, 973–982.
37. Yastrebov, V.A.; Anciaux, G.; Molinari, J.F. Contact between representative rough surfaces. *Phys. Rev. E* **2012**, *86*, 035601.
38. Krick, B.; Vail, J.; Persson, B.; Sawyer, W. Optical In Situ Micro Tribometer for Analysis of Real Contact Area for Contact Mechanics, Adhesion, and Sliding Experiments. *Tribol. Lett.* **2012**, *45*, 185–194.
39. Selvadurai, P.A.; Glaser, S.D. Direct measurement of contact area and seismic stress along a sliding interface. In Proceedings of the 46th US Rock Mechanics/Geomechanics Symposium, Chicago, IL, USA, 24–27 June 2012.
40. Langers, S.; Weatherley, D.; Olsen-Kettle, L.; Finzi, Y. Stress heterogeneities in earthquake rupture experiments with material contrasts. *J. Mech. Phys. Solids* **2013**, *61*, 742–761.
41. Archard, J.F. Single Contacts and Multiple Encounters. *J. Appl. Geophys.* **1961**, *32*, 1420–1425.
42. McLaskey, G.C.; Glaser, S.D. Micromechanics of asperity rupture during laboratory stick slip experiments. *Geophys. Res. Lett.* **2011**, *38*, doi:10.1029/2011GL047507.
43. Selvadurai, P.A.; Glaser, S.D. Asperity formations and their relation to seismicity on a planar fault. Unpublished work, 2015.

12

Diamond Nanowires: A Novel Platform for Electrochemistry and Matrix-Free Mass Spectrometry

Sabine Szunerits *, Yannick Coffinier and Rabah Boukherroub

Institute of Electronics, Microelectronics and Nanotechnology (IEMN), UMR-CNRS 8520, Université Lille 1, Avenue Poincaré—BP 60069, 59655 Villeneuve d'Ascq, France; E-Mails: yannick.coffinier@iri.univ-lille1.fr (Y.C.); rabah.boukherroub@univ-lille1.fr (R.B.)

* Author to whom correspondence should be addressed; E-Mail: sabine.szunerits@iri.univ-lille1.fr

Academic Editor: Stephen Holler

Abstract: Over the last decades, carbon-based nanostructures have generated a huge interest from both fundamental and technological viewpoints owing to their physicochemical characteristics, markedly different from their corresponding bulk states. Among these nanostructured materials, carbon nanotubes (CNTs), and more recently graphene and its derivatives, hold a central position. The large amount of work devoted to these materials is driven not only by their unique mechanical and electrical properties, but also by the advances made in synthetic methods to produce these materials in large quantities with reasonably controllable morphologies. While much less studied than CNTs and graphene, diamond nanowires, the diamond analogue of CNTs, hold promise for several important applications. Diamond nanowires display several advantages such as chemical inertness, high mechanical strength, high thermal and electrical conductivity, together with proven biocompatibility and existence of various strategies to functionalize their surface. The unique physicochemical properties of diamond nanowires have generated wide interest for their use as fillers in nanocomposites, as light detectors and emitters, as substrates for nanoelectronic devices, as tips for scanning probe microscopy as well as for sensing applications. In the past few years, studies on boron-doped diamond nanowires (BDD NWs) focused on increasing their electrochemical active surface area to achieve higher sensitivity and selectivity compared to planar diamond interfaces. The first part of the present review article will cover the promising applications of BDD NWS for label-free sensing. Then, the potential use of diamond nanowires as inorganic substrates for matrix-free laser desorption/ionization mass spectrometry, a powerful label-free approach for quantification and identification of small compounds, will be discussed.

Keywords: diamond nanowires; diamond nanostructures; synthetic methods; electrochemical sensing; mass spectrometry; SALDI

---

1. Introduction

Diamond, a natural as well as a synthetic material, has captured researchers' attention since decades. From any list summarizing the specific material properties, diamond is often at the extreme [1]: crystalline diamond shows the highest atomic density of any bulk crystal, the highest bulk modulus and the highest thermal conductivity. Diamond, a wide band gap semiconductor, is optically transparent from the far infrared to the ultraviolet, making it an ideal candidate for optical applications. During the growth of diamond films using chemical vapor deposition (CVD) systems, dopants and impurities can be readily incorporated into the material, allowing for tuning its optical and electrical properties. Substantial progress has been made in this area using boron for *p*-type doping and heavily boron-doped diamond (BDD) films (B-doping levels > 10^{20} cm^{-3}) are now produced routinely for electrochemical investigations. The interest of BDD films for electrochemical sensing arises from the wide potential window and negligible capacitive current achieved as well as their stability required for use in *in vitro* biosensing applications [2–7]. Figure 1 summarizes some examples of BDD applications in electrochemical sensing.

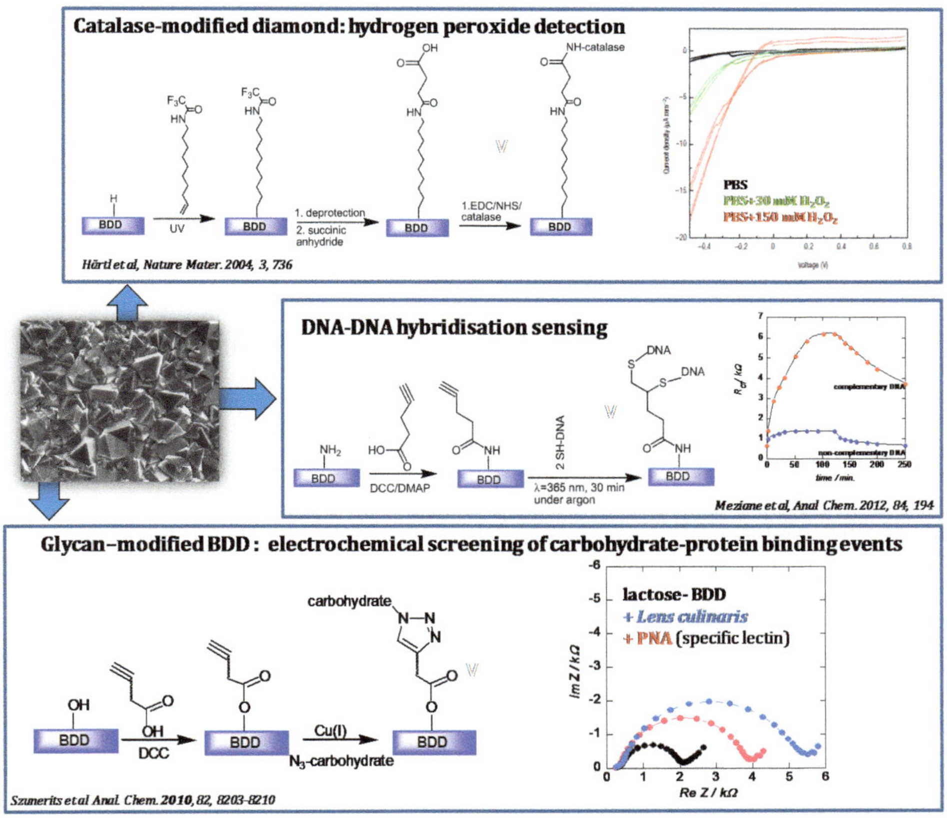

Figure 1. Selected examples of electrochemical sensing using BDD electrodes.

The attractiveness of diamond is that different morphologies and forms can be obtained from this sp^3 hybridized material (Figure 2). Indeed, modulation of the growth parameters results in microcrystalline to ultrananocrystalline CVD diamond films. Ultrananocrystalline films have the advantage of smooth surfaces, lower strain and improved fracture resistance. Such films are characterized by diamond domains that are ≈10 nm or less in size with thin sp^2-bonded boundaries.

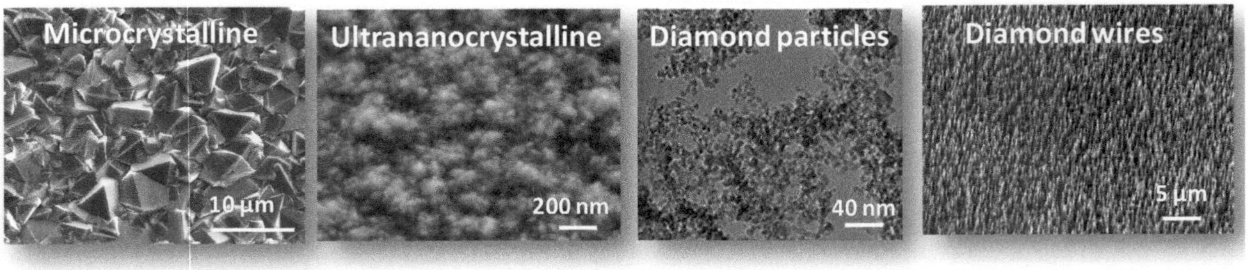

Figure 2. Different morphologies and forms of diamond.

Nanoscale diamond particles (also termed nanodiamonds, NDs) represent another interesting form widely explored for applications in drug delivery or medical diagnostics. The first records of the production of NDs date back to the 1960s, when a group of Soviet scientists discovered single crystals of cubic diamond particles in soot produced by detonating an oxygen-deficient TNT/hexogen composition in inert media without using any extra carbon source [8]. Since then, NDs have been rapidly gaining popularity in bioimaging and drug delivery applications, since colloidal suspensions of individual diamond particles with diameter of 4–10 nm are commercially available.

Diamond in the form of nanowires has to be added to this list. The use of diamond nanowires is believed to address positively issues related to improving the overall performance of sensors, including sensitivity and selectivity [9–12]. For a long time, the routine use of diamond nanowires was restricted, as no viable methods for their fabrication were available. The first attempt to synthesize diamond nanowires dates back to 1968 using a radiation heating unit developed from a super-high pressure Xenon lamp [13]. Only limited progress in this direction was made due to the difficulty of controlling the dimension of the diamond filament and lack of characterization facilities at that time, restricting further investigation of such diamond-based whiskers.

The intent of this review article is to make the reader more familiar with recent developments for the preparation of diamond nanostructures and their use for label-free sensing applications. The different synthetic methods can be classified into two main approaches: "top-down" and "bottom-up" approaches. Due to their high surface area, boron-doped diamond nanowires (BDD NWs) represent an interesting platform for electrochemical sensing as compared to planar BDD electrodes. This will be demonstrated in various examples of electrochemical sensing of different chemical/biological species using BDD NWs. Finally, the potential applications of diamond nanowires as inorganic matrix for surface-assisted laser desorption/ionization mass spectrometry (SALDI-MS) will be discussed in details.

2. Synthetic Routes of Diamond Nanowires

Reports on the fabrication of diamond structured surfaces with diameters as small as 25 μm and hundreds of microns in length date back to the 1960s [13]. However, it was only around the beginning of the 21st century that further attempts for the synthesis of diamond nanostructures were undertaken. The main approaches for the successful fabrication of diamond nanostructures are generally based on "top-down" and "bottom-up" processes.

2.1. Top-Down Approach

One of the initial attempts for a top-down synthesis of diamond nanostructures was reported by Shiomi [14]. Reactive ion etching (RIE) with oxygen plasma of CVD diamond film coated with a 400 nm thick aluminum layer resulted in columnar structures of approximately 300 nm in length and 10 nm in diameter (Figure 3A) [14]. The plasma-assisted RIE technology has since then been widely investigated for the top-down fabrication of diamond nanowires and pillars. Masuda *et al.* proposed, for example, the use of porous anodic aluminum oxide masks for the formation of diamond honeycomb films via oxygen plasma etching of CVD diamond films through the holes of porous alumina films (Figure 3B) [15]. They showed that the etching rate of alumina, compared to that of the diamond film, is negligible, resulting in the formation of honeycomb structures with high aspect ratios. Uniform holes with an average diameter of 70 nm and spacing of 100 nm with uniform depth of ~0.6 μm have been etched perpendicular to the diamond film surface, yielding an aspect ratio of ~9. An important fact is that this lithographic process was carried out in non-contact mode: the mask is merely placed on the substrate and therefore does not adhere to the substrate surface, unlike the situation with the resist-type masks commonly used in lithography. This makes the approach fast and easy to perform on different diamond interfaces.

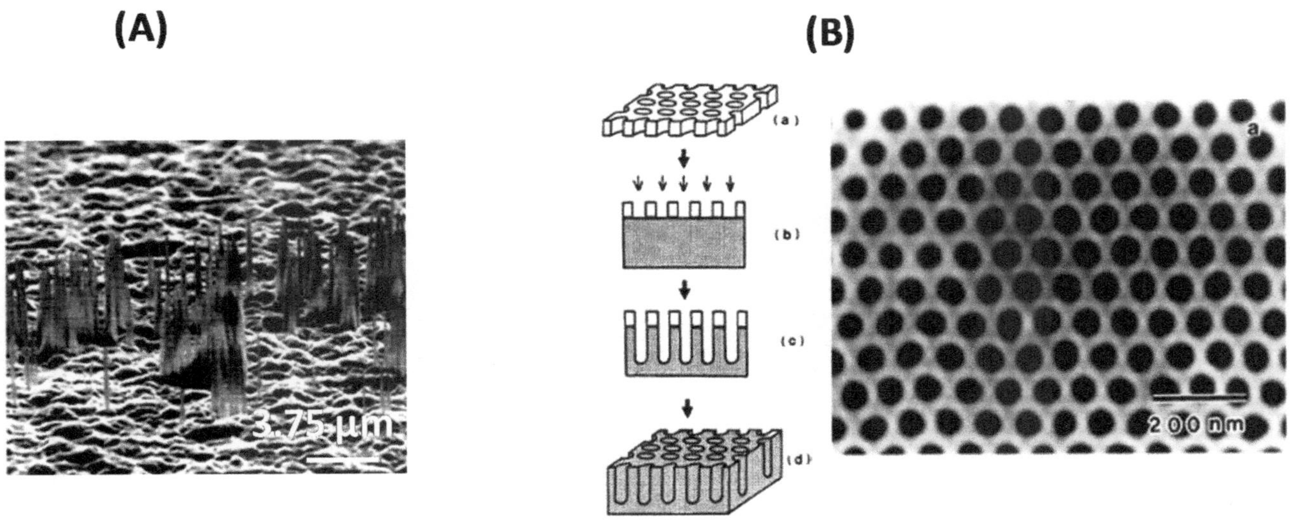

Figure 3. (**A**) SEM images of diamond nanowiskers formed on as-grown diamond films coated with 400 nm thick Al layer using reactive ion etching (RIE) with oxygen plasma (reprint with permission from [16]); (**B**) Formation of honeycomb diamond films (reprint with permission from [15]).

Beside such masks, arrays of nanoparticles, seeded onto CVD grown diamond films, have been investigated by several groups as attractive alternatives. Aluminum [17], SiO_2 [18], gold [19], as well as diamond nanoparticles [11,20] proved to be useful etching masks (Figure 4A). Okuyama *et al.* used RIE with oxygen plasma through a two-dimensionally ordered SiO_2 particle array to form diamond cylinders [18]. The diameter and the length of the cylinders depend on the etching time and vary between 0.6–1 μm in diameter and 3–4 μm in length (Figure 4B) [18]. High-density and uniform diamond nanopillar arrays were obtained by Zou *et al.* by employing bias-assisted RIE in a hydrogen/argon plasma using gold nanoislands of 150 nm in diameter as etching masks [19]. The gold islands protect the underlying diamond from etching and sputtering; nanopillars with gold clusters at the tip are produced. Gold nanoparticles were indeed found to be one of the most suitable seeding masks as they are easy to disperse, resulting in single nanoparticles on the surface requiring no further processing step. The etch rate of the gold nanoparticles mask is 25 nm/min [21]. This yields good etch selectivity and diamond wires of 900 nm in height, and diameters from 275 nm at bottom to 310 nm at the top can be produced [21].

The use of nanodiamond (ND) particles as a hard mask for RIE of diamond was examined by research groups at the AIST (Japan) and IAF (Germany) [11,20,22,23]. Vertically aligned diamond nanowires were obtained using RIE in an O_2/CF_4 (97/3%) gas mixture for etching times of 10 s (Figure 4C). The length of the wires was limited by simultaneous etching of the ND particles mask with an etching rate of 10 $Å \cdot s^{-1}$.

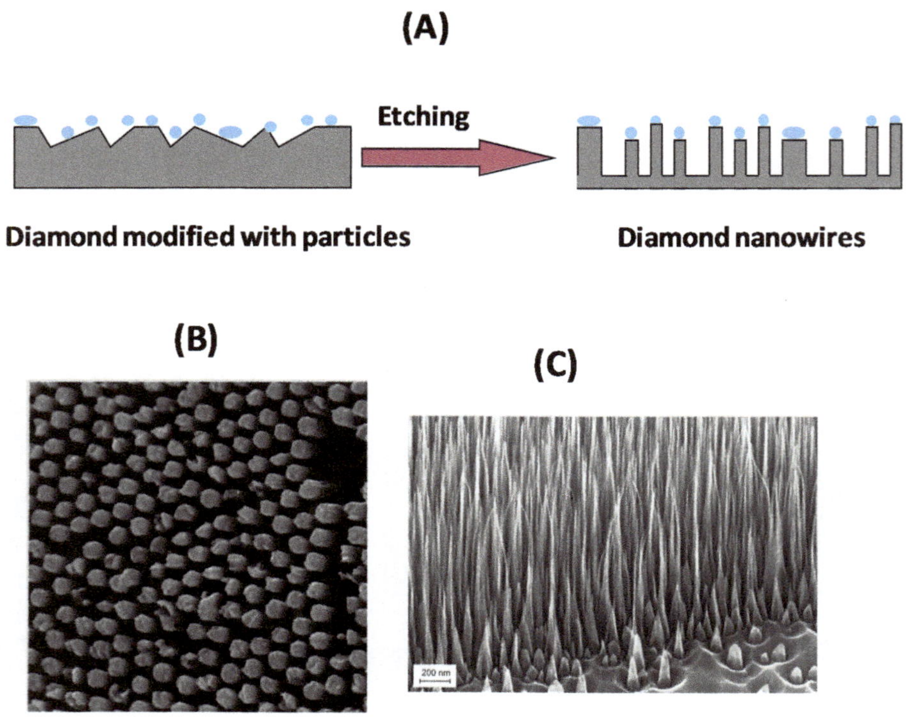

Figure 4. (**A**) Top-down etching process of CVD diamond using seed particles; (**B**) SEM image of diamond cylinders obtained using oxygen reactive ion etching (RIE) for 60 min through a 1 μM SiO_2 particle array (reprint with permission from [18]); (**C**) SEM image of vertically aligned diamond nanowires using diamond nanoparticles as masks (with courtesy of C. Nebel).

Mask-less top-down approaches have been recently proposed as alternatives [16,24–27] (Figure 5A). Such methods have the intrinsic advantage of being simple and straightforward, not requiring complicated processing steps such as mask deposition or template removal. Our group demonstrated that diamond nanowires can be easily prepared from highly boron-doped microcrystalline diamond thin films by RIE in an oxygen plasma (Figure 5A) [12,25–27]. The resulting nanowires are 1.4 ± 0.1 μm long with a tip and base radius of r_{tip} = 10 ± 5 nm and r_{base} = 40 ± 5 nm, respectively. The nanowires are about 140 times longer than aligned diamond nanowires prepared using diamond nanoparticles as a hard mask. X-ray photoelectron spectroscopy (XPS) analysis of the chemical composition of the diamond nanowires revealed that next to C_{1s} at 285 eV and O_{1s} at 532 eV, additional peaks at 402, 104 and 169 eV due respectively to N_{1s}, Si_{2p} and S_{2p} are present in the spectrum (Figure 5B). The latter elements are believed to originate from surface contamination during the RIE process. Indeed, the presence of SiO_x shell around the BDD NWs was confirmed by HR-TEM analysis (Figure 5B). SiO_x was most likely deposited during the etching process due to sputtering of the substrate holder or the silicon wafer onto which the diamond film was deposited. A similar behavior was observed by Baik *et al.* when Mo sample holder was used [16]. The SiO_x deposits can be easily removed through immersion in HF aqueous solutions.

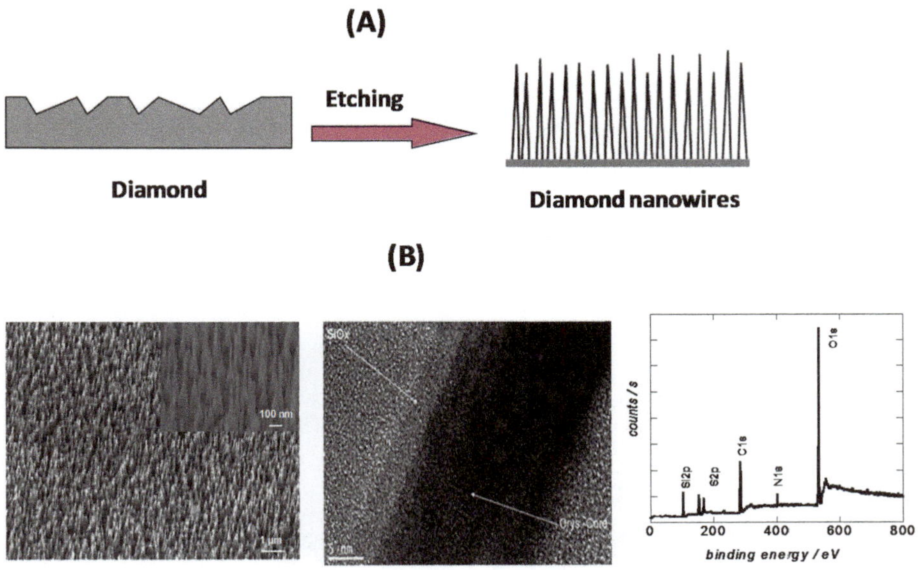

Figure 5. (**A**) Top-down etching process of CVD diamond without mask; (**B**) SEM and HR-TEM images, and XPS survey spectrum of boron-doped diamond nanowires synthesized through maskless technique (reprint with permission from [26]).

2.2. Bottom-Up Approach

As the formation of diamond nanowires by the top-down approach is often accompanied with surface damage, whose impact is generally greater at the nanoscale due to the large surface-to-volume ratio, bottom-up procedures were also investigated in the literature. One of the first bottom-up approaches was described by Masuda *et al.* (Figure 6A) [28]. In this technique, microcrystalline diamond nanocylinders were grown on anodic oxide templates using microwave plasma-assisted CVD and 50 nm nanodiamond particles as seeds. The density of the obtained wires was as high as 4.6 × 10^8 cylinders/cm^2 with a wire length of about 5 μm and 300 nm in diameter (Figure 6A). Diamond nanorods of 8–10 nm in diameter and up to 200 nm in length coated with an amorphous carbon layer were grown along the (110) direction upon applying a prolonged hydrogen plasma to multi-walled carbon nanotubes (MWCNTs) (Figure 6B) [29]. The authors suggested that initial diamond nuclei can be formed at defect sites of the MWCNTs due to the presence of hydrogen. At high temperature (1000 K) in the presence of hydrogen, MWCNTs transform to amorphous material, where the nucleation of diamond phase is facilitated. The addition of N_2 into the growth mixture of ultrananocrystalline diamond was reported by Vlasov and co-workers to change the surface morphology to wire-like structures [30]. The addition of 25% of N_2 to the $Ar/CH_4/H_2$ gas mixture resulted in the formation of hybrid diamond-graphite nanowires with lengths up to a few hundred nanometers. This hybrid material consists of a single crystalline diamond core of 5–6 nm in diameter oriented along the (110) principal axis and graphitic shells of different thicknesses covering the core. Using a mixture of N_2 and CH_4 allowed also the growth of ultrathin diamond nanorods by microwave-assisted CVD [31]. The resulting nanorods exhibited a diameter as small as 2.1 nm, which is not only smaller than any other reported diamond nanostructures but also smaller than the theoretical value of energetically stable diamond nanorods. More recently, the synthesis of straight, thin and long diamond nanowires using atmospheric-pressure chemical vapor deposition was proposed [32,33]. The diamond nanowires showed a uniform diameter of 60–90 nm with over tens of micrometers in length. Spectroscopic analysis provided information that these nanowires are diamond with high crystallinity and high structural uniformity.

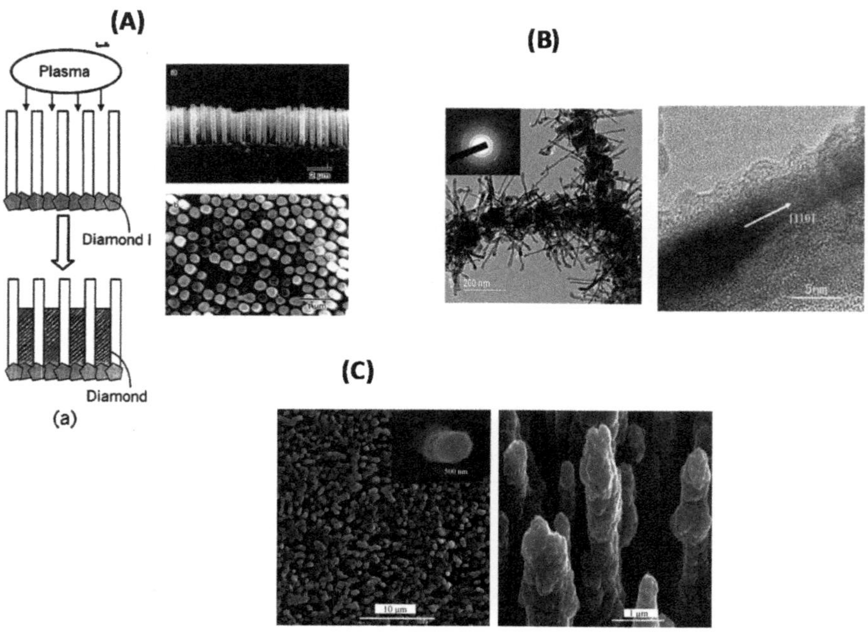

Figure 6. (**A**) Schematic diagram of the bottom-up fabrication of cylindrical diamond wires in porous alumina template (**left**) and the corresponding SEM images (reprint with permission from [28]); (**B**) TEM (**left**) and HR-TEM (**right**) images of diamond rods grown on carbon nanotubes (reprint with permission from [29]); (**C**) SEM image of diamond coated silicon nanowires (reprint with permission from [10]).

Post-coating of preformed silicon nanowires by diamond thin films has attracted considerable interest in the past five years for the preparation of core-shell nanowires [10,34,35]. Boron-doped diamond nanoforest electrode could be fabricated on pre-formed silicon nanowires as illustrated in Figure 6C [10].

Silicon nanowires (Si NWs) were formed according to a procedure reported by Peng *et al.* [36], and subsequently post-coated with boron-doped diamond thin films deposited by high frequency CVD technology. The coverage of the nanocrystalline diamond film is complete and continuous along the whole length (5 μm) of the Si NWs as seen in the SEM images in Figure 6C. A similar strategy was employed by Gao and co-workers to coat Si NWs with a 100 nm layer of nanocrystalline diamond by microwave-enhanced CVD [34]. The grafting of negatively-charged diamond nanoparticles over cationic polymer-coated nanometric patterns was proposed by Girard and co-workers as a bottom-up strategy towards 3D diamond nanostructures [35].

3. Applications of Diamond Nanowires

In general, the synthesis of diamond nanostructures has been advanced to a high level in a very short time span. The access to such nanostructures allowed finally the use of diamond nanowires for different applications ranging from solid-state electron emitters, high performance nano-electromechanical switches, probes for scanning probe microscopy and photonic systems, to the formation of superhydrophobic and oleophobic interfaces [15,21,26,31,33,37] (Figure 7). The use of boron-doped diamond nanowires has in particular found interest for electrochemical sensing [10–12,25,38–43] and as surface-assisted laser desorption/ionization (SALDI) matrix [27].

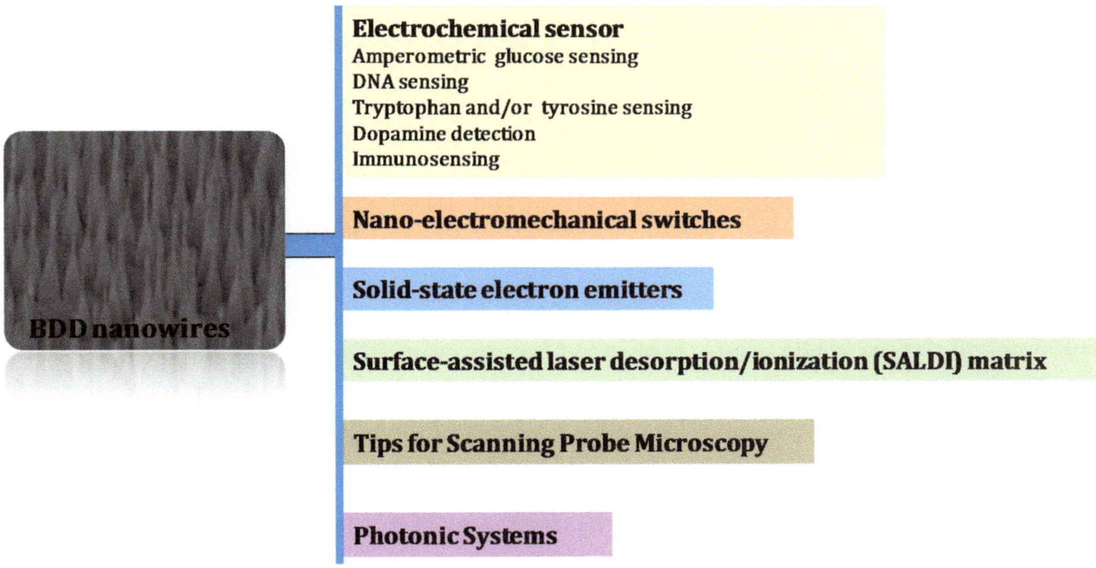

Figure 7. Applications of diamond nanowires.

3.1. Diamond Nanowires for Electrochemical Sensing

Diamond nanowires are among the fairly new but promising materials for chemical and biochemical sensing. The common theme of diamond sensors is that they convert biological or chemical information into an electrical signal, which can be measured accurately using a panel of electrochemical methods (e.g., cyclic voltammetry, differential pulse voltammetry, electrochemical impedance spectroscopy, *etc.*). As the technology required to create electrochemical biosensors is much cheaper than that required for fluorescence-based sensors, electrochemical sensors are dominating the analytical field. Their label-free character adds to their general value. While the conversion of a biological interaction to an electrical signal is attractive for sensors that are in continuous use or need to withstand harsh environments, so far, electrochemical sensors are in general several orders of magnitude less sensitive than the best fluorescence-based detection sensors.

Yang *et al.* were the first to demonstrate that the detection limit of electrochemical biosensors can be markedly improved if vertically aligned diamond nanowires are used [11]. DNA sensors were prepared through immobilization of single strand DNA probes onto diamond nanowires pre-functionalized with amine-terminated phenyl groups in an electrochemical functionalization step (Figure 8) [23,44]. The enhanced electrical field at the very end of the diamond tips resulted in a preferential DNA alignment at the tip rather than at the walls of the wires, increasing the probes' accessibility for interaction. This gave rise to optimized hybridization kinetics of complementary DNA (cDNA) and high sensitivity with a detection limit of 2 pM for cDNA as well as single-base mismatch discrimination.

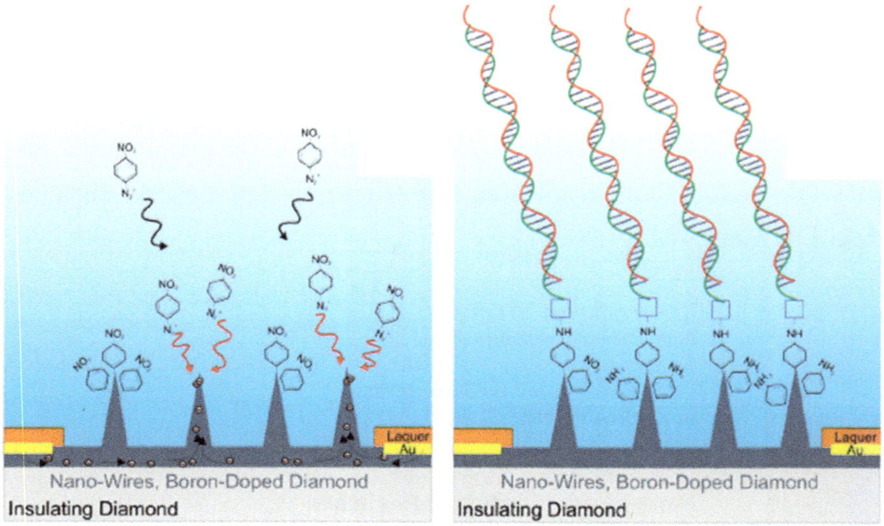

Figure 8. Diamond nanowires for DNA sensing: Preferential linking of phenyl aryl and DNA molecules to the tip of the wires (reprint with permission from [23,44]).

Diamond nanowires electrodes allow also the direct electrochemical detection of glucose under strong basic conditions [10,12,45]. While almost no visible anodic peak for glucose oxidation was observed during the positive potential scan on a planar BDD electrode, a well-defined current response for glucose was obtained, for example, on BDD NWs electrodes of ≈3 μm in length with a diameter ranging from 10–50 nm [12]. The detection limit of this sensor was 60 μM (Figure 9). Such an improvement in glucose oxidation suggests that the Faradaic current of glucose oxidation depends strongly on the surface structure and porosity of the electrode, and the accessible surface area. The results reported by Nebel and co-workers showed that decoration of diamond nanopillar electrodes with Ni-nanoparticles improves the sensitivity of the sensor for glucose detection with a detection limit of 10 μM [45]. Diamond coated Si nanowires were investigated by Luo *et al.* for glucose sensing with an estimated detection limit of 0.2 μM and a sensitivity of 8.1 $\mu A \cdot mM^{-1} \cdot cm^{-2}$ [10].

Diamond nanowires are also adapted for the sensitive electrochemical detection of aromatic amino acids such as tryptophan and tyrosine, two important precursors of adrenaline, dopamine or melatonine [12,25]. A detection limit of 5×10^{-7} M was recorded for tryptophan on BDD NWs electrodes [25], being significantly lower than on planar microcrystalline BDD (1×10^{-5} M) [46]. The simultaneous detection of tryptophan and tyrosine by differential pulse voltammetry is also possible on BDD NWs electrodes, when the amount of tryptophan present in the mixture is not exceeding tryptophan/tyrosine ≤ 0.5 [41]. The oxidation of other small molecules such as dopamine, uric acid and ascorbic acid was reported by Shalini *et al.* using nitrogen-doped diamond nanowire electrodes [47].

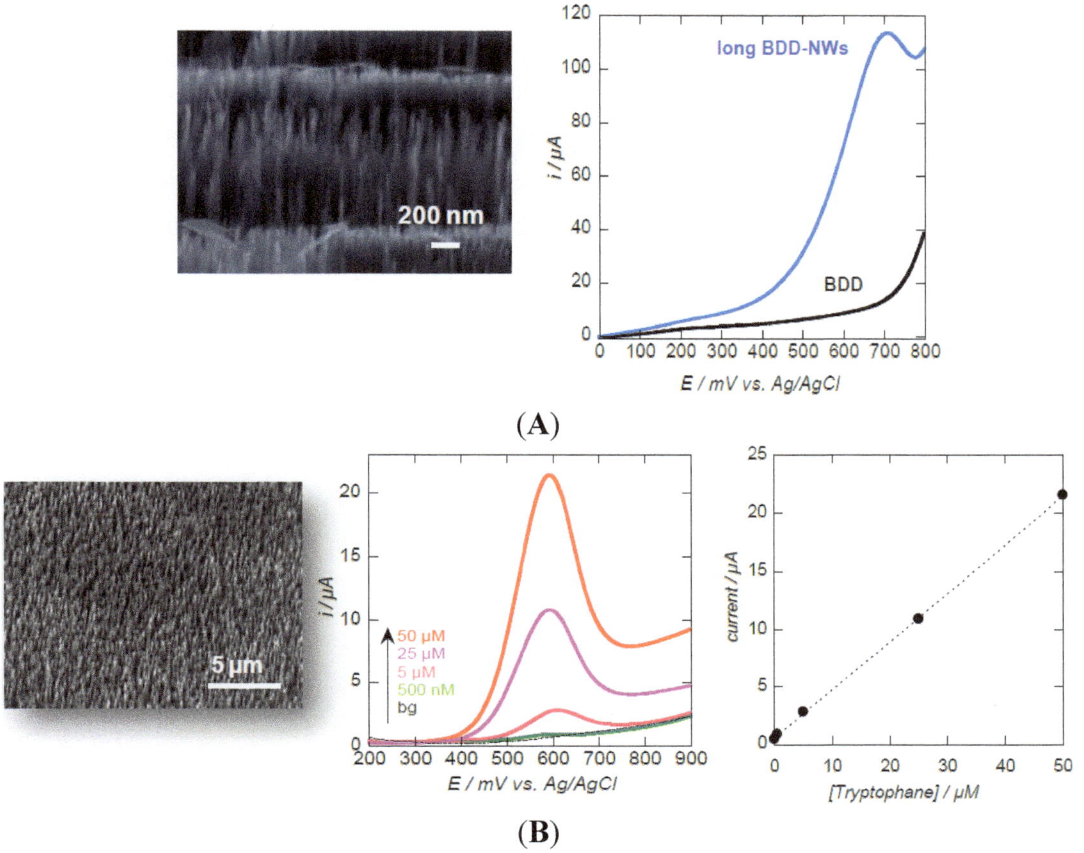

Figure 9. Diamond nanowires for enzyme-free glucose and tryptophan sensing: (**A**) SEM image of long diamond nanowires produced through mask-less RIE of CVD diamond films and linear sweep voltammogram recorded in 2 mM glucose solution (0.1 M NaOH) for BDD (black) and long BDD NWs (blue) (reprint with permission from [12]); (**B**) SEM image of short diamond nanowires together with differential pulse voltammograms for different concentrations of tryptophan and the corresponding calibration curve (reprint with permission from [25]).

Electrochemical immunosensors based on chemically modified BDD NWs electrodes have been lately developed by our group [38–40]. Diamond nanowires immunosensors were constructed by coating diamond nanowires with functional conducting polymer films (e.g., carboxylic acid-terminated poly(pyrrole), copper ion (Cu^{2+}) chelation followed by linkage of histidine-tagged peptides [39]) (Figure 10A) or by electrochemical deposition of nickel nanoparticles (Ni-NPs) onto diamond nanowires followed by immobilization of biotin-tagged anti-IgG (Figure 10B) [38,40]. Post-coating of diamond nanowires with polymer films can be achieved by amperometrically biasing diamond nanowire electrodes at 1.2 V *vs.* Ag/AgCl in 3-(pyrrole) carboxylic acid solution [39]. Fine-tuning the charge allowed coating the wires rather than the formation of polymer films (PPA) in solution. Figure 10A shows SEM images of BDD NWs coated with carboxylic acid-terminated polypyrrole (PPA-BDD NWs) by varying the deposition time. At very low deposition charge (2 mC·cm^{-2}), the polymer started to form preferentially at the defect sites of the interface *i.e.*, in-between the BDD NWs. Increasing the deposition charge to 11 mC·cm^{-2} resulted in polymer coated BDD NWs, while large deposition charges (23 mC·cm^{-2} and higher) led to a loss of the wire structure in favor of continuous film formation. The available carboxylic groups of the poly(pyrrole) coated wire electrode allows their coordination with copper ion (Cu^{2+}), known to be specific binding sites for His-tagged analytes. Indeed, the affinity constant (K_A) of His-Tag-des-Arg^6-Bradykinine peptide to Cu^{2+} coordinated carboxyl-terminated diamond wires was determined as $K_A = (1.15 \pm 0.5) \times 10^6$ M^{-1}, higher than that determined in the absence of Cu^{2+} ($K_A = (0.31 \pm 0.5) \times 10^6$ M^{-1}). Concentrations as low as 10 nM resulted in R_{CT} shift of 50 ± 22 Ω on these interfaces, while on a planar BDD interface modified with carboxylic groups and chelated with Cu^{2+}, His-tagged peptide concentrations had to exceed 100 nM to cause a comparable shift.

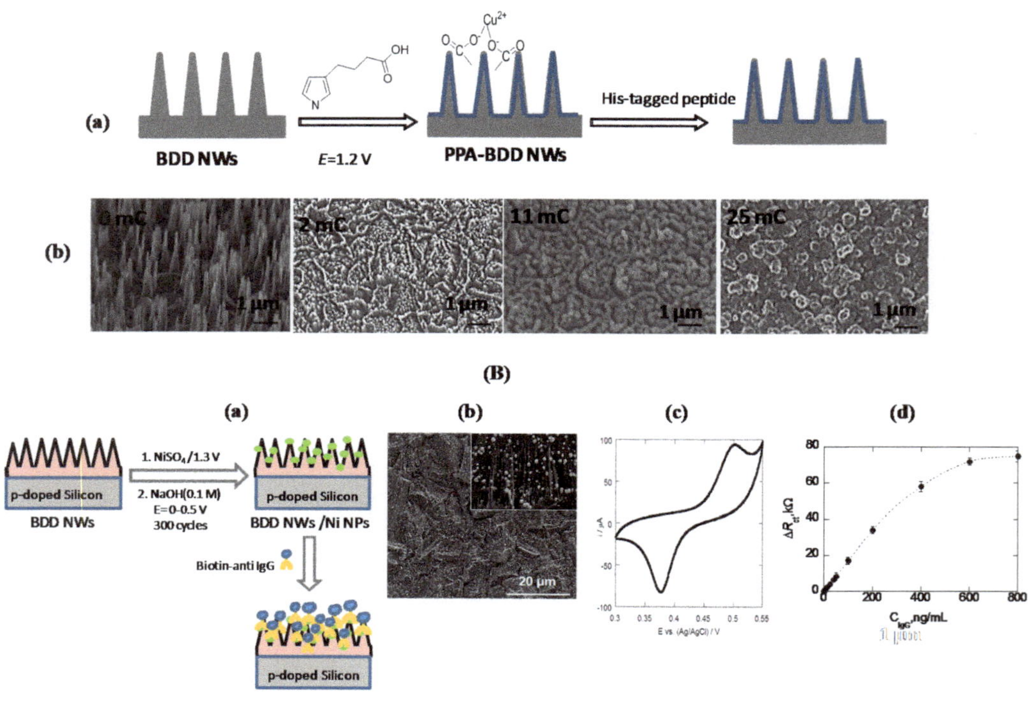

Figure 10. Diamond nanowires for immunosensing. (**A**) (**a**) fabrication method of polymer coated BDD NWs (PPA BDD NWs) with chelated Cu^{2+} ions and subsequently modified with histidine-terminated target; (**b**) SEM images of BDD NWs after electrochemical deposition of carboxylic acid-terminated poly(pyrrole) (100 mM) in TBATFB (0.1 M)/ acetonitrile solution at E = +1.2 V for different deposition charges (2, 11, 23 mC·cm^{-2}) (reprint with permission from [39]); (**B**) Fabrication method of Ni NPs modified diamond nanowires (**a**), SEM image of Ni NPs modified diamond nanowires (**b**), CV of Ni NPs-BDD NWs in 0.1 M NaOH (**c**), Calibration curve for IgG (**d**) (reprint with permission from [40]).

The other approach consists of the deposition of nickel nanoparticles. Figure 10B(b) shows a representative SEM image of Ni NPs modified diamond nanowires (BDD NWs/Ni NPs) revealing particles of 20 ± 5 nm in diameter and a particle density of 150 Ni NPs/µm^2. The oxidation state of the deposited nickel was confirmed by the presence of a quasi-reversible redox peak at ≈0.47 V/Ag/AgCl in the cyclic voltammogram of the BDD NWs/Ni NPs electrodes in PBS (Figure 10B(c) assigned to $Ni(OH)_2/NiOOH$. The hydrated nickel hydroxide can be present in two crystalline forms: the hydrated α-$Ni(OH)_2$ and the anhydrous β-$Ni(OH)_2$, the latter being the more stable and preferentially formed by cycling in NaOH (0.1 M) [48]. Biotinylated anti-IgG was specifically linked to the Ni^{2+} particles and the change in the interfacial properties upon binding of different concentrations of IgG was detected using electrochemical impedance spectroscopy. A linear relation between ΔR_{ct} and IgG concentration in the range of 0.3–400 ng/mL was recorded (Figure 10A(d)) with a correlation coefficient of $r = 0.9996$ according to ΔR_{et} (kΩ) = 0.02 + 0.0451 × [IgG]. The detection limit of IgG was determined to be ≈0.3 ng/mL from five blank noise signals (95% confidence level). The advantage of this approach is the possibility of a controlled immobilization of biotinylated antibodies and antigens with the omission of avidin layers, influencing the electrochemical behavior of the electrical interface. As the affinity of the biotin-tag to the Ni NPs is weaker than avidin-biotin interaction, an easy regeneration of the interface through a simple ethylenediaminetetraacetic acid (EDTA) wash was achieved. This could allow in the future the immobilization of different antibodies using the same interface.

3.2. Diamond Nanowires for Matrix-Free Mass Spectrometry

Mass spectrometry (MS) is widely accepted as a ‘gold-standard’ method for the identification of chemicals or biological products. It is, nowadays, applied in highly diversified domains like those directly or indirectly tied to healthcare or regulation-driven demand such as drug development, diagnostics, food and environmental safety testing. Among the different methodologies, matrix-assisted laser desorption/ionization mass spectrometry (MALDI-MS), first introduced in 1988 by Hillenkampf and Karas [49], constitutes one of the soft ionization techniques that provides the nondestructive vaporization and ionization of analytes using UV-absorbing organic matrices. The utility of MALDI-MS for protein and peptide analysis lies in its ability to provide high accurate molecular-weight information on intact molecules. Acquiring optimum MALDI data depends, however, not only on the functional and structural properties of the analyte itself but also largely on the choice of suitable matrices. As the matrix is the medium by which the analyte is transported to the gaseous phase and provides the conditions that makes ionization possible, the matrix and the sample-matrix preparation procedure greatly influence the quality of MALDI mass spectra. Organic matrices such as 2,5-dihydroxylbenzoic acid (DHB), α-cyano-4-hydroxycinnamic acid (HCCA) or sinapic acid (SA) are commonly used. However, the choice of the matrix remains an empirical issue. While MALDI-MS has been successfully used to analyze large molecules [50–52], it has rarely been applied to low-molecular weight compounds (<500 Da) as a large number of matrix ions appear in the low-mass range.

Tanaka and co-workers proposed the use of a suspension of cobalt nanoparticles of 30 nm in size mixed with glycerol as inorganic matrix for the laser desorption/ionization of analytes [53]. Since then, several alternative inorganic matrices have been proposed as assisting material [54] in a process that was named surface-assisted laser desorption/ionization mass spectrometry (SALDI-MS) by Sunner *et al.* [55]. The basic principle of SALDI-MS analysis is schematically outlined in Figure 11A. Using inorganic species as the assisting material in MALDI is an alternative approach to avoid the problems arising in conventional MALDI analysis using organic compounds as matrix systems [56]. Nanostructured diamond-like carbon (DLC) coated targets were proposed by Bonn and co-workers as versatile platform for the analysis of amino acids, carbohydrates, lipids, peptides and other metabolites using laser desorption/ionization mass spectrometry [57]. A nanodiamond MALDI-MS support was demonstrated by Wei *et al.* to improve the ionization efficiency of samples [58]. The use of boron-doped diamond nanowires as an inorganic matrix for the D/I of peptides and small molecules, and their analysis by mass spectrometry with a very high sensitivity has been demonstrated by us [27]. To minimize droplet spreading on the matrix surface, the nanowires were chemically functionalized with octadecyltrichlorosilane (OTS) to reach a final water contact angle of 120 °C (Figure 11B). The sub-bandgap absorption under UV laser irradiation and the heat confinement inside the nanowires allowed LDI-MS of various compounds, most likely via a thermal mechanism (Figure 11B). One of the most used compounds to assess the LDI-MS performances of SALDI surfaces is verapamil, a calcium channel blocker. Detection of verapamil on silicon nanowires, prepared by the Metal-Assisted Chemical Etching (MACE) method, has shown a detection limit value of 5 fmol [59]. Walker *et al.* explored silicon nanopost arrays (NAPA) in combination with LDI-MS. The high ionization efficiencies enabled detection of ultratrace amounts of analytes (800 zmol of verapamil) within a dynamic range spanning up to four orders of magnitude [60]. For comparison, the detection limit value of verapamil on diamond nanowires was 200 zeptomoles, which is slightly better. Impressive results were obtained by Northen *et al.* by nanostructure-initiator mass spectrometry (NIMS). They demonstrated a detection limit value for verapamil of 700 ymol [61]. Although this technique is not a "matrix" *sensu stricto* (as MALDI), NIMS takes advantage of an "initiator" compound (e.g., bis(tridecafluoro-1,1,2,2-tetrahydrooctyl)tetramethyldisiloxane), which assists desorption and/or ionization of analyte molecules. Even though the limit of detection (LoD) that is needed depends on the targeted application, most of the time, the sensitivity race is not scientifically or even technically relevant. Other parameters should be taken into account for performant SALDI surfaces such as laser fluence, dynamic range, versatility, peptide discrimination and post-translational modification (PTMs) detection (proteomics field), cost, reproducibility.

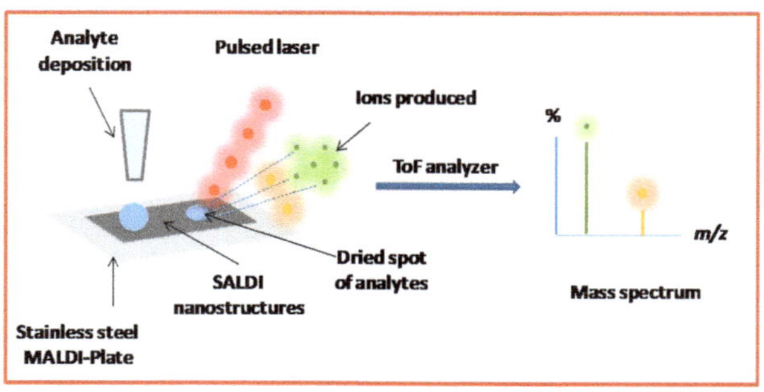

(A)

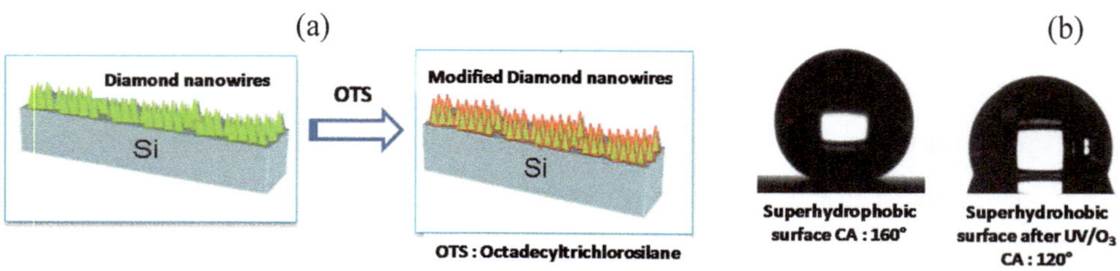

(B)

Figure 11. *Cont.*

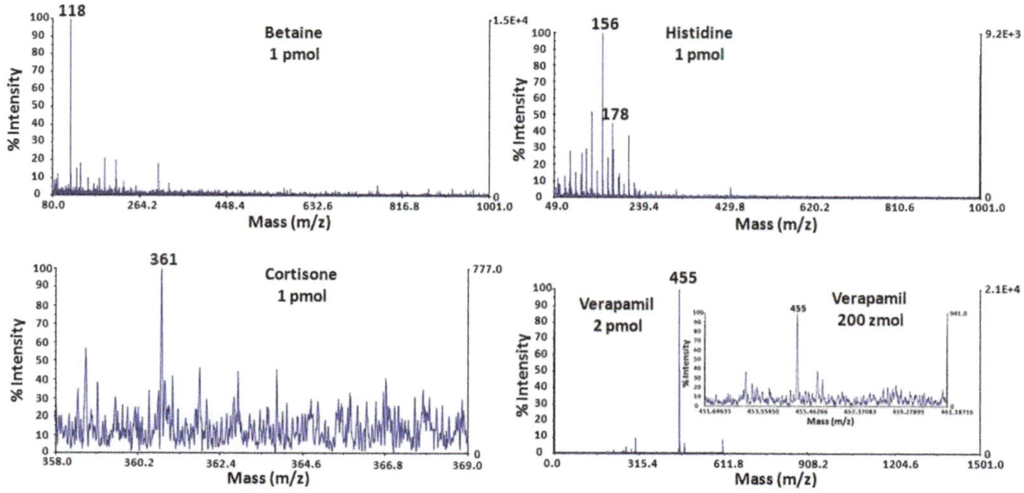

(C)

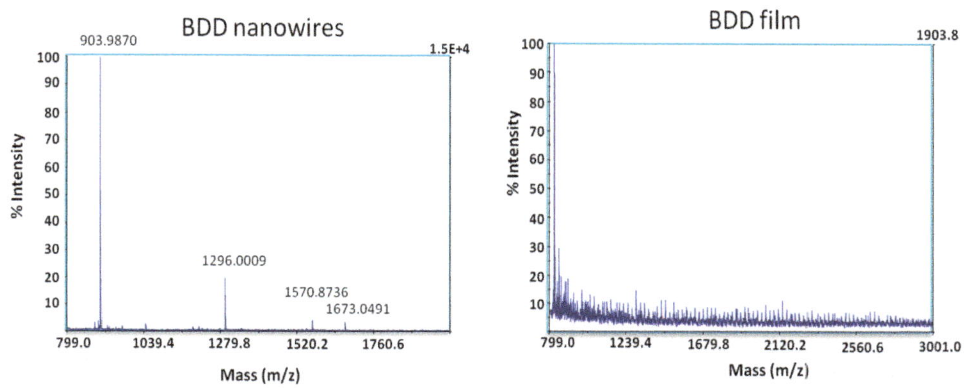

(D)

Figure 11. Diamond nanowires as inorganic matrices for SALDI. (**A**) Schematic presentation of the principle of surface-assisted laser desorption/ionization mass spectrometry (SALDI-MS) with analyte deposition on SALDI based nanostructures such as diamond nanowires; (**B**) OTS modified diamond nanowires (**a**) and contact angle values before and after UV/O_3 treatment (**b**); (**C**) LDI-MS spectra on BDD NWs of various compounds (Histidine, *m*/*z* 156, 1 pmol; Betaine m/z 118, 1 pmol; Cortisone m/z 361, 1 pmol and verapamil m/z 455, 2 pmol and 200 zmol (inset)); (**D**) LDI-MS detection of a peptide mixture ([Des-Arg[1]]-bradykinin m/z 904 (50 fmol/μL), angiotensin I *m*/*z* 1296 (50 fmol/μL), [Glu[1]]-fibrinopeptide B m/z 1570 (50 fmol/μL), neurotensin *m*/*z* 1673 (10 fmol/μL)) on BDD NWs and BDD films (reprint with permission from [27]).

The investigated diamond nanowires, undoped (UDD NWs) or boron-doped (BDD NWs), display antireflective properties permitting photons absorption at $\lambda = 337$ nm. However, on the mass spectrum of a peptide mixture obtained on BDD NWs interface, all peptides have been detected with relatively high signal intensity (Figure 11D). On the contrary, the same experiment performed on a planar nanocrystalline BDD, *i.e.*, the same interface without any RIE step process etching, no peaks were observed. Thus, the absence of peaks in the MS spectrum (Figure 11D) clearly indicates that the presence of nanostructures on the BDD substrate is mandatory for achieving the laser desorption/ionization (LDI) of biomolecules [27].

3.3. Future Trend: Coupled Electrochemistry-Mass Spectrometry Analysis on Diamond Nanowires

Due to the great performance of BDD nanowires for electrochemical sensing and SALDI-MS, their combined use might provide a powerful quantitative and qualitative analysis platform. Electrochemistry coupled with mass spectrometry (EC/MS) allowed the identification of products or/and intermediates of electrochemical reactions, which is not only useful for elucidation of redox reaction mechanisms but also leads to many valuable bioanalytical applications [62,63]. The versatility of EC/MS stems from two facts. MS can serve as a sensitive and general detector for electrochemical compounds and can provide molecular weight information about an analyte of interest. In addition, tandem MS analysis can be used for structural determination based on ion dissociation. Electrochemical conversion, on the other hand, can improve analyte ionization and provides desired modification to the analyte prior to MS analysis. Attracted by the complementary nature of these two techniques, the coupling of EC and MS appears perfect and appealing. For more than four decades, researchers have been engaged in coupling these two techniques [64–67].

A key challenge of the coupling is how to interface an electrochemical cell online with a mass spectrometer for efficiently ionizing the electrolyzed samples. Therefore, selecting an appropriate ionization method is critical. In this regard, the evolution of the coupled EC/MS technique followed the advances of ionization methods in the field of mass spectrometry. In 1971, the first EC/MS device was introduced by Bruckenstein and Gadde for *in situ* mass spectrometric determination of volatile electrode reaction products [64]. In the experiment, a Teflon membrane was placed between the porous electrode and the mass spectrometer ionization chamber so that volatile reaction products could penetrate through the membrane and be ionized by electron impact (EI) without the interference of the solvent. In 1984, differential electrochemical mass spectrometry (DEMS) was established for the online MS detection of volatile electrochemical products in real time. The total response time was less than 1 s [68]. Furthermore in DEMS, the product MS signal intensity was proportional to the Faradaic current as only the volatile compounds produced were transferred to the ionization chamber, from which quantitation could be achieved. Then, several methods of ionization have been used including thermospray (TS), fast atom bombardment (FAB), inductively coupled plasma (ICP), chemical ionization (CI), atmospheric pressure chemical ionization (APCI), atmospheric pressure photoionization (APPI) and electrospray ionization (ESI). This former combination (EC/ESI-MS) was the most widely used for many applications [62]. Desorption electrospray ionization (DESI) or nanoDESI and direct analysis in real time (DART) are other recent methods to perform ambient soft ionization with little or no sample preparation [69]. Applications of EC/MS are various and numerous and among them we can cite mechanistic elucidation of electrochemical reactions [70–73], mimicking of metabolic pathways [65], tagging of protein/peptide thiol groups using various electrochemical generated species to electrochemically enhance MS signals [74,75], pre-concentrate target via electrochemical deposition [66], and following protein/peptide cleavage and online MS analysis that is very notably useful in the proteomics field [76,77].

4. Conclusions

From the above discussion, it becomes clear that a large amount of effort has been devoted to the synthesis of diamond nanostructures to a point where they can be considered for device-oriented applications. The discoveries and research undertaken in the last years hope to trigger the development of diamond nanowire sensors for clinical diagnostic, environmental sensing and other applications at the interface between biology, physics and chemistry. However, the full spectrum of such nanostructures for other technological applications cannot be overseen. Diamond coated silicon nanowires have been lately investigated for supercapacitor applications [34], bringing diamond nanostructures to the field of energy. A full and detailed understanding of the electrical and electrochemical properties of a single diamond nanowire might be of ultimate importance in the near future to foster further such developments.

The use of diamond nanowires is believed to have great potential for EC/MS. Indeed, coupling EC/MS using diamond nanowires will permit performing both electrochemistry and MS detection of ionized compounds achieved by either DESI (or nanoDESI) or the LDI ionization process.

Surface functionalization is required for almost any kind of sensing applications. Currently, the reported surface functionalization schemes of diamond wires are limited to some examples. Widening this area is thus one aspect that should be undertaken by research groups working in this field. The formation of superhydrophobic and oleophobic interfaces has, for example, been demonstrated to have impact on cell and bacteria adhesion [26]. An important aspect will be the determination of the influence of diamond doping levels and even wire length on SALDI results. A better understanding will help to optimize the technique and achieve highly reproducible and accurate results. This will make the approach of high interest for any laboratory.

Acknowledgments

Financial support from the Centre National de la Recherche Scientifique (CNRS), the Université Lille 1, the Nord Pas de Calais region and the Institut Universitaire de France (IUF) is gratefully acknowledged.

Conflicts of Interest

The authors declare no conflict of interest.

References

1. Nemanich, R.J.; Carlisle, J.A.; Hirata, A.; Haenen, K. CVD diamond—Research, applications, and challenges. *MRS Bull.* **2014**, *39*, 490–548.
2. Fortin, E.; Chane-Tune, J.; Mailley, P.; Szunerits, S.; Marcus, B.; Petit, J.P.; Mermoux, M.; Vieil, E. Nucleosides and ODN electrochemical detection onto boron-doped diamond electrodes. *Bioelectrochemistry* **2004**, *63*, 303–306.
3. McCreery, R.L. Advanced carbon electrode materials for molecular electrochemistry. *Chem. Rev.* **2008**, *108*, 2646–2687.

4. Hartl, A.; Schmich, E.; Garrido, J.A.; Hernando, J.; Catharino, S.C.R.; Walter, S.; Feulber, P.; Kromka, A.; Steinmuller, D.; Stutzmann, M. Protein-modified nanocrystalline diamond thin films for biosensor applications. *Nat. Mater.* **2004**, *3*, 736–742.
5. Ferro, S.; de Battist, A. The 5-V window of polarizability of fluorinated diamond electrodes in aqueous solutions. *Anal. Chem.* **2003**, *75*, 7040–7042.
6. Meziane, D.; Barras, A.; Kromka, A.; Houdkova, J.; Boukherroub, R.; Szunerits, S. Thiol-yne reaction on boron-doped diamond electrodes. *Anal. Chem.* **2012**, *84*, 194–200.
7. Szunerits, S.; Niedziółka-Jönsson, J.; Boukherroub, R.; Woisel, P.; Baumann, J.S.; Siriwardena, A. Label-free detection of lectins on carbohydrate-modified boron-doped diamond Surfaces. *Anal. Chem.* **2010**, *82*, 8203–8210.
8. Danilenko, V.V. On the history of the discovery of nanodiamond synthesis. *Phys. Solid State* **2004**, *46*, 595–599.
9. Szunerits, S.; Coffinier, Y.; Galopin, E.; Brenner, J.; Boukherroub, R. Preparation of boron-doped diamond nanowires and their application for sensitive electrochemical detection of tryptophan. *Electrochem. Commun.* **2010**, *12*, 438–441.
10. Luo, D.; Wu, L.; Zhi, J. Fabrication of boron-doped diamond nanorod forest electrodes and their application in nonenzymatic amperometric glucose sensing. *ACS Nano* **2009**, *3*, 2121–2128.
11. Yang, N.; Uetsuka, H.; Osawa, E.; Nebel, C.E. Vertically aligned diamond nanowires for DNA sensing. *Angew. Chem. Int. Ed.* **2008**, *47*, 5183–5185.
12. Wang, Q.; Subramanian, P.; Li, M.; Yeap, W.S.; Haenen, K.; Coffinier, Y.; Boukherroub, R.; Szunerits, S. Non-enzymatic glucose sensing on long and short diamond nanowires electrodes. *Electrochem. Commun.* **2013**, *34*, 286–290.
13. Derjaguin, B.V.; Fedoseev, D.V.; Lukyanovich, V.M.; Spitzin, B.V.; Ryabov, V.A.; Lavrentyev, A.V. Filamentary diamond crystals. *J. Cryst. Growth* **1968**, *2*, 380–384.
14. Shiomi, H. Reactive ion etching of diamond in O_2 and CF_4 plasma and fabrication of porous diamond for field emitter cathodes. *Jpn. J. Appl. Phys.* **1997**, *36*, 7745–7748.
15. Masuda, H.; Watanaba, M.; Yasui, K.; Tryk, D.; Rao, T.; Fujishima, A. Fabrication of a nanostructured diamond honeycomb film. *Adv. Mater.* **2000**, *12*, 444–447.
16. Baik, E.S.; Baik, Y.J.; Jeaon, D. Aligned diamond nanowhiskers. *J. Mater. Res.* **2000**, *15*, 923–926.
17. Ando, Y.; Nishibayashi, Y.; Sawaben, A. 'Nano-rods' of single crystalline diamond. *Diamond Relat. Mater.* **2004**, *13*, 633–637.
18. Okuyama, S.; Matsushita, S.I.; Fujishima, A. Periodic submicrocylinder diamond surfaces using two-dimensional fine particle arrays. *Langmuir* **2002**, *18*, 8282–8287.
19. Zou, Y.S.; Yang, T.; Zhang, W.J.; Chong, Y.M.; He, B.; Bello, I.; Lee, S.T. Fabrication of diamond nanopillar and their arrays. *Appl. Phys. Lett.* **2008**, *92*, 053105.
20. Yang, N.; Uetsuka, H.; Osawa, E.; Nebel, C.E. Vertically aligned nanowires from boron-doped diamond. *Nano Lett.* **2008**, *8*, 3572–3576.
21. Hausmann, B.J.M.; Khan, M.; Zhang, Y.; Bainec, T.M.; Martinick, K.; McCutcheon, M.; Hemmer, P.; Loncar, M. Fabrication of diamond nanowires for quantum information processing applications. *Diamond Relat. Mater.* **2010**, *19*, 621–629.

22. Smirnov, W.; Kriele, A.; Yang, N.; Nebel, C.F. Aligned diamond nano-wires: Fabrication and characterisation for advanced applications in bio and electrochemistry. *Diamond Relat. Mater.* **2009**, *18*, 186–189.
23. Nebel, C.E.; Yang, N.; Uetsuka, H.; Osawa, E.; Tokuda, N.; Williams, O. Diamond nano-wires, a new approach towards next generation electrochemical gene sensor platforms. *Diamond Relat. Mater.* **2009**, *18*, 910–917.
24. Zheng, W.W.; Hsieh, Y.H.; Chiu, Y.C.; Cai, S.J.; Cheng, C.L.; Chen, C. Organic functionalization of ultradispersed nanodiamond: Synthesis and applications. *J. Mater. Chem.* **2009**, *19*, 8432–8441.
25. Marcon, L.; Riquet, F.; Vicogne, D.; Szunerits, S.; Bodart, J.F.; Boukherroub, R. Cellular and *in vivo* toxicity of functionalized nanodiamond in Xenopus embryos. *J. Mater. Chem.* **2010**, *20*, 8064–8069.
26. Coffinier, Y.; Galopin, E.; Szunerits, S.; Boukherroub, R. Preparation of superhydrophobic and oleophobic diamond nanograss array. *J. Mater. Chem.* **2010**, *20*, 10671–10675.
27. Coffinier, Y.; Szunerits, S.; Drobecq, H.; Melnyk, O.; Boukherroub, R. Diamond nanowires for highly sensitive matrix-free mass spectrometry analysis of small molecules. *Nanoscale* **2012**, *4*, 231–238.
28. Masuda, H.; Yanagishita, T.; Yasui, K.; Nishio, K.; Yagi, I.; Rao, N.; Fujishima, A. Synthesis of well-aligned diamond nanocylinders. *Adv. Mater.* **2001**, *13*, 247–249.
29. Sun, L.T.; Gond, J.; Zhu, D.Z.; Zhu, Z.Y.; He, S. Diamond nanorods from carbon nanotubes. *Adv. Mater.* **2004**, *16*, 1849–1853.
30. Vlasov, I.I.; Lebedev, O.I.; Ralchenko, V.G.; Goovaerts, E.; Bertoni, G.; Tendeloo, G.V.; Konov, V.I. Hybrid diamond-graphite nanowires produced by microwave plasma chemical vapor deposition. *Adv. Mater.* **2007**, *19*, 4058–4062.
31. Shang, N.; Papakonstantinou, P.; Wang, P.; Zakharov, A.; Palnitkar, U.; Lin, I.N.; Chu, M.; Stamboulis, A. Self-assembled growth, microstructure, and field-emission high-performance of ultrathin diamond nanorods. *ACS Nano* **2009**, *3*, 1032–1038.
32. Hsu, C.H.; Cloutier, S.G.; Palefsky, S.; Xu, J. Synthesis of diamond nanowires using atmopsheric-pressure chemical vapor deposition. *Nano Lett.* **2010**, *10*, 3272–3276.
33. Hsu, C.H.; Xu, J. Diamond nanowire-a challenge from extremes. *Nanoscale* **2012**, *4*, 5293–5299.
34. Gao, F.; Lewes-Malandrakis, G.; Wolfer, M.T.; Muller-Sebert, W.; Gentile, P.; Aradilla, D.; Schubert, T.; Nebel, C.E. Diamond-coated silicon wires for supercapacitor applications in ionic liquids. *Diamond Relat. Mater.* **2015**, *51*, 1–6.
35. Girard, H.A.; Scorsone, E.; Saada, S.; Gesset, C.; Arnault, J.C.; Perruchas, S.; Rousseau, L.; David, S.; Pichot, V.; Spitzer, D.; Berganzo, P. Electrostatic grafting of diamond nanoparticles towards 3D diamond nanostructures. *Diamond Relat. Mater.* **2012**, *23*, 83–87.
36. Peng, K.Q.; Yan, Y.J.; Gao, S.P.; Zhu, J. Synthesis of large-area silicon nanowire arrays via self-assembling nanoelectrochemistry. *Adv. Mater.* **2002**, *14*, 1164–1167.
37. Babinec, T.M.; Hausmann, B.J.M.; Khan, M.; Zhang, Y.; Maze, J.R.; Hemmer, P.R.; Loncar, M. A diamond nanowire single-photon source. *Nature Nanotechnol.* **2010**, *5*, 195–199.
38. Subramanian, P.; Foord, J.; Steinmueller, D.; Coffinier, Y.; Boukherroub, R.; Szunerits, S. Diamond nanowires decorated with metallic nanoparticles: A novel electrical interface for the immobilization of histidinylated biomolecules. *Electrochim. Acta* **2013**, *110*, 4–8.

39. Subramanian, P.; Mazurenko, I.; Zaitsev, V.; Coffinier, Y.; Boukherroub, R.; Szunerits, S. Diamond nanowires modified with poly[3-(pyrrolyl)carboxylic acid] for the immobilization of histidine-tagged peptides. *Analyst* **2014**, *139*, 4343–4349.
40. Subramanian, P.; Motorina, A.; Yeap, W.S.; Haenen, K.; Coffinier, Y.; Zaitsev, V.; Niedziolka-Jonsson, J.; Boukherroub, R.; Szunerits, S. Impedimetric immunosensor based on diamond nanowires decorated with nickel nanoparticles. *Analyst* **2014**, *139*, 1726–1731.
41. Wang, Q.; Vasilescu, A.; Subramanian, P.; Vezeanu, A.; Andrei, V.; Coffinier, Y.; Li, M.; Boukherroub, R.; Szunerits, S. Simultaneous electrochemical detection of tryptophan and tyrosine using boron-doped diamond and diamond nanowires electrodes. *Electrochem. Commun.* **2013**, *35*, 84–87.
42. Yang, N.; Uetsuka, H.; Nebel, C.E. Biofunctionalization of vertically aligned diamond nanowires. *Adv. Funct. Mater.* **2009**, *19*, 887–893.
43. Yang, N.; Uetsuka, H.; Williams, O.A.; Osawa, E.; Tokuda, N.; Nebel, C.E. Vertically aligned diamond nanowires: Fabrication, characterization, and application for DNA sensing. *Phys. Stat. Sol. A* **2009**, *206*, 2048–2056.
44. Uetsuka, H.; Shin, D.; Tokuda, N.; Saeki, K.; Nebel, C.E. Electrochemical grafting of boron-doped single-crystalline chemical vapor deposition diamond with nitrophenyl molecules. *Langmuir* **2007**, *23*, 3466–3472.
45. Yang, N.; Smirnov, W.; Nebel, C.E. Three-dimensional electrochemical reactions on tip-coated diamond nanowires with nickel nanoparticles. *Electrochem. Commun.* **2013**, *27*, 89–91.
46. Zhao, G.; Qi, Y.; Tian, Y. Simultaneous and direct determination of tryptophan and tyrosine at boron-doped diamond electrode. *Electroanalysis* **2006**, *18*, 830–834.
47. Shalini, J.; Sankaran, K.J.; Dong, C.L.; Lee, C.Y.; Tai, N.H.; Lin, I.N. *In situ* detection of dopamine using nitrogen incorporated diamond nanowire electrode. *Nanoscale* **2013**, *5*, 1159–1167.
48. Toghill, K.E.; Xiao, L.; Stradiotto, N.R.; Compton, R.G. The determination of methanol using an electrolytically fabricated nickel microparticle modified boron-doped diamond electrode. *Electroanalysis* **2010**, *22*, 491–500.
49. Karas, M.; Hillenkamp, F. Laser desorption ionization of proteins with molecular masses exceeding 10,000 daltons. *Anal. Chem.* **1988**, *60*, 259–280.
50. Hanton, S.D. Mass spectrometry of polymers and polymer surfaces. *Chem. Rev.* **2001**, *101*, 527–569.
51. Knochenmuss, R.; Zenobi, R. MALDI ionization: The role of in-plume processes. *Chem. Rev.* **2003**, *103*, 441–452.
52. Li, L. *MALDI Mass Spectrometry for Synthetic Polymer Analysis (Chemical Analysis: A Series of Monographs on Analytical Chemistry and Its Applications)*; Wiley-VCH: Weinheim, Germany, 2009.
53. Tanaka, K.; Waki, H.; Ido, Y.; Akita, S.; Yoshida, Y.; Yoshida, T. Protein and polymer analyses up to m/z 100,000 by laser ionization time-of-flight mass spectrometry. *Rapid Commun. Mass Spectrom* **1988**, *2*, 151–153.
54. Arakawa, R.; Kawasaki, H. Functionalized nanoparticles and nanostructured surfaces for surface-assisted laser desorption/ionization mass spectrometry. *Anal. Sci.* **2010**, *26*, 1229–1240.

55. Sunner, J.; Dratz, E.; Chen, Y.C. Graphite surface-assisted laser desorption/ionization time-of-flight mass spectrometry of peptides and proteins from liquid solutions. *Anal. Chem.* **1995**, *67*, 4335–4342.
56. Qiao, L.; Liu, B.H.; Girault, H.H. Nanomaterial-assisted laser desorption ionization for mass spectrometry-based biomedical analysis. *Nanomedicine* **2010**, *5*, 1641–1652.
57. Najam-ul-Haq, M.; Rainer, M.; Huck, C.W.; Stecher, G.; Feuerstein, I.; Steinmueller, D.E.A. Chemically modified nano crystalline diamond layer as material enhanced laser desorption ionisation (MELDI) surface in protein profiling. *Curr. Nanosci.* **2006**, *2*, 1–7.
58. Wei, L.M.; Xue, Y.; Zhou, X.W.; Jin, H.; Shi, Q.; Lu, H.L.; Yang, P.Y. Nanodiamond MALDI support for enhancing the credibility of identifying proteins. *Talanta* **2008**, *74*, 1363–1370.
59. Piret, G.; Drobecq, H.; Coffinier, Y.; Melnyk, O.; Boukherroub, R. Matrix-free laser desorption/ionization mass spectrometry on silicon nanowire arrays prepared by chemical etching of crystalline silicon. *Langmuir* **2010**, *26*, 1354–1361.
60. Walker, B.N.; Stolee, J.A.; Vertes, A. Nanophotonic ionization for ultratrace and single-cell analysis by mass spectrometry. *Anal. Chem.* **2012**, *84*, 7756–7762.
61. Northen, T.R.; Yanes, O.; Northen, M.T.; Marrinucci, D.; Uritboonthai, W.; Apon, J.; Golledge, S.L.; Nordstrom, A.; Siuzdak, G. Clathrate nanostructures for mass spectrometry. *Nature* **2007**, *449*, 1033–1036.
62. Liu, P.; Lu, M.; Zheng, Q.; Zhang, Y.; Dewald, H.D.; Chen, H. Recent advances of electrochemical mass spectrometry. *Analyst* **2013**, *128*, 5519–5539.
63. Abonnenc, M.; Qiao, L.; Liu, B.H.; Girault, H.H. Electrocheical aspects of electrosptray and laser desorption/ionization for mass spectometry. *Annu. Rev. Anal. Chem.* **2010**, *3*, 231–254.
64. Bruckenstein, S.; Gadde, R.R. Use of a porous electrode for *in situ* mass spectrometric determination of volatile electrode reaction products. *J. Am. Chem. Soc.* **1971**, *93*, 793–794.
65. Jahn, S.; Karst, U. Electrochemistry coupled to (liquid chromatography/) mass spectrometry—Current state and future perspectives. *J. Chromatogr. A* **2012**, *1259*, 16–49.
66. Gutkin, V.; Gunand, J.; Lev, O. Electrochemical deposition-stripping analysis of molecules and proteins by online electrochemical flow cell/mass spectrometry. *Anal. Chem.* **2009**, *81*, 8396–8404.
67. Van Berkel, G.J.; Kertesz, V. Electrochemically initiated tagging of thiols using an electrospray ionization based liquid microjunction surface sampling probe two-electrode cell. *Rapid Commun. Mass Spectrom.* **2009**, *23*, 1380–1386.
68. Wolter, O.; Heitbaum, J. Differentiel electrochemical mass spectroscopy (DEMS)—A new method for the study of electrode processes. *Ber. Bunsenges. Phys. Chem.* **1984**, *88*, 2–6.
69. Takats, Z.; Wiseman, J.M.; Gologan, B.; Cooks, R.G. Mass spectrometry sampling under ambient conditions with desorption electrospray ionization. *Science* **2004**, *306*, 471–473.
70. Jurva, U.; Bissel, P.; Isin, E.M.; Igarashi, K.; Kuttab, S.; Castagnoli, N. Model electrochemical-mass spectrometric studies of the cytochrome P450-catalyzed oxidations of cyclic tertiary allylamines. *J. Am. Chem. Soc.* **2005**, *127*, 12368–12377.
71. Lu, W.; Xu, X.; Cole, R.B. On-line linear sweep voltammetry−electrospray mass spectrometry. *Anal. Chem.* **1997**, *69*, 2478–2484.

72. Zhang, T.; Palii, S.P.; Eyler , J.R.; Brajter-Toth, A. Enhancement of ionization efficiency by electrochemical reaction products in on-line electrochemistry/electrospray ionization Fourier transform ion cyclotron resonance mass spectrometry *Anal. Chem.* **2002**, *74*, 1097–1103.
73. Zettersten, C.; Sjöberg, P.J.R.; Nyholm, L. Oxidation of 4-chloroaniline studied by on-line electrochemistry electrospray ionization mass spectrometry. *Anal.Chem.* **2009**, *81*, 5180–5187.
74. Roussel, C.; Rohner, T.C.; Jensen, H.; Girault, H.H. Mechanistic aspects of on-line electrochemical tagging of free L-cysteine residues during electrospray ionisation for mass spectrometry in protein analysis. *ChemPhysChem* **2003**, *4*, 200–206.
75. Rohner, T.C.; Rossier, J.S.; Girault, H.H. On-line electrochemcial tagging of cysteines in proteins during nanospray. *Electrochem. Commun.* **2002**, *74*, 695–700.
76. Roeser, J.; Permentier, H.P.; Bruins, A.P.; Bischoff, R. Electrochemical oxidation and cleavage of tyrosine- and tryptophan-containing tripeptides. *Anal. Chem.* **2010**, *820*, 7556–7565.
77. Roeser, J.; Permentier, H.P.; Bruins, A.P.; Bischoff, R. Oxidative protein labeling in mass-spectrometry-based proteomics. *Anal. Bioanal. Chem.* **2010**, *82*, 7556–7565.

13

Accuracy Analysis of a Multi-View Stereo Approach for Phenotyping of Tomato Plants at the Organ Level

Johann Christian Rose *, Stefan Paulus and Heiner Kuhlmann

Department of Geodesy, Institute of Geodesy and Geoinformation, University of Bonn, Nussallee 17, 53115 Bonn, Germany; E-Mails: paulus@igg.uni-bonn.de (S.P.); heiner.kuhlmann@uni-bonn.de (H.K.)

* Author to whom correspondence should be addressed; E-Mail: rose@igg.uni-bonn.de

Academic Editor: Vittorio M.N. Passaro

Abstract: Accessing a plant's 3D geometry has become of significant importance for phenotyping during the last few years. Close-up laser scanning is an established method to acquire 3D plant shapes in real time with high detail, but it is stationary and has high investment costs. 3D reconstruction from images using structure from motion (SfM) and multi-view stereo (MVS) is a flexible cost-effective method, but requires post-processing procedures. The aim of this study is to evaluate the potential measuring accuracy of an SfM- and MVS-based photogrammetric method for the task of organ-level plant phenotyping. For this, reference data are provided by a high-accuracy close-up laser scanner. Using both methods, point clouds of several tomato plants were reconstructed at six following days. The parameters leaf area, main stem height and convex hull of the complete plant were extracted from the 3D point clouds and compared to the reference data regarding accuracy and correlation. These parameters were chosen regarding the demands of current phenotyping scenarios. The study shows that the photogrammetric approach is highly suitable for the presented monitoring scenario, yielding high correlations to the reference measurements. This cost-effective 3D reconstruction method depicts an alternative to an expensive laser scanner in the studied scenarios with potential for automated procedures.

Keywords: SfM and MVS photogrammetry; close-up laser scanning; plant phenotyping; organ-level parameterization

---

1. Introduction

With the rise of phenotyping, the demand for access to the plants' 3D shape has become of significant importance [1]. Various methods using laser scanning [2], time of flight cameras [3,4] or structured light approaches [5], opening the door to 3D phenotyping, have been published. Close-up laser scanning has shown its advantage of high accuracy and high resolution [6] combined with direct access to the point cloud, but requires huge investment costs (100 k€) in advance. Using the 3D shape, descriptive parameters, like the plant volume or the geometry of single organs, can be extracted. A detailed parameterization and monitoring of growth or reactions to environmental changes are possible and can be tracked at the organ level. Furthermore, the effect of occlusion, a huge challenge in 2D phenotyping approaches [7,8], can be reduced tremendously.

In recent years, with the boost in computational power and the widespread availability of digital cameras, photogrammetric approaches, like structure from motion (SfM) [9] and multi-view stereo (MVS) [10], have found their way into phenotyping. These technologies offer non-invasive and non-destructive ways to measure a huge variety of plant characteristic traits through the creation of a full 3D point cloud of the plant. Simultaneously, they hold potential for future automatic measuring processes.

Ivanov *et al.* [11] were the first to use a photogrammetric approach to build a 3D point cloud of such a complex medium as a plant. They used a fixed base stereovision system pointing towards nadir, which was mounted 8.5 m above the ground to create a 3D model of the canopy of maize plants with a height of 2.5 m. From the model, they were able to derive the leaf position and orientation, as well as the leaf area distribution. This approach served as a stepping stone for future advances in the field of plant phenotyping with photogrammetric methods.

Aguilar *et al.* [12] used the close-range photogrammetric package, PhotoModeler Pro5 (Eos System Inc., Vancouver, BC, Canada), to measure the volume of several large tomato plants, as it is directly related to the Leaf Area Index (LAI), which represents the ground area covered by the plant canopy projected on the ground. The LAI itself is used for determining the optimal plant volume for pesticide application, as there is an ideal volume for spraying [13]. Around 400 target points of different colors attached to a plastic net were placed on each of the tomato plants, simulating the plants' enclosed surface. Afterwards, the target points were manually marked in PhotoModeler, and their 3D coordinates were calculated. Using these 3D coordinates, polygon meshes were computed, and from these, ultimately, the plant volume was calculated. A correlation of $R^2 = 0.75$ between the measured volumes and manually-derived reference volumes was found, whereby the correlation between LAI and manual volume was found to be $R^2 = 0.817$.

Santos *et al.* [14] used a consumer-grade camera mounted on a tripod to take images of basil and Ixora plants. The images were processed in an SfM approach called Bundler [15] to compute the camera's intrinsic parameters and relative poses [16] in order to compute sparse 3D point clouds of the plants. A patch-based MVS (PMVS) approach [17] was utilized to densify the sparse point cloud using the relative camera poses from the SfM model. Leafs and nodes were automatically segmented and classified afterwards [18]. They could show that these methods were able to effectively reconstruct the full three-dimensional shape of plants with a sparse canopy.

Paulus *et al.* [5] investigated the applicability of several low-cost 3D imaging systems for the task of plant phenotyping. The Microsoft Kinect and the DAVIDlaser scanner were compared to the high-precision close-up Perceptron v5 laser scanner, which served as a reference method. From the volumetric shape of sugar beet taproots, their leaves and the shape of wheat ears, plant parameters were extracted and compared to the reference measurements. They showed that low-cost sensors were as suitable as high-cost sensors, like the Perceptron v5, depending on the parameter of interest and the required level of detail.

Jay *et al.* [19] used SIFT keypoints to find homologous points between overlapping image regions. They applied an SfM approach named MicMac (MicMac 2007, IGN, Paris, France) to compute 3D point clouds from sunflowers, Savoy cabbages, cauliflowers and Brussels sprouts organized in crop rows. Their motivation was to assess the applicability of the image-based SfM approach for high-throughput phenotyping in outdoor scenarios where a variety of uncontrolled factors, like leaf movement through wind and inhomogeneous lightening conditions, complicate the data acquisition. Image acquisition followed the classical remote sensing method in that the camera is pointed towards nadir and linearly moved along a translation axis alongside the crop rows. Images are taken regularly after a certain baseline length between successive images has been reached. After the point cloud construction, plant and background were separated using criteria, like height and the Excess Green Index. Afterwards, plant height and leaf area were measured and compared to reference measurements, yielding high correlations for every species.

As the application of the SfM and MVS 3D reconstruction concept for plant phenotyping grows, it is important to gain knowledge about the potential measuring accuracy for parameter extraction. Accuracy evaluation then requires a high number of reference measurements, which possess higher accuracy and resolution than the SfM/MVS measurements. In the present study, the commercial photogrammetric software, Pix4DMapper (Pix4D SA, 1015 Lausanne, Switzerland), based on SfM and MVS techniques, was used to non-invasively and non-destructively monitor the growth process of a set of tomato plants for a six-day period. Colored 3D point clouds of the plants were reconstructed solely from image sets taken with the DSLR camera Canon EOS 450D. From the point clouds, several plant parameters at the organ level important for phenotyping were extracted. For accuracy and error evaluation, reference data for the same tomato plants were created using the high accuracy close-up triangulation line scanner, Perceptron v5 (Perceptron Scan Works V5, Perceptron Inc., Plymouth, MI, USA), combined with an articulated measuring arm (Romer Infinite 2.0 (1.4 m), Hexagon Metrology Services Ltd., London, U.K.). Reconstructing highly detailed and accurate 3D point clouds, the Perceptron has been found to be a high-precision non-invasive phenotyping tool [6]. This is the first time a photogrammetric method has been compared to a reference method at this high level of detail and accuracy. The applicability of the photogrammetric approach is evaluated using the means of linear regression and the RMSE (root-mean-square-error) and MAPE (mean-absolute-percentage-error) indicators.

2. Method

2.1. Experimental Setup

Five three-week-old tomato plants were monitored over a period of 6 days, resulting in 30 point clouds per method. The BBCHscale [20] value, determining the developmental stage of the plants, was 14/15 at the beginning of the monitoring and reached 17/18 at the end of monitoring.

The plants were stored in a greenhouse for optimal nourishment. In each case, the plant itself grew in a nourishment cube of a standardized size (100 × 100 × 100 mm) made from rock wool. To reduce plant movement due to wind and to avoid measurement uncertainties due to high temperature and humidity, the plants were moved to a prepared measurement table outside the greenhouse. Direct sunlight radiation was minimized by window coverages, and the temperature was modest at about 20–25 °C. A joint in the table ensured an identical positioning and orientation of the plant throughout the monitoring period.

2.2. Reference Data through High-Accuracy Laser Scanning

The reference data were acquired using the articulated measuring arm Romer Infinite 2.0 combined with the close-up triangulation line scanner Perceptron v5. The system provides a point-to-point resolution of 14 μm with an accuracy of 45 μm. The measuring field has a depth of 110 mm and a mean width of 105 mm. This scanning combination has shown its applicability for 3D imaging of various plants, like grapevine [21], barley [6] and sugar beet [5]. The scanning field was manually moved over the plant (Figure 1A), resulting in a huge amount of single 3D scan lines. These were automatically combined to an occlusion-free and very dense point cloud of the plant. A point cloud consisted of 3–10 million points, dependent on the size and developmental stage of the plant. Data acquisition took about 10–15 min per plant.

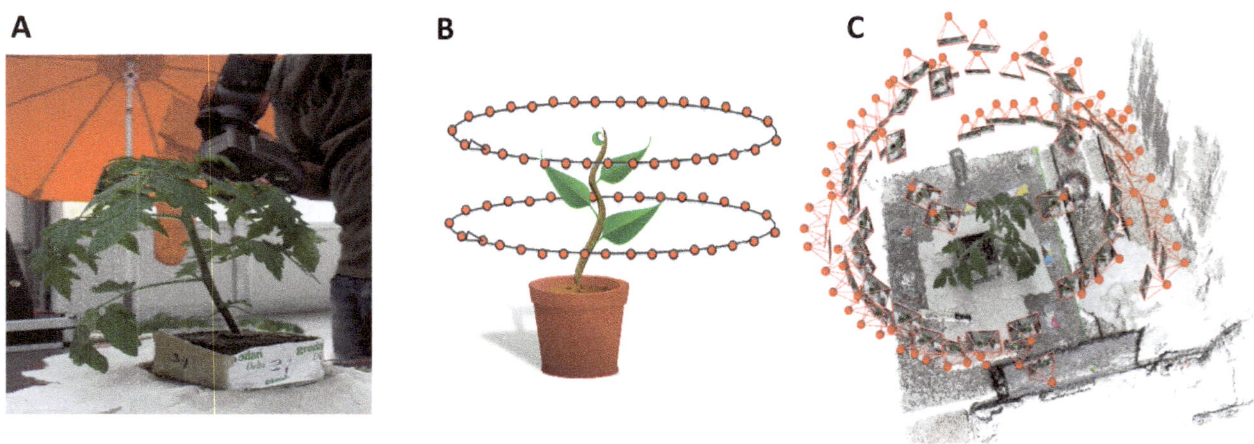

Figure 1. Both reconstruction methods: scanning the plant with the Perceptron v5 (**A**); the paths of the camera during image acquisition (**B**); the photogrammetric 3D reconstruction of the plant and the environment (**C**). Estimated camera positions are marked as red dots with attached image thumbnails.

2.3. Structure from Motion and Multi-View Stereo Using Pix4DMapper

The camera used for the image acquisition was a Canon EOS 450D (Canon Inc., Tokyo, Japan). It has a CMOS sensor with a pixel count of 12.2 MP and a dimension of 22.2 mm × 14.8 mm. The focal length is about 18 mm. The photogrammetric software Pix4DMapper was used to derive the 3D point clouds of the tomato plants. Pix4DMapper is based on SfM and MVS techniques employing keypoint matching [22,23] and bundle block adjustment. Processing runs semi-automatically with a working capacity of up to 1000 images. The PC used for processing worked with a Win7 64-bit system with 8 GB of memory and 8 cores of 3.7 GHz.

The quality and detail of the point cloud (Figure 1C) is determined by image quality and image content. To produce sharp edges without any image artifacts, the aperture, focus and exposure time were set in a balanced way to attain the largest depth of field possible without introducing motion blur. Room lights were switched on to reach an optimal image brightness. The heterogeneous texture of image content is an important prerequisite for keypoint detection and correct keypoint matching, the success of which determines the quality of the camera calibration and relative image orientation. Several texture-rich objects were placed around the table to increase the heterogeneity of the scene.

The images were taken by hand, *i.e.*, without a tripod, from a standing position using an image dimension of 3088 × 2056 pixels. To compensate for occlusions, two camera angles and heights were utilized. The camera was first held at a tilt of about 45° downwards from the horizontal at a height of about 50 cm over the top point of the plant. Focus was manually set and kept fixed. The depth of focus was increased by a small aperture to take sharp images. The user moved around the plant in a circular fashion, taking a new image every time the image content between the former and the current image had reached an overlap of about 70%. The whole time, the tilt of the camera and the initial distance of 50 cm were maintained as well as possible. The image overlap was estimated using the camera's viewfinder. Once the full circle had been closed, a second image set was taken in the same way with a different angle and height. The camera angle was horizontal and at a height at about the middle height level of the plant (Figure 1B). Again, the full circle around the plant was closed.

In this way, about 40–70 images per plant were recorded. During the monitoring phase, the plants grew about 10–15 cm in height while their canopy developed significantly, as well. Image acquisition took about 2–3 min per plant, regardless of the plants' growth stadium, as the amount of images necessary to image the whole plant did not change drastically. This corresponds to a mean acquisition speed of about one image every three seconds. The image sets were separately processed with Pix4DMapper, yielding colored 3D point clouds of the plant and close-by objects (Figure 1C). A point cloud consisted of about 15 k points at the beginning and of about 55 k points at the end of monitoring.

2.4. Point Cloud Preparation

Ahead of the parameter derivation, the point clouds of both sensors have to be processed. The preparation procedures and the methods of parameter derivation will be described in the following chapter. The point clouds were processed using Geomagic Studio 12 (Raindrop Geomagic Inc., Morrisville, NC, USA).

The laser scanning point clouds were cleaned from outliers manually. The point clouds were rasterized, resulting in a point-to-point distance of 0.5 mm to achieve a homogenous point distribution. Irrelevant objects were manually removed.

Noise reduction and surface smoothing filters were applied to the Pix4dMapper point clouds in a way that geometrical features remained sharp and features of smaller scale, like tiny branches or leaves, were maintained. The point clouds were visually controlled for missing parts (e.g., from occlusions) after computation. Figure 2 shows a laser and a Pix4Dmapper point cloud in their entirety before preparation.

Figure 2. (**A**) A shaded and artificially-colored laser scanning point cloud; (**B**) a photogrammetric point cloud in real colors. The difference in point density is clearly visible.

Remaining outliers and false object points clearly distinguished through their texture (white and grey points at leaf and stem borders; Figure 3) were manually removed. Irrelevant objects visible in the point clouds, e.g., the table and nourishment cube, were manually removed until only the plant itself remained.

Initially, the Pix4DMapper point clouds are arbitrarily scaled. They were thus manually scaled to units of millimeters. The metric scaling factor was derived through the known value of a geometrical feature in the point cloud, e.g., the length of an object, and its real millimeter value. We either used the measuring table's length or width in millimeters, as at least one side was always fully visible in the point cloud. The scale factor is the ratio of the feature in millimeters and in the pixel system of the original point cloud. The original point cloud coordinates were multiplied with the deduced scale factor to reach a scaling in millimeters. In this way, an individual scale factor was determined for every point cloud.

Figure 3. Close-up view of a photogrammetric point cloud before (**A**,**B**) and after (**C**) preparation (leaves only). The point size was enhanced for visualization.

2.5. Parameter Extraction

Four parameters were extracted from the point clouds: single leaf areas, cumulated leaf area, main stem height and volume of the convex hull of the whole plant. Stem height and leaf areas were derived using Geomagic, while the convex hull was calculated using MATLAB 2009b routines (The MathWorks Inc., Natick, MA, USA). The parameter extraction was identical for both methods.

Main Stem Height: The height of the main stem is an important indicator for the growth response of plants after fertilizer application [24] or under varying CO_2 rates in the environment [25]. In this study, the beginning of the main stem was defined as the point where the stem emerges from the plant cube. The end was defined as the point of intersection of the first lateral branch after the two cotyledon leaves—the first leaves the seedling develops after emerging from the soil for nourishment—and the stem. For height determination, points covering the area from the beginning to the end of the main stem were selected, and a least-square approximation for a cylinder was applied, as it represents a standard approach in phenotyping [26,27]. The height of the estimated cylinder was taken as the main stem height.

Leaf area: The analysis of the leaf area is important in determining the plant's developmental stage [28], chlorophyll production [29] and health status. Furthermore, it serves as an indicator for mechanical tasks, such as pesticide application [12]. Stem and branch points were manually cut from the plant model, so that only leaf points remained. These were meshed, whereby the plausibility of the meshed surface was visually verified by its smoothness. The sum of the meshed areas constitutes the cumulated leaf area of the plant. Furthermore, 26 single leaf areas from two different plants were extracted to assess how accurately the approach is able to measure single leaves. Attention was focused on selecting leaves of different sizes (small, medium, large), forms and crookedness to assess its general applicability.

Convex hull: The convex hull is defined as the shape of an object created through joining its outermost points. From 2D images, the convex hull is often used to estimate the surface coverage of plants [30,31]. In 3D, it is used in the close-up range, e.g., in [32] to determine the soil exploration extent of root systems. It further serves as an estimator for tree canopy biomass [33] and has been used as an effective indicator for plant drought on barley plants [6].

Tables 1 and 2 illustrate the influence of the cleaning process on the final results. Especially, the leaf area estimation is affected by false object points on leaf borders.

Table 1. Example of the influence of cleaning on the cumulated leaf area.

	Cumulated Leaf Area (mm^2)		
	Pre-Cleaning	**After Cleaning**	**Difference (%)**
Pix4DMapper	50,795.9	48,468.1	**4.6**
Laser (Reference)	48,661.2	48,404.1	**0.5**

Table 2. Example of the influence of cleaning on the convex hull.

	Convex Hull (dm^3)		
	Pre-Cleaning	**After Cleaning**	**Difference (%)**
Pix4DMapper	5.301	5.217	**1.6**
Laser (Reference)	5.131	5.125	**0.1**

2.6. Estimation of Accuracy and Error Distribution

Accuracy assessment is done by calculating the correlation coefficient R^2 between laser and photogrammetric measurements through the means of linear regression. The RMSE (root-mean-square-error) and MAPE (mean-absolute-percentage-error) indicators are used for error estimation. They are calculated using Equations (1) and (2), whereby ref designates the reference measurement from the Perceptron v5 and *act* refers to the actual measurement extracted from the photogrammetric point clouds:

$$RMSE = \sqrt{mean(ref - act)^2} \tag{1}$$

$$MAPE = mean\left(\left|\frac{ref - act}{ref} * 100\right|\right) \tag{2}$$

3. Results

All plants were reconstructed in their entirety through the usage of the two complementary perspective angles during data acquisition (Figure 2B; Figure 3A) with only very filigree parts missing in some cases (Figure 4C). Fine leaf veins are clearly distinguished texturally from the rest of the leaf.

At the leaf and branch borders, triangulation errors were noticed (grey points at leaf and branch borders; Figure 3B). All parameters extracted from the Pix4DMapper point clouds yielded high correlations of $R^2 \geq 0.96$ to the parameters from the reference point clouds. The parameter single leaf area (Figure 5A) yielded a correlation $R^2 = 0.99$ to the reference measurements. The RMSE reached 58.49 mm^2, while the MAPE lay at 6.41%. The cumulated leaf area (Figure 5B) yielded a correlation of $R^2 = 0.99$. The RSME for the cumulated leaf area reached 1679 mm^2. The MAPE lay at 2.26%. A correlation of $R^2 = 0.96$ was reached for the main stem height (Figure 5C) with an RMSE of about 1.39 mm and an MAPE lying at 1.87%. The convex hull (Figure 5D) yielded a correlation of $R^2 = 0.99$, with an RMSE reaching 0.03 dm^3 and an MAPE lying at 4.14%.

The relative error between reference and actual measurements mostly remained below 10% for all parameters. Relative errors over 10% were present only for single leaf area measurements (see Section 4). Figure 6 shows a histogram of the relative error distributions of all parameters.

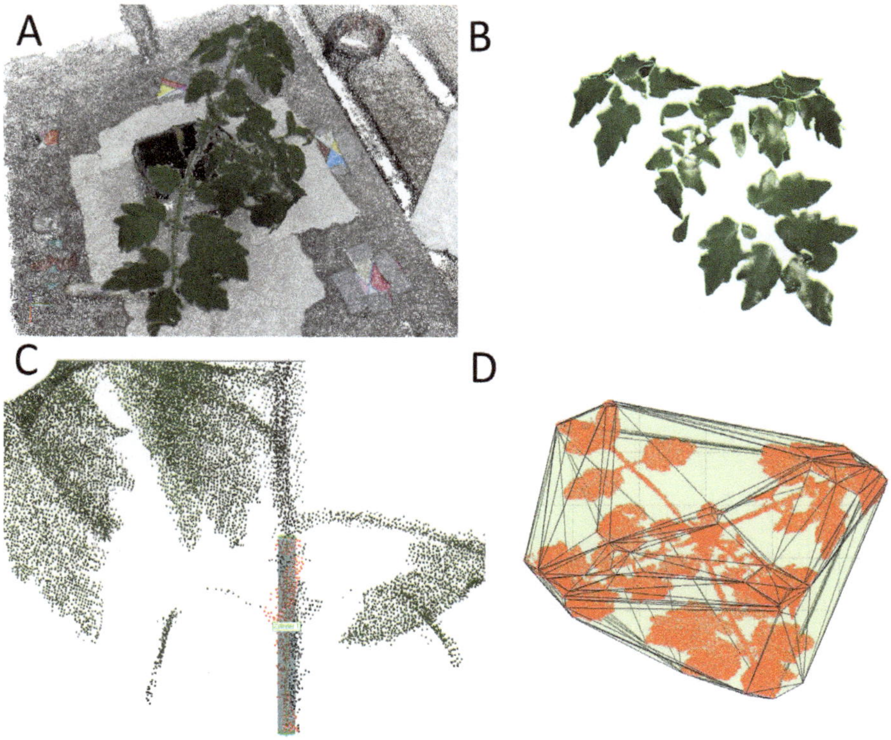

Figure 4. (**A**) The whole plant; (**B**) meshed leaves separated from stems and branches; (**C**) the measuring of the stem height; and (**D**) the computed 3D convex hull encompassing the plant.

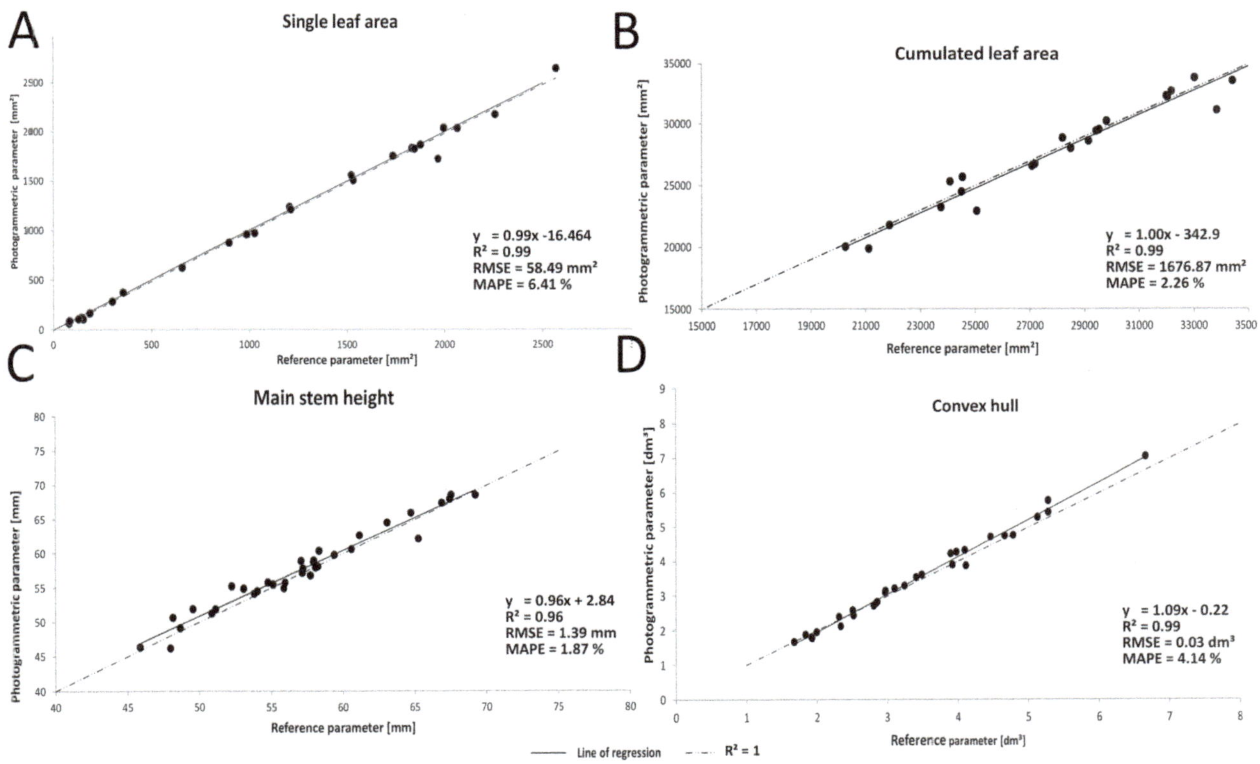

Figure 5. Regression analysis, RSME and MAPE results for the extracted parameters. (**A**) Single leaf area; (**B**) cumulated leaf area; (**C**) main stem height; (**D**) convex hull.

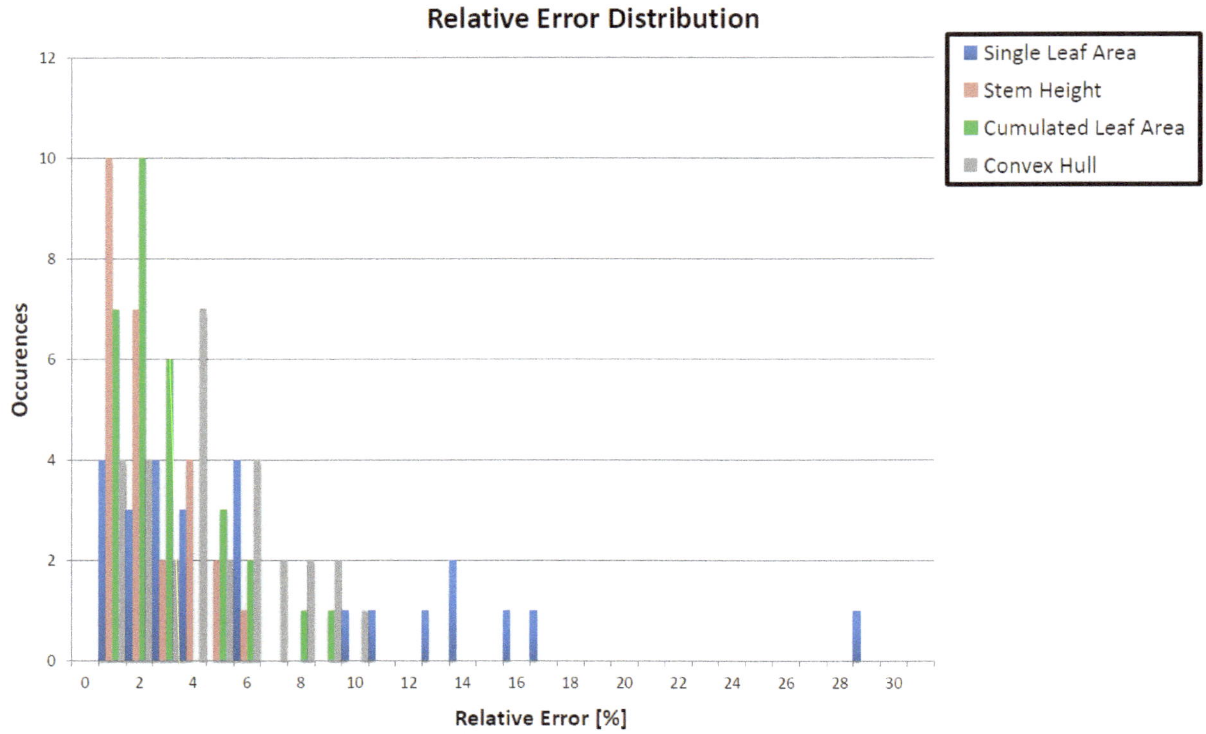

Figure 6. Histogram of the parameters' relative error distribution.

4. Discussion

The results show that the SfM-/MVS-based method is well suited for determining the parameters inspected here. All extracted parameters highly correlate with the reference measurements, yielding a coefficient of R^2 lying between 0.96–0.99. The MAPE lay under 7% for all parameters, thus lying under the limit of tolerance of 10%. This error limit was defined in [34] as the error already inherent between manual measurements. Furthermore, according to [34], detecting morphological changes in the plant between different measuring dates with a MAPE under this limit is still possible.

Measuring crooked and curled leaves is possible. Reconstructing small leaves in their completeness depends on their visibility in multiple images, as well as the level of detail and sharpness of the images. Additionally, small leaves are more sensible to filtering and, thus, more prone to loss of true area. This is seen in Figure 6, where the single leaf area alone from all parameters bears relative errors over 10% up to 30%. The level of detail can be increased by lowering the distance of the camera, while sharpness can be achieved by a wide depth of field. The plausibility of the meshed area was accounted for through visual inspection, but should be replaced by mathematical smoothness constraints. Holes in meshed areas can be filled manually and automatically if deemed plausible.

The stem height was measured manually using a cylindrical best-fit primitive, which allowed measuring even when the stem was only partly reconstructed. The stems beginning and end are defined by visual cues alone. A mathematical definition of the stems' beginning and end is needed to make measuring more objective. Furthermore, as the stem shape usually does not grow completely straight, the cylindrical form can only approximate its shape and height. More precise ways of parameterization need to be investigated, e.g., through the approximation of multiple adjacent cylinders or through circle approximation.

The convex hull is computed using the full plant without being separated into single plant organs first. As such, the convex hull is effortlessly extractable through automatic methods to monitor growth processes and reaction to environmental conditions, such as drought [6].

Data acquisition was done manually with a user taking images by hand without the use of a tripod, making the approach very adaptable and space-saving. Expert knowledge for the data acquisition is not necessary. The SfM approach calculates the cameras intrinsic parameters automatically and, in this way, avoids the long camera calibration phase before acquisition often necessary for stereo systems, as used in [35]. It is a very quick in the field method taking 2–3 min for data acquisition per plant. The Perceptron v5 is superior in point resolution and accuracy, but the time needed for data acquisition increases with plant size and complexity due to the limited working distance of 110 mm. The here depicted 3D reconstruction method is a cost-effective method amounting to 6500 € for the commercial Pix4DMapper software and to about 2500 € for a high-end camera. In comparison, the laser scanning system amounts to about 100 k€. We stress that while a specific commercial software was used for the point cloud derivation, at least similar if not equal results could probably be achieved by using alternative solutions, as demonstrated in [10], where the general accuracy and completeness of six MVS algorithms was investigated on several challenging 3D objects. Alternative software, like Bundler/PMVS (see e.g., [14]) or Agisoft Photoscan [36] are potential candidates. Likewise, open source software, like CloudCompare [37] and Octave [38] pose well-suited alternatives for parameter computation, point removal and scaling procedures.

A difficulty is the post-processing character, as the completeness and the quality of the point cloud can only be assessed after computation. During the data acquisition, it is therefore important to take images with sufficient content overlap and from complementary perspectives to avoid occlusion and to reconstruct the plant completely. False white and grey object points predominantly exist on leaf and stem borders. These are probably due to false matches caused by the homogeneous texture of the white foil of the nourishment cube and white paper towels placed underneath the cube. These could be removed through color filtering.

The amount of time invested per plant in post-processing procedures amounts to 10 to 20 min for point cloud generation, 5 min for manual scaling and 5–10 min for error removal and separation into plant organs. Data acquisition using the Perceptron laser scanner amounted to 10–15 min in contrast. However, as the working distance of the scanner is only 110 mm, the time needed for a full scanning will grow with the plant's size and canopy complexity. Images, on the other hand, capture more parts of the plant in a shorter period of time. There will be a plant size for which scanning the object is no longer feasible, and SfM/MVS methods require less time for data acquisition and 3D reconstruction. In [18,21], the feasibility of automatic geometry-based segmentation and classification algorithms for plant organ parameterization was already demonstrated. These could further be improved through the integration of the color information of the photogrammetric point clouds. In combination with methods for automatic coordinate system transformation of the point cloud (e.g., image geotagging, usage of measuring arms) and color filters for error removal, these approaches could reduce the need for supervised post-processing and enhance the amount of throughput that phenotyping often requires [1]. The color information should further be utilized for detecting plant reactions to environmental stress factors, like drought.

Future works should focus on extracting additional morphological parameters, like LAI, leaf length/inclination and stem diameter, which can probably be extracted from the point clouds with a similar accuracy. The limit of detectable details should be studied, as well. While only one kind of plant was used, the morphology of the plants changed drastically over the period of monitoring. In this period, they grew from a relatively open and simple canopy with some larger leaves, into a complex and denser canopy with varying leaf sizes and branch thicknesses. Our results demonstrate that the reconstruction is possible even under changing plant morphologies. The approach should nevertheless be tested on plants with even more complex and denser morphologies.

5. Conclusions

We show that the here presented photogrammetric approach depicts a reliable tool for high accuracy phenotyping at the organ level. A correlation of $R^2 \geq 0.96$ to high accuracy reference data was reached. The data acquisition is very simple and can be handled by non-trained personal, while the approaches flexibility holds potential for application in a variety of scenarios. It thus poses a suitable alternative to expensive and stationary tools for the studied scenarios.

Acknowledgments

The authors want to express their gratitude to Jan Dupuis and Florian Zimmermann for assisting in the data acquisition.

Author Contributions

Johann Christian Rose, Stefan Paulus and Heiner Kuhlmann designed the study. Johann Christian Rose and Stefan Paulus interpreted the data, carried out the parameter extraction and drafted the manuscript.

Conflicts of Interest

The authors declare no conflict of interest.

References

1. Dhondt, S.; Wuyts, N.; Inzé, D. Cell to whole-plant phenotyping: The best is yet to come. *Trends Plant Sci.* **2013**, *18*, 433–444.
2. Dornbusch, T.; Lorrain, S.; Kuznetsov, D.; Fortier, A.; Liechti, R.; Xenarios, I.; Fankhauser, C. Measuring the diurnal pattern of leaf hyponasty and growth in Arabidopsis a novel phenotyping approach using laser scanning. *Funct. Plant Biol.* **2012**, *39*, 860–869.
3. Kraft, M.; Regina, S.; Freitas, D.; Munack, A. Test of a 3D time of flight camera for shape measurements of plants. In Proceedings of the CIGR Workshop on Image Analysis in Agriculture, Budapest, Hungary, 26–27 August 2010; pp. 26–27.

4. Kizma, W.; Foix, S.; Alenyà, G. Plant leaf analalysis using Time of Flight camera under sun, shadow and room conditions. In Proceedings of the IEEE International Symposium on Robotic and Sensors Environments, Magdeburg, Germany, 16–18 November 2012.
5. Paulus, S.; Behmann, J.; Mahlein, A.K.; Plümer, L.; Kuhlmann, H. Low-cost 3D systems: Suitable tools for plant phenotyping. *Sensors* **2014**, *14*, 3001–3018.
6. Paulus, S.; Schumann, H.; Kuhlmann, H.; Léon, J. High-precision laser scanning system for capturing 3D plant architecture and analysing growth of cereal plants. *Biosyst. Eng.* **2014**, *121*, 1–11.
7. Tsaftaris, S.; Noutsos, C. Plant phenotyping with low cost digital cameras and image analytics. In *Environmental Science and Engineering (Subseries: Environmental Science)* Springer: Berlin/Heidelberg, Germany, 2009; pp. 238–251.
8. Arvidsson, S.; Pérez-Rodríguez, P.; Mueller-Roeber, B. A growth phenotyping pipeline for Arabidopsis thaliana integrating image analysis and rosette area modeling for robust quantification of genotype effects. *New Phytol.* **2011**, *191*, 895–907.
9. Hartley, R.I.; Zisserman, A. *Multiple View Geometry in Computer Vision*, 2nd ed.; Cambridge University Press: Cambridge, UK, 2004.
10. Seitz, S.M.; Curless, B.; Diebel, J.; Scharstein, D.; Szeliski, R. A comparison and evaluation of multi-view stereo reconstruction algorithms. *Proc. IEEE Conf. Compu. Vis. Pattern Recogn.* **2006**, *1*, 519–528.
11. Ivanov, N.; Boissard, P.; Chapron, M.; Andrieu, B. Computer stereo plotting for 3-D reconstruction of a maize canopy. *Agric. For. Meteorol.* **1995**, *75*, 85–103.
12. Aguilar, M.; Pozo, J.; Aguilar, F.; Sanchez-Hermosilla, J.; Páez, F.; Negreirosb, J. 3D Surface Modelling of Tomato Plants Using Close-Range. *Int. Arch. Photogramm. Remote Sens. Spat. Inf. Sci.* **2008**, *XXXVII*, 139–144.
13. Zhu, H.; Dorner, J.; Rowland, D.; Derksen, R.; Ozkan, H. Spray Penetration into Peanut Canopies with Hydraulic Nozzle Tips. *Biosyst. Eng.* **2004**, *87*, 275–283.
14. Santos, T.; Oliveira, A. Image-based 3D digitizing for plant architecture analysis and phenotyping. In Proceedings of the Workshop on Industry Applications (WGARI) in SIBGRAPI 2012 (XXV Conference on Graphics, Patterns and Images), Minas Gerais, Brazil, 22–25 August 2012.
15. Snavely, N.; Seitz, S.M.; Szeliski, R. Photo tourism: Exploring photo collections in 3D. In Proceedings of the SIGGRAPH Conference, Boston, MA, USA, 30 July–3 August 2006; pp. 835–846.
16. Nistér, D. An efficient solution to the five-point relative pose problem. *IEEE Trans. Pattern Anal. Mach. Intell.* **2004**, *26*, 756–770.
17. Furukawa, Y.; Ponce, J. Accurate, dense, and robust multiview stereopsis. *IEEE Trans. Pattern Anal. Mach. Intell.* **2010**, *32*, 1362–1376.
18. Santos, T.; Ueda, J. Automatic 3D plant reconstruction from photographies, segmentation and classification of leaves and internodes using clustering. In Proceedings of the International Conference on Function-Structural Plant Models, Saariselkä, Finland, 9–14 June 2013; pp. 95–97.

19. Jay, S.; Rabatel, G.; Gorrettta, N. In-Field Crop Row Stereo-Reconsruction for Plant Phenotyping. In Proceedings of the Second International Conference on Robotics and associated High-Technologies and Equipment for Agriculture and forestry (RHEA), Madrid, Spain, 21–23 May 2014; pp. 21–23.
20. Feller, C.; Bleiholder, H.; Buhr, L.; Hack, H.; Hess, M.; Klose, R.; Meier, U.; Stauss, R.; Van den Boom, T.; Weber, E. Phänologische Entwicklungsstadien von Gemüsepflanzen: II. Fruchtgemüse und Hülsenfrüchte, 1995 b. *Nachrichtenbl. Deut. Pflanzenschutzd.* **1995**, 217–232. In Germany.
21. Paulus, S.; Dupuis, J.; Mahlein, A.K.; Kuhlmann, H. Surface feature based classiffication of plant organs from 3D laserscanned point clouds for plant phenotyping. *BMC Bioinform.* **2013**, *14*, 238, doi:10.1186/1471-2105-14-238.
22. Strecha, C.; Bronstein, A.M.; Bronstein, M.M.; Fua, P. LDAHash: Improved Matching with Smaller Descriptors. *IEEE Trans. Pattern Anal. Mach. Intell.* **2012**, *34*, 66–78.
23. Küng, O.; Strecha, C.; Beyeler, A.; Zufferey, J.C.; Floreano, D.; Fua, P.; Gervaix, F. The Accuracy of Automatic Photogrammetric Techniques on Ultra-Light Uav Imagery. *ISPRS* **2012**, *XXXVIII-1/C22*, 125–130.
24. Rui, Y.K.; Peng, Y.F.; Wang, Z.R.; Shen. J.B. Stem perimeter, height and biomass of maize (*Zea mays* L.) grown under different N fertilization regimes in Beijing, China. *Int. J. Plant Prod.* **2009**, *3*, 85–90.
25. Thornley, J.H.M. Modelling Stem Height and Diameter Growth in Plants. *Ann. Bot.* **1999**, *84*, 195–205.
26. Paulus, S.; Dupuis, J.; Riedel, S.; Kuhlmann, H. Automated analysis of barley organs using 3D laser scanning: An approach for high throughput phenotyping. *Sensors* **2014**, *14*, 12670–12686.
27. Frasson, R.P.D.M.; Krajewski, W.F. Three-dimensional digital model of a maize plant. *Agric. For. Meteorol.* **2010**, *150*, 478–488.
28. Granier, C.; Tardieu, F. Multi-scale phenotyping of leaf expansion in response to environmental changes: The whole is more than the sum of parts. *Plant Cell Environ.* **2009**, *32*, 1175–1184.
29. Gonzalez, N.; Vanhaeren, H.; Inzé, D. Leaf size control: Complex coordination of cell division and expansion. *Trends Plant Sci.* **2012**, *17*, 332–340.
30. Jansen, M.; Gilmer, F.; Biskup, B.; Nagel, K.; Rascher, U.; Fischbach, A.; Briem, S.; Dreissen, G.; Tittmann, S.; Braun, S.; *et al.* Simultaneous phenotyping of leaf growth and chlorophyll fluorescence via GROWSCREEN FLUORO allows detection of stress tolerance in Arabidopsis thaliana and other rosette plants. *Funct. Plant Biol.* **2009**, *36*, 902–914.
31. Rascher, U.; Blossfeld, S.; Fiorani, F.; Jahnke, S.; Jansen, M.; Kuhn, A.J.; Matsubara, S.; Märtin, L.L.; Merchant, A.; Metzner, R.; *et al.* Non-invasive approaches for phenotyping of enhanced performance traits in bean. *Funct. Plant Biol.* **2011**, *38*, 968–983.
32. Topp, C.N.; Iyer-Pascuzzi, A.S.; Anderson, J.T.; Lee, C.R.; Zurek, P.R.; Symonova, O.; Zheng, Y.; Bucksch, A.; Mileyko, Y.; Galkovskyi, T.; *et al.* 3D phenotyping and quantitative trait locus mapping identify core regions of the rice genome controlling root architecture. *Proc. Natl. Acad. Sci. USA* **2013**, *110*, E1695–E1704.
33. Korhonen, L.; Vauhkonen, J.; Virolainen, A.; Hovi, A.; Korpela, I. Estimation of tree crown volume from airborne lidar data using computational geometry. *Int. J. Remote Sens.* **2013**, *34*, 7236–7248.

34. Paproki, A.; Sirault, X.; Berry, S.; Furbank, R.; Fripp, J. A novel mesh processing based technique for 3D plant analysis. *BMC Plant Biol.* **2012**, 1–13, doi:10.1186/1471-2229-12-63.
35. Biskup, B.; Scharr, H.; Schurr, U.; Rascher, U. A stereo imaging system for measuring structural parameters of plant canopies. *Plant Cell Environ.* **2007**, *30*, 1299–1308.
36. Agisoft Photoscan. Available online: http://www.agisoft.com/ (accessed on 23 April 2015).
37. CloudCompare. Available online: http://www.danielgm.net/cc/ (accessed on 23 April 2015).
38. Octave. Available online: https://www.gnu.org/software/octave/ (accessed on 23 April 2015).

Permissions

All chapters in this book were first published in Sensors, by MDPI; hereby published with permission under the Creative Commons Attribution License or equivalent. Every chapter published in this book has been scrutinized by our experts. Their significance has been extensively debated. The topics covered herein carry significant findings which will fuel the growth of the discipline. They may even be implemented as practical applications or may be referred to as a beginning point for another development.

The contributors of this book come from diverse backgrounds, making this book a truly international effort. This book will bring forth new frontiers with its revolutionizing research information and detailed analysis of the nascent developments around the world.

We would like to thank all the contributing authors for lending their expertise to make the book truly unique. They have played a crucial role in the development of this book. Without their invaluable contributions this book wouldn't have been possible. They have made vital efforts to compile up to date information on the varied aspects of this subject to make this book a valuable addition to the collection of many professionals and students.

This book was conceptualized with the vision of imparting up-to-date information and advanced data in this field. To ensure the same, a matchless editorial board was set up. Every individual on the board went through rigorous rounds of assessment to prove their worth. After which they invested a large part of their time researching and compiling the most relevant data for our readers.

The editorial board has been involved in producing this book since its inception. They have spent rigorous hours researching and exploring the diverse topics which have resulted in the successful publishing of this book. They have passed on their knowledge of decades through this book. To expedite this challenging task, the publisher supported the team at every step. A small team of assistant editors was also appointed to further simplify the editing procedure and attain best results for the readers.

Apart from the editorial board, the designing team has also invested a significant amount of their time in understanding the subject and creating the most relevant covers. They scrutinized every image to scout for the most suitable representation of the subject and create an appropriate cover for the book.

The publishing team has been an ardent support to the editorial, designing and production team. Their endless efforts to recruit the best for this project, has resulted in the accomplishment of this book. They are a veteran in the field of academics and their pool of knowledge is as vast as their experience in printing. Their expertise and guidance has proved useful at every step. Their uncompromising quality standards have made this book an exceptional effort. Their encouragement from time to time has been an inspiration for everyone.

The publisher and the editorial board hope that this book will prove to be a valuable piece of knowledge for researchers, students, practitioners and scholars across the globe.

List of Contributors

Yongbin Qi
School of Electrical and Electronic Engineering, Nanyang Technological University, 50 Nanyang Avenue, 639798 Singapore

Cheong Boon Soh
School of Electrical and Electronic Engineering, Nanyang Technological University, 50 Nanyang Avenue, 639798 Singapore

Erry Gunawan
School of Electrical and Electronic Engineering, Nanyang Technological University, 50 Nanyang Avenue, 639798 Singapore

Kay-Soon Low
School of Electrical and Electronic Engineering, Nanyang Technological University, 50 Nanyang Avenue, 639798 Singapore

Rijil Thomas
School of Electrical and Electronic Engineering, Nanyang Technological University, 50 Nanyang Avenue, 639798 Singapore

Natasa Reljin
Department of Biomedical Engineering, University of Connecticut, 260 Glenbrook Road, Storrs, CT 06269, USA

Bersain A. Reyes
Department of Biomedical Engineering, University of Connecticut, 260 Glenbrook Road, Storrs, CT 06269, USA

Ki H. Chon
Department of Biomedical Engineering, University of Connecticut, 260 Glenbrook Road, Storrs, CT 06269, USA

Jiangping Chen
State Key Laboratory of Chemical Engineering, Tianjin Key Laboratory of Membrane Science and Desalination Technology, School of Chemical Engineering and Technology, Tianjin University, Tianjin 300072, China

Se Shi
State Key Laboratory of Chemical Engineering, Tianjin Key Laboratory of Membrane Science and Desalination Technology, School of Chemical Engineering and Technology, Tianjin University, Tianjin 300072, China

Rongxin Su
State Key Laboratory of Chemical Engineering, Tianjin Key Laboratory of Membrane Science and Desalination Technology, School of Chemical Engineering and Technology, Tianjin University, Tianjin 300072, China
Collaborative Innovation Center of Chemical Science and Engineering (Tianjin), Tianjin 300072, China

Wei Qi
State Key Laboratory of Chemical Engineering, Tianjin Key Laboratory of Membrane Science and Desalination Technology, School of Chemical Engineering and Technology, Tianjin University, Tianjin 300072, China
Collaborative Innovation Center of Chemical Science and Engineering (Tianjin), Tianjin 300072, China

Renliang Huang
School of Environmental Science and Engineering, Tianjin University, Tianjin 300072

Mengfan Wang
State Key Laboratory of Chemical Engineering, Tianjin Key Laboratory of Membrane Science and Desalination Technology, School of Chemical Engineering and Technology, Tianjin University, Tianjin 300072, China

Libing Wang
State Key Laboratory of Chemical Engineering, Tianjin Key Laboratory of Membrane Science and Desalination Technology, School of Chemical Engineering and Technology, Tianjin University, Tianjin 300072, China

Zhimin He
State Key Laboratory of Chemical Engineering, Tianjin Key Laboratory of Membrane Science and Desalination Technology, School of Chemical Engineering and Technology, Tianjin University, Tianjin 300072, China

Ernst Niederleithinger
BAM Federal Institute for Materials Research and Testing, Berlin 12200, Germany

Julia Wolf
BAM Federal Institute for Materials Research and Testing, Berlin 12200, Germany

Frank Mielentz
BAM Federal Institute for Materials Research and Testing, Berlin 12200, Germany

Herbert Wiggenhauser
BAM Federal Institute for Materials Research and Testing, Berlin 12200, Germany

Stephan Pirskawetz
BAM Federal Institute for Materials Research and Testing, Berlin 12200, Germany

Victor Hugo C. de Albuquerque
Programa de Pós-Graduação em Informática Aplicada, Universidade de Fortaleza, Fortaleza, Ceará 60811-905, Brazil

Cleisson V. Barbosa
Programa de Pós-Graduação em Informática Aplicada, Universidade de Fortaleza, Fortaleza, Ceará 60811-905, Brazil

Cleiton C. Silva
Departamento de Engenharia Metalúrgica e de Materiais, Universidade Federal do Ceará, Fortaleza, Ceará 60455-900, Brazil

Elineudo P. Moura
Departamento de Engenharia Metalúrgica e de Materiais, Universidade Federal do Ceará, Fortaleza, Ceará 60455-900, Brazil

Pedro P. Rebouças Filho
Programa de Pós-Graduação em Ciências da Computação, Instituto Federal de Educação,
Ciência e Tecnologia do Ceará, Fortaleza, Ceará 61939-140, Brazil

João P. Papa
Departamento de Ciência da Computação, Universidade Estadual Paulista, Bauru, São Paulo 17033-360, Brazil

João Manuel R. S. Tavares
Instituto de Ciência e Inovação em Engenharia Mecânica e Engenharia Industrial,
Departamento de Engenharia Mecânica, Faculdade de Engenharia, Universidade do Porto, Porto 4200-465, Portugal

Wann-Yun Shieh
Department of Computer Science and Information Engineering, Chang Gung University, No. 259, Wen-Hwa 1st Road, Kwei-Shan, Tao-Yuan 333, Taiwan

Chin-Man Wang
Department of Physical Medicine and Rehabilitation, Chang Gung Memorial Hospital, Taoyuan 333
Department of Medicine, Chang Gung University, No. 5, Fu-Hsing Street, Kwei Shan, Tao-Yuan 333, Taiwan

Chia-Shuo Chang
Department of Computer Science and Information Engineering, Chang Gung University, No. 259, Wen-Hwa 1st Road, Kwei-Shan, Tao-Yuan 333, Taiwan

Giampiero de Cesare
Department of Information Engineering, Electronics and Telecommunications, "La Sapienza" University of Rome, via Eudossiana 18, 00184 Rome, Italy

Augusto Nascetti
Department of Astronautics, Electrical and Energetic Engineering, "La Sapienza" University of Rome, via Eudossiana 18, 00184 Rome, Italy

Domenico Caputo
Department of Information Engineering, Electronics and Telecommunications, "La Sapienza" University of Rome, via Eudossiana 18, 00184 Rome, Italy

Jonathan Lueke
Department of Mechanical Engineering, University of Alberta, University of Alberta, Edmonton, AB T6G 2G8, Canada

Ahmed Badr
Department of Electrical and Computer, University of Alberta, University of Alberta, Edmonton, AB T6G 2V4, Canada

Edmond Lou
Department of Electrical and Computer, University of Alberta, University of Alberta, Edmonton, AB T6G 2V4, Canada

Walied A. Moussa
Department of Mechanical Engineering, University of Alberta, University of Alberta, Edmonton, AB T6G 2G8, Canada

Intae Kim
Department of Mechanical Engineering, Pohang University of Science and Technology, 77 Cheongam-Ro, Nam-Gu, Pohang, Gyeongbuk 790-784, Korea

Geon Hwee Kim
Department of Mechanical Engineering, Pohang University of Science and Technology, 77 Cheongam-Ro, Nam-Gu, Pohang, Gyeongbuk 790-784, Korea

Chang Sup Kim
School of Biotechnology and Graduate School of Biochemistry, Yeungnam Univerisity, 280 Daehak-Ro, Gyeongsan, Gyeongbuk 712-749, Korea

Hyung Joon Cha
Department of Chemical Engineering, Pohang University of Science and Technology, 77 Cheongam-Ro, Nam-Gu, Pohang, Gyeongbuk 790-784, Korea

Geunbae Lim
Department of Mechanical Engineering, Pohang University of Science and Technology, 77 Cheongam-Ro, Nam-Gu, Pohang, Gyeongbuk 790-784, Korea

Gungun Lin
Institute for Integrative Nanosciences, IFW Dresden, Helmholtzstr. 20, 01069 Dresden, Germany
Material Systems for Nanoelectronics, Chemnitz University of Technology, Reichenhainerstr. 70, 09107 Chemnitz, Germany

Denys Makarov
Institute for Integrative Nanosciences, IFW Dresden, Helmholtzstr. 20, 01069 Dresden, Germany

Oliver G. Schmidt
Institute for Integrative Nanosciences, IFW Dresden, Helmholtzstr. 20, 01069 Dresden, Germany
Material Systems for Nanoelectronics, Chemnitz University of Technology, Reichenhainerstr. 70, 09107 Chemnitz, Germany

Paul A. Selvadurai
University of California, Berkeley, CA 94720, USA

Steven D. Glaser
University of California, Berkeley, CA 94720, USA

Sabine Szunerits
Institute of Electronics, Microelectronics and Nanotechnology (IEMN), UMR-CNRS 8520,
Université Lille 1, Avenue Poincaré—BP 60069, 59655 Villeneuve d'Ascq, France

Yannick Coffinier
Institute of Electronics, Microelectronics and Nanotechnology (IEMN), UMR-CNRS 8520,
Université Lille 1, Avenue Poincaré—BP 60069, 59655 Villeneuve d'Ascq, France

Rabah Boukherroub
Institute of Electronics, Microelectronics and Nanotechnology (IEMN), UMR-CNRS 8520,
Université Lille 1, Avenue Poincaré—BP 60069, 59655 Villeneuve d'Ascq, France

Johann Christian Rose
Department of Geodesy, Institute of Geodesy and Geoinformation, University of Bonn, Nussallee 17, 53115 Bonn, Germany

Stefan Paulus
Department of Geodesy, Institute of Geodesy and Geoinformation, University of Bonn, Nussallee 17, 53115 Bonn, Germany

Heiner Kuhlmann
Department of Geodesy, Institute of Geodesy and Geoinformation, University of Bonn, Nussallee 17, 53115 Bonn, Germany